Solution of Problems in Structures

Solution of Problems in Structures: A Problem-based Textbook

W. T. MARSHALL, B.Sc., Ph.D., F.I.C.E., F.I.Struct.E.
Regius Professor of Civil Engineering,
University of Glasgow

METRIC EDITION

Pitman Publishing

First Metric edition 1971
Reprinted (*with corrections*) 1975

SIR ISAAC PITMAN AND SONS LTD.
Pitman House, 39 Parker Street, Kingsway, London WC2B 5PB
P.O. Box 46038, Banda Street, Nairobi, Kenya

PITMAN PUBLISHING PTY. LTD.
Pitman House, 158 Bouverie Street, Carlton, Victoria 3053, Australia

PITMAN PUBLISHING CORPORATION
6 East 43rd Street, New York, N.Y. 10017, U.S.A.

SIR ISAAC PITMAN (CANADA) LTD.
495 Wellington Street West, Toronto 135, Canada

THE COPP CLARK PUBLISHING COMPANY
517 Wellington Street West, Toronto 135, Canada

ISBN: 0 273 36099 X

Text set in 10 pt. Monotype Modern, printed by letterpress, and bound in Great Britain at The Pitman Press, Bath.
(T.1347:73)

Preface to Metric Edition

The original motive for writing this little book was to produce something which would be useful to students. If it is to continue to fulfil this, then it must obviously be written in those units which the students are using. This has necessitated changing the Imperial units in which it was originally written to the S.I. units which are now standard practice.

The examples taken from examination papers were in Imperial units. The change to S.I. has been done to keep the loads and dimensions approximately the same as before, but the questions as they appear in the book are not in the form in which they originally were as examination questions. In making the change, the nearest whole number, instead of an accurate conversion, has been taken.

It is hoped that the book in this form will continue to be useful to students.

Glasgow, April 1971 W. T. MARSHALL

Preface to First Edition

"What we want, Sir, is a book with loads of examples in it." These words were spoken to me by a student about ten years ago and they have been a challenge to me ever since. This book, with its examples on loads, is the answer.

There are so many good books on the Theory of Structures that to add another to the catalogue would be a waste of time and effort. These books deal mostly with the fundamental principles which are of course absolutely essential. In the mind of the student, however, degree examinations hold the largest place and he knows that in these degree examinations he will not be asked to prove, say, the first theorem of Castigliano but to answer some question which an examiner has thoughtfully composed on that theorem.

In this book therefore only passing reference is made to the fundamentals. What is given is the solution of various problems based on these fundamentals together with a number of examples for the student to work through for himself.

Many of these examples have been taken from University examination papers and I am grateful to the Universities of Oxford, Cambridge, St. Andrews, Glasgow, Aberdeen, London and Durham, all of whom have given permission for questions to be reproduced from their degree papers.

Every care has been taken to ensure that the answers given are correct, but as there are more than 350 problems in the book it is possible that a slip may occasionally be found. In the checking of the examples I have been helped by many old students, particularly by Messrs. A. L. Florence, A. Gibb and J. Hughes, of St. Andrews, and W. M. Jenkins and J. G. S. Smith of Glasgow University. To each of them my thanks are due.

Glasgow, Sept. 1957 W. T. MARSHALL

Contents

Chapter 1

Statically Determinate Frames

A STATICALLY DETERMINATE FRAME is a frame in which the forces in the members can be obtained by the application of the equations of static equilibrium. For a plane frame there are three such equations and for a space frame, six.

The first consideration is whether any given frame is statically determinate. A *framework* is defined as an assemblage of bars which is able to resist geometrical distortion under any system of applied loads. There are three equations of static equilibrium and the simplest plane frame to satisfy both conditions, namely (i) that it can resist geometrical distortion and (ii) that it can be solved by the application of the equations of static equilibrium in a triangle as shown in Fig. 1.1. This simple frame can be extended and still satisfy both conditions provided that each additional joint (or panel point) is tied to the existing frame by two members only.

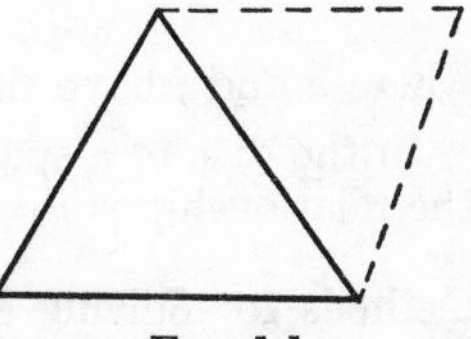

FIG. 1.1

Number of Bars in a Statically Determinate Frame

The relationship between the number of bars and joints for a plane frame can be expressed as

$$n = 2j - 3$$

where n = number of bars
j = number of joints

In the case of frames which have pinned supports such supports are equivalent to an existing frame and the relationship between the number of bars and joints is given by

$$n = 2j_f$$

where n = number of bars
j_f = number of free joints, i.e. joints outside the pinned supports

It must first be established by the application of the above rule together with the fundamental principle of the triangle and the two bracing members for each panel point, that the frame is statically determinate. It should be remembered, however, that the rule applies only if each member of the frame is able to take both tension and compression. Some frames contain members which can only take tension forces—such frames are in general statically determinate and the loads in the members are obtained as described in the section on counter-bracing (*see* p. 41).

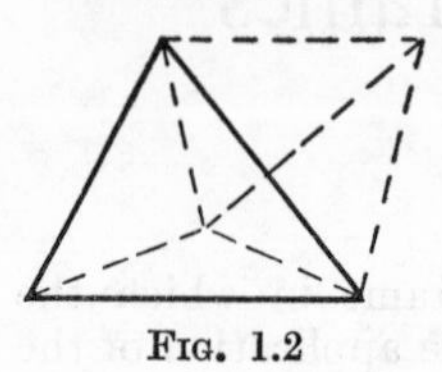

FIG. 1.2

The simplest *space frame* to satisfy the two conditions previously mentioned is the tetrahedron which can be extended by tying each fresh joint by three members to the initial frame as shown in Fig. 1.2. Thus the relationship between the number of bars and joints in the case of a space frame can be expressed as

$$n = 3j - 6$$

where n and j have the same meanings as before.

In the case of a space frame which is pinned to three supports the relationship is $n = 3j_f$.

Methods for Solving Statically Determinate Frames

Any method which employs the equations of static equilibrium can be used for the solution of a statically determinate frame. These equations are three in number, namely, two resolutions along perpendicular axes, and one moment, in the case of plane frames, and six for space frames. The six for a space frame consist of three resolutions along axes at right angles and moments about three points.

The main methods used for the solution of statically determinate frames are—

1. The stress diagram.
2. The method of sections.
3. The method of inspection or resolution at joints.
4. The method of tension coefficients.

The first method, which incidentally would be more correctly called the load diagram, is dealt with fully in books on graphic statics or in the relevant sections of the larger books on the Theory of Structures. It is not proposed to deal with it in this book.

The method of sections consists of applying the equations of static equilibrium to a section of the frame which cuts through those members whose loads have to be found.

For example, assume it is required to find the forces in the members X, Y and Z in the frame shown in Fig. 1.3. A section is cut along the line C–C and the left-hand portion of the truss considered. This is in equilibrium under the action of the external load W_1, the reaction R_A and the forces in the three members X, Y and Z.

By taking moments about point 1, the force in the member X is obtained. Knowing this the forces can now be resolved vertically giving Y. Finally, resolving horizontally gives Z.

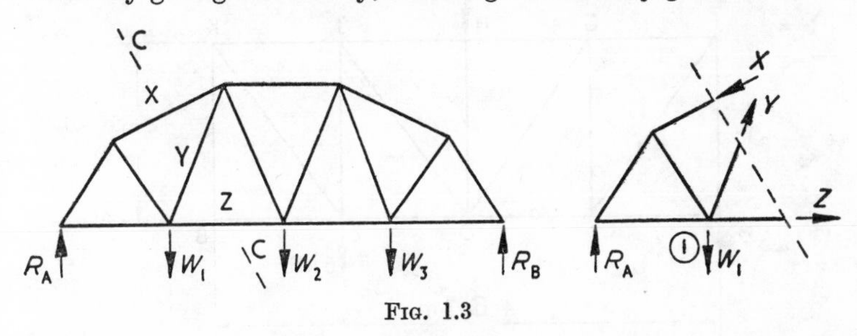

FIG. 1.3

This method is lengthy since a number of sections must be drawn to analyse a complete frame. No examples will be given on it but it will be incorporated into the next method.

Method of Inspection or Resolution at Joints

When drawing a stress diagram the student has to remember that he can only consider the equilibrium of a joint at which there are not more than two unknown forces. If there are only two unknowns then the force polygon can be completed and the unknown forces obtained. These two unknown forces could, however, have been obtained by considering the equilibrium of the components in two perpendicular directions of the forces acting on the joint.

This principle is the basis of the method of resolution at the joints. Any joint at which there are not more than two unknown forces is selected and the forces on the joint resolved into two components at right angles. Since the joint is in equilibrium the sum of the components in each direction must be zero. This fact yields two equations from which the two unknown forces can be obtained.

If the angles of the members of the frame have trigonometrical properties which are well known it will be possible, after some practice, to form the equations and solve them mentally so that, virtually, the forces may be obtained by inspection.

Careful attention should be given to this method for there are many examples in redundant frame analysis which must be started by writing down the forces in the members of the frame (when the redundant members have been removed) and an initial error in writing down the forces can waste much time.

Example 1.1. Determine by inspection the forces in the members of the frame shown in Fig. 1.4 (a), which carries the loads shown.

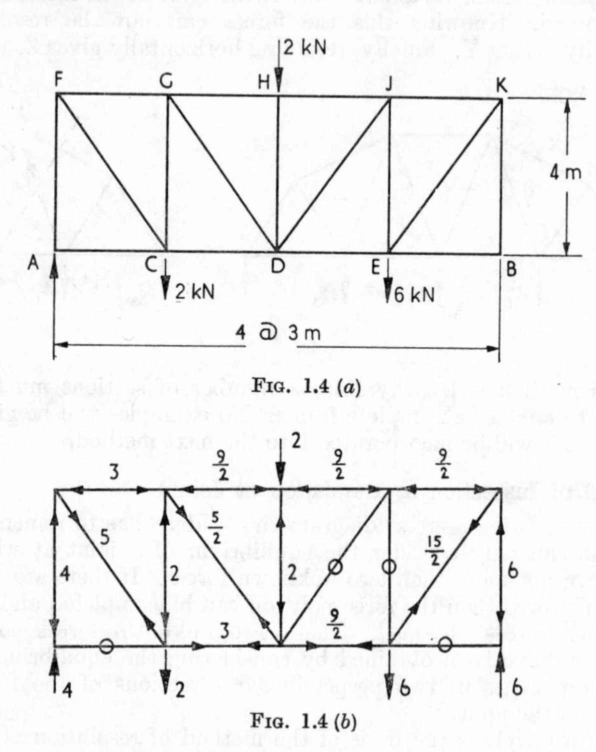

FIG. 1.4 (a)

FIG. 1.4 (b)

Solution. The reactions must first be calculated.
Taking moments about B gives

$$R_A \times 12 = 2 \times 9 + 2 \times 6 + 6 \times 3$$

giving $\qquad R_A = 4 \text{ kN}$

hence $\qquad R_B = 6 \text{ kN}$

The work will be given in full but a number of the steps are simple and the student should be able to perform them mentally.

Joint A. Resolving horizontally $AC = 0$

Resolving vertically $AF = R_A = 4 \text{ kN}$

Joint F. Resolving vertically $4/5 \times FC = AF = 4$

$\therefore \quad FC = 5$

Resolving horizontally $FG = 3/5 \times FC = 3$

Joint C. Resolving vertically $GC = 4/5 \times FC - 2$

$= 4 - 2 = 2$

Resolving horizontally $CD = 3/5 \times FC = 3$

Joint G. Resolving vertically $4/5 \times GD = 2 \qquad GD = 5/2$

Resolving horizontally $GH = 3/5 \times GD + FG$

$= 3/2 + 3 = 9/2$

Joint H. Resolving vertically $HD = 2$

Resolving horizontally $HJ = HG = 9/2$

Joint D. Resolving vertically $4/5 \times DJ = 2 - 4/5 \times GD$

$= 2 - 2 = 0$

Resolving horizontally $DE = DC + 3/5 \times GD$

$= 3 + 3/5 \times 5/2$

$= 9/2$

Joint J. Resolving vertically $JE = 4/5 \times DJ = 0$

Resolving horizontally $JK = JH = 9/2$

Joint E. Resolving vertically $4/5 \times EK = 6$

$\therefore \quad EK = 15/2$

Resolving horizontally $BE = ED - 3/5 \times EK$

$= 9/2 - 3/5 \times 15/2$

$= 0$

Joint K. Resolving vertically $KB = 4/5 \times KE = 6$

Joint B. The force in KB is equal to the reaction at B thus providing a check that the work is correct.

The forces in kN are shown in Fig. 1.4 (*b*).

Example 1.2. The truss shown in Fig. 1.5 (*a*) is supported at A and F and carries loads of 10 kN and 5 kN at C and D respectively. If the forces in the members CD and JK are numerically equal, determine the value of the forces in all the members of the truss.

Solution. In this frame there are 12 joints, hence the number of members for a statically determinate frame is $2 \times 12 - 3 = 21$. A count, however, shows that there are 22. In order to solve the frame a further equation must be provided and this is the reason why the question states that the forces in CD and JK are equal.

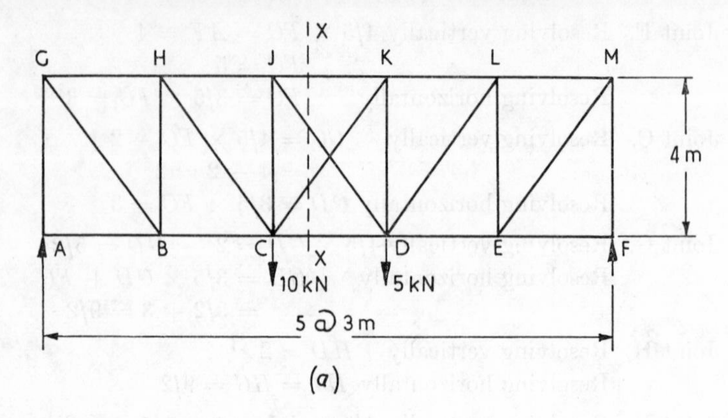

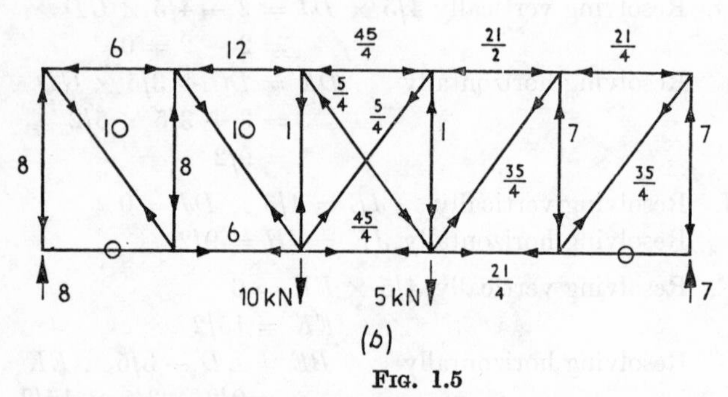

FIG. 1.5

The reactions are first calculated. Taking moments about F gives

$$15 \times R_A = 10 \times 9 + 5 \times 6$$

Thus

$$R_A = 8 \text{ kN} \quad \text{and} \quad R_F = 7 \text{ kN}$$

Cut a section through X–X. This passes through four members JK, CD, JD and CK. Taking moments, however, about the intersection of JD and CK and considering loads to the right gives

$$2JK + 2CD = 7 \times 7{\cdot}5 - 5 \times 1{\cdot}5 = 45$$

Thus

$$JK = 45/4 = CD$$

As with the previous example the work will be given in full, but a number of the steps can be carried out mentally.

Joint A. Resolving horizontally $AB = 0$
Resolving vertically $AG = 8$

Joint G. Resolving vertically $BG \times 4/5 = 8$

hence $BG = 10$

Resolving horizontally $GH = 3/5 \times BG = 6$

Joint B. Resolving vertically $BH = 4/5 \times BG = 8$

Resolving horizontally $BC = 3/5 \times BG = 6$

Joint H. Resolving vertically $4/5 \times HC = HB = 8$

hence $HC = 10$

Resolving horizontally $HJ = HG + 3/5 \times HC$

$= 6 + 6 = 12$

Joint C. Resolving horizontally

$$3/5 \times CK = CB + 3/5 \times CH - CD$$
$$= 6 + 6 - 45/4 = 3/4$$

hence $CK = 5/4$

Resolving vertically

$$JK = 10 - 4/5 \times HC - 4/5 \times CK$$
$$= 10 - 8 - 1 = 1$$

Joint J. Resolving vertically $4/5 \times JD = JC = 1$

hence $JD = 5/4$

Joint K. Resolving vertically $KD = 4/5 \times KC$

$= 4/5 \times 5/4 = 1$

Resolving horizontally $KL = JK - 3/5 \times KC$

$= 45/4 - 3/4 = 21/2$

Joint D. Resolving vertically

$$4/5 \times DL = KD + 5 + 4/5 \times JD = 1 + 5 + 1$$

hence $DL = 35/4$

Resolving horizontally

$$DE = DC - 3/5 \times JD - 3/5 \times DL$$
$$= 45/4 - 3/4 - 21/4 = 21/4$$

Joint L. Resolving vertically $LE = 4/5 \times DL = 7$

Resolving horizontally $LM = KL - 3/5 \times DL$

$= 21/2 - 21/4 = 21/4$

Joint E. Resolving vertically $4/5 \times EM = 7$

hence $EM = 35/4$

Resolving horizontally $EF = DE - 3/5 \times EM$

$= 21/4 - 21/4 = 0$

Joint M. Resolving vertically $MF = 4/5 \times EM = 7$

At joint F, MF as calculated from previous stressing is equal to the calculated vertical reaction thus giving a check on the work. The forces in kN are shown in Fig. 1.5 (*b*).

Examples for Practice on Method of Inspection

1. The frame shown in Fig. 1.6 is supported at B and E and carries loads of 3 kN and 6 kN at A and F. Determine by

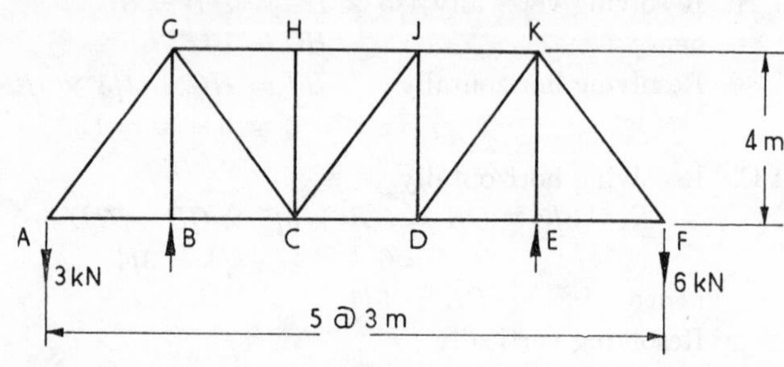

Fig. 1.6

inspection the loads in the members: (i) when the configuration is as shown, (ii) when members CG, CJ and DK are replaced by BH, DH and EJ.

Answer. (i) $AB = -9/4$, $BC = -9/4$, $CD = -15/4$, $DE = -9/2$, $EF = -9/2$, $AG = +15/4$, $GH = +12/4$, $HJ = +12/4$, $JK = +15/4$, $KF = +15/2$, $BG = -2$, $GC = -5/4$, $HC = 0, CJ = +5/4$, $JD = -1$, $DK = +5/4$, $KE = -7$.

(ii) $AB = -9/4$, $BC = -12/4$, $CD = -12/4$, $DE = -15/4$, $EF = -9/2$, $AG = +15/4$, $GH = +9/4$, $HJ = +15/4$, $JK = +9/2$, $KF = +15/2$, $BG = -3$, $BH = +5/4$, $HC = 0$, $HD = -5/4$, $JD = +1$, $JE = -5/4$, $KE = -6$.

2. Determine by inspection the forces in the members of the frame shown in Fig. 1.7. when it carries a load of 1 kN vertically at F.

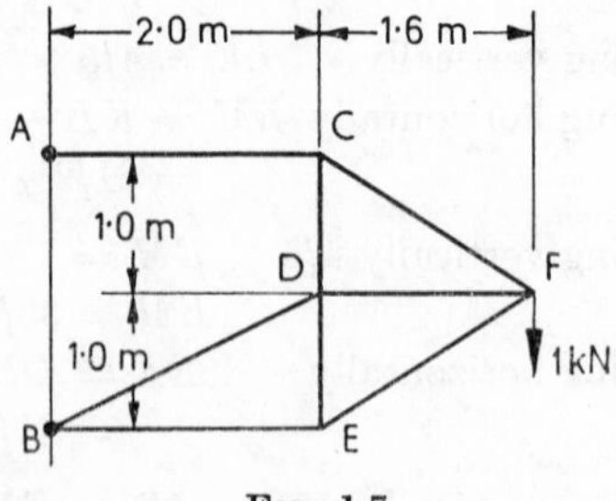

Fig. 1.7

Answer. $AC = +1{\cdot}8$, $CF = +2{\cdot}12$, $FE = +0{\cdot}236$, $BE = +0{\cdot}2$, $CD = -1{\cdot}125$, $DE = -0{\cdot}127$, $DB = -2{\cdot}23$, $DF = -2{\cdot}00$.

3. The frame shown in Fig. 1.8 is built up from 45° triangles and is supported at A and B and carries a load W as shown.

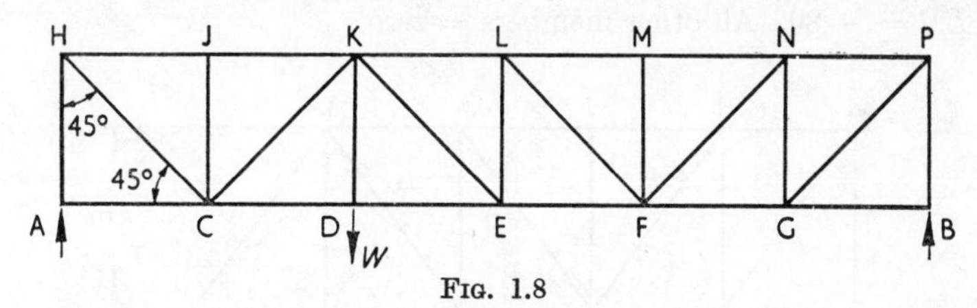

Fig. 1.8

Determine the forces in all the members by the method of inspection and check the values for KL and LE by the method of sections.

Answer. $AH = -2/3$, $HJ = -2/3$, $JK = -2/3$, $KL = -1$, $LM = -2/3$, $MN = -2/3$, $NP = -1/3$, $PB = -1/3$, $CD = +4/3$, $DE = +4/3$, $EF = +1$, $FG = +1/3$, $CH = +2\sqrt{2}/3$, $CK = -2\sqrt{2}/3$, $KD = +1$, $KE = -\sqrt{2}/3$, $LE = +1/3$, $LF = -\sqrt{2}/3$, $FN = +\sqrt{2}/3$, $NG = -1/3$, $GP = +\sqrt{2}/3$. All other members = zero. All forces expressed in terms of W.

4. Determine by inspection the forces in the frame shown in Fig. 1.9 if the tension in HD is twice the compression in JC.

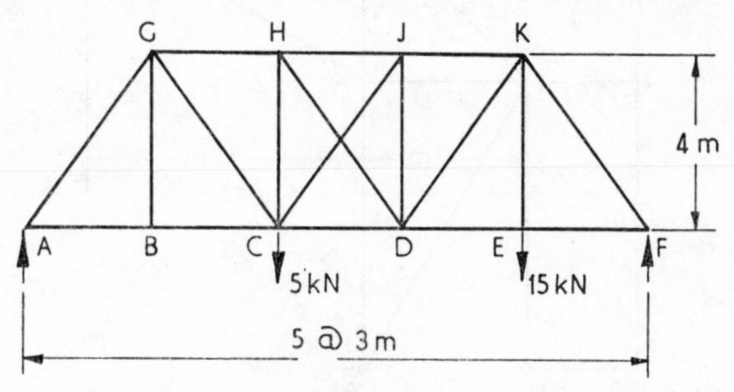

Fig. 1.9

Answer. $AB = +9/2$, $BC = +9/2$, $CD = +37/4$, $DE = +21/2$, $EF = +21/2$, $AG = -15/2$, $GH = -9$, $HJ = -19/2$, $JK = -39/4$, $KF = -35/2$, $GB = 0$, $GC = +15/2$, $HC = -2/3$, $HD = +5/6$, $CJ = -5/12$, $JD = +1/3$, $KD = -5/4$, $KE = +15$.

5. The frame shown in Fig. 1.10 is supported at A and B and carries loads of 20 kN and 60 kN at C and F respectively. Determine the forces in the members of the frame.

Answer. $DE = -30$, $EB = -45$, $BF = -45$, $FL = +75$, $KL = +30$, $JK = +15$, $HJ = +15$, $HC = +20$, $HD = -25$, $DK = +25$, $KE = -20$, $LE = +25$, $LB = -80$. All other members = zero.

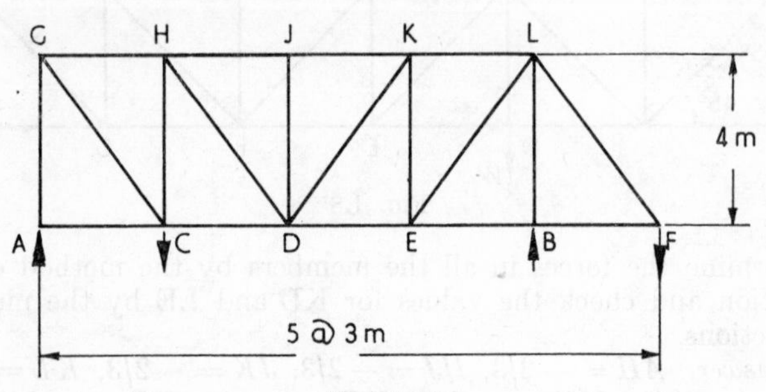

FIG. 1.10

6. The frame shown in Fig. 1.11 is pinned to the supports A and B. Determine the forces when it carries a load W acting vertically at F.

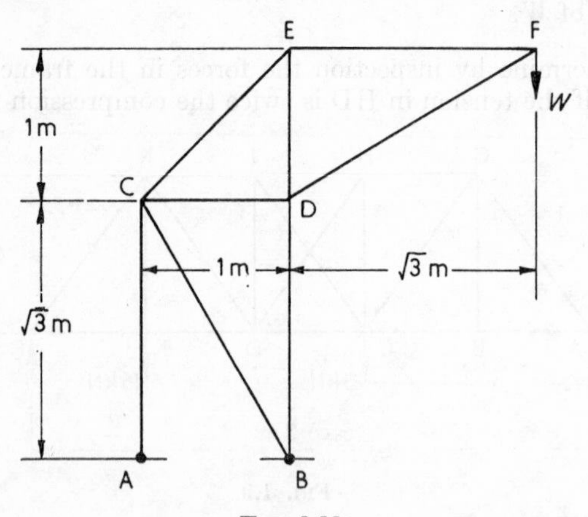

FIG. 1.11

Answer. $AC = +\sqrt{3}W$, $CE = +\sqrt{6}W$, $EF = +\sqrt{3}W$, $BD = -(\sqrt{3}+1)W$, $CD = -\sqrt{3}W$, $DE = -\sqrt{3}W$, $DF = -2W$, $CB = 0$.

7. Determine the forces in the frame shown in Fig. 1.11 when the load carried is 10 kN acting horizontally at F.

Answer. $AC = +\ 27{\cdot}32$, $CE = +\ 14{\cdot}14$, $EF = +\ 10$, $BD = -\ 10$, $DE = -\ 10$, $CB = -\ 20$. (All other members zero.)

8. Determine the forces in the members of the frame shown in Fig. 1.12 which is pinned to the supports A and D.

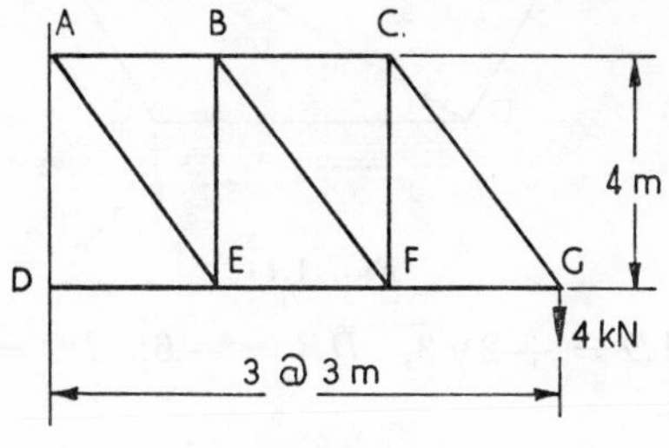

FIG. 1.12

Answer. $AB = +\ 6$, $BC = +\ 3$, $CG = +\ 5$, $DE = -\ 9$, $EF = -\ 6$, $FG = -\ 3$, $AE = +\ 5$, $BE = -\ 4$, $BF = +\ 5$, $CF = -\ 4$.

9. Determine the forces in the frame shown in Fig. 1.13 which is pinned to the supports D and C.

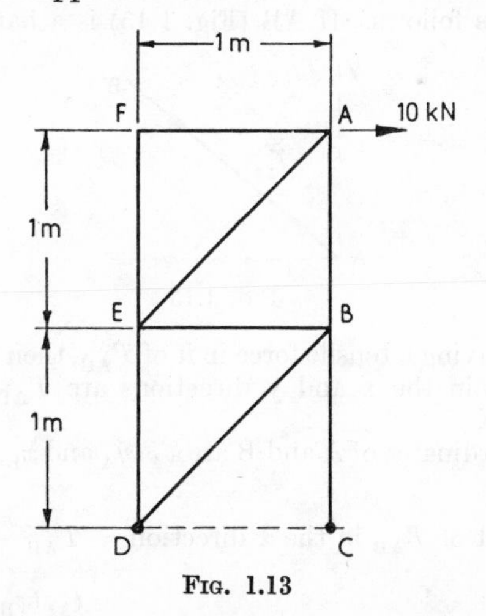

FIG. 1.13

Answer. $DE = +\ 10$, $EA = +\ 14{\cdot}14$, $EB = -\ 10$, $DB = +\ 14{\cdot}14$, $AB = -\ 10$, $BC = -\ 20$, $AF = FE = 0$.

10. The frame shown in Fig. 1.14 is pinned at A and B. Determine the forces in the members if DB bisects the angle ABC.

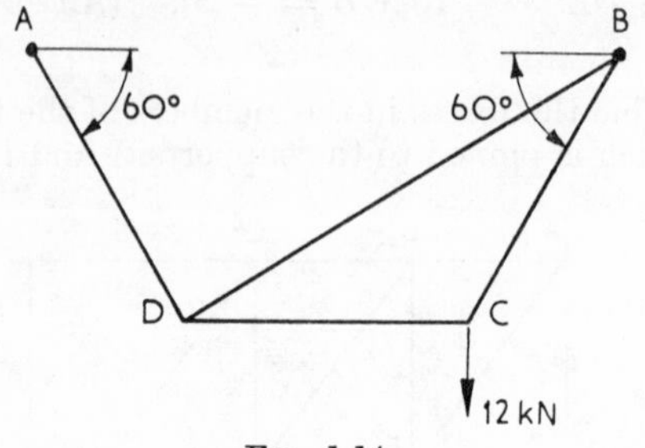

FIG. 1.14

Answer. $AD = +\,2\sqrt{3}$, $DB = -\,6$, $DC = +\,4\sqrt{3}$, $CB = +\,8\sqrt{3}$.

The Method of Tension Coefficients

This method of analysis can be applied to both plane and space frames. In the former case it does not generally give the answer as quickly as a stress diagram or the method of inspection, but in the case of space frames it is the most useful method at the student's disposal.

The fundamental principles of the method in the case of plane frames are as follows. If AB (Fig. 1.15) is a bar of length L_{AB}

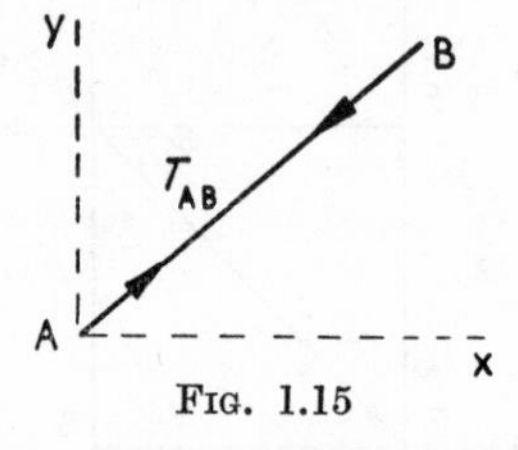

FIG. 1.15

in a frame, having a tensile force in it of T_{AB}, then the components of this force in the x and y directions are T_{AB} cos BAX and T_{AB} sin BAX.

If the co-ordinates of A and B are x_A, y_A and x_B, y_B respectively then

$$\text{Component of } T_{AB} \text{ in the x direction} = T_{AB}\frac{x_B - x_A}{L_{AB}}$$

$$= t_{AB}(x_B - x_A)$$

where $t_{AB} = \dfrac{T_{AB}}{L_{AB}}$ and is known as the tension coefficient of the bar AB.

Similarly the component in the y direction $= t_{AB}(y_B - y_A)$.

If at the joint A in the frame there are a number of bars AB, AC, . . . AN and external loads X_A, Y_A acting in the x and y directions, then, since the joint is in equilibrium the sum of the components of the external and internal forces must be zero in each of these directions.

Expressing these relationships symbolically gives the equations

$$t_{AB}(x_B - x_A) + t_{AC}(x_C - x_A) \ldots + t_{AN}(x_N - x_A) + X_A = 0 \quad (1)$$

$$t_{AB}(y_B - y_A) + t_{AC}(y_C - y_A) \ldots + t_{AN}(y_N - y_A) + Y_A = 0 \quad (2)$$

A similar pair of equations can be formed for each joint in the frame giving in all $2j$ equations in the case of a frame having j joints. These equations will contain the tension coefficients as unknowns and if the frame has n members then there are n unknown tension coefficients. But for a plane frame $n = 2j - 3$, hence there are three superfluous equations. These can be used either to determine the reactions or to check the values of the tension coefficients obtained from the previous equations.

In the case of a space frame each joint has three co-ordinates and the forces have components in three directions, x, y and z. Thus if there are j joints in a space frame the consideration of the equilibrium in the three directions produces $3j$ equations containing "n" unknown tension coefficients. But $n = 3j - 6$, hence there are six superfluous equations which can be used either to determine the reactions or to check the values of the tension coefficients.

Having found the tension coefficients t_{AB} the force in the bar is the product $t_{AB}L_{AB}$.

This is known as the method of "tension coefficients" and the equations of equilibrium are built up by assuming that the bars are in tension. A bar that is in compression has a negative tension coefficient.

The procedure in using the method is as follows—

1. Take positive directions for x, y and z.
2. Assume that all members are in tension.
3. Write down equations for each joint in the frame. The terms are positive or negative according as they tend to move the joints in the positive or negative directions of x, y or z. The student should note particularly that the whole build-up of equations contains terms such as $t_{AB}(x_B - x_A)$ and $t_{AB}(x_A - x_B)$. Thus a simple check on the accuracy of the build-up is to note that a positive coefficient to one of the unknowns must be accompanied by an equal negative one.
4. Solve equations for t_{AB}, etc.
5. Check values for t_{AB} from equations of static equilibrium.
6. Calculate $T_{AB} = L_{AB}t_{AB}$.

Example 1.3. Use the method of tension coefficients to determine the forces in the members of the frame shown in Fig. 1.16 which is pinned at A and rests on a roller bearing at B.

Solution. The three superfluous tension coefficients' equations can be used to determine the unknown reactions H_A, V_A and V_B.

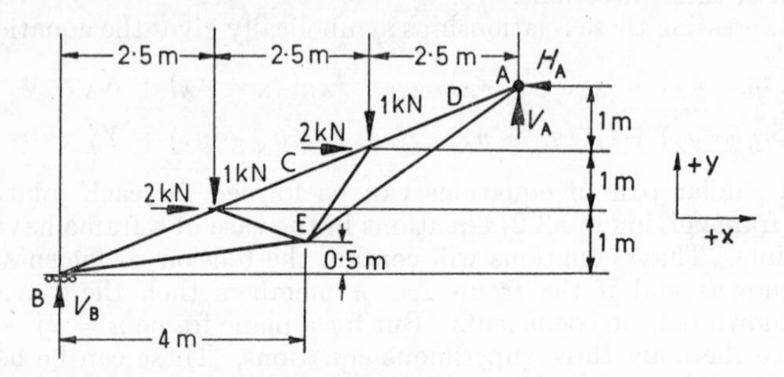

Fig. 1.16

These, however, will be determined from static equilibrium giving the extra equations as a check.

In writing down the equations of static equilibrium the author recommends that the student does not write t_{AB} but simply AB. It is the suffix AB which is important and if the expression t_{AB} is used the important suffixes are lost in a maze of "t"s.

The reactions are first obtained. The pin has horizontal and vertical components.

Resolving horizontally gives $H_A = 4$

Moments about A give

$$7{\cdot}5V_B = 1 \times 5 + 1 \times 2{\cdot}5 + 2 \times 2 + 2 \times 1$$

$$= 13{\cdot}5$$

$$\therefore \qquad V_B = 1{\cdot}8$$

and $\qquad V_A = 0{\cdot}2$

The algebra for the solution of the equations is not given but the procedure is as follows. Joint A which gives two simultaneous equations in AD and AE is first solved; the value obtained for AD is then substituted in the equations for joint D giving two simultaneous equations in CD and DE. Joint B gives two simultaneous equations in BC and BE; the value obtained for BC is substituted in the equations for joint C producing two simultaneous equations in CD and CE. Joint E is used as a check.

Joint	Direction	Equations	Tension Coefficient
A	x	$-2{\cdot}5AD - 3{\cdot}5AE - 4 = 0$	$AD = -3{\cdot}90$
	y	$-1{\cdot}0AD - 2{\cdot}5AE + 0{\cdot}2 = 0$	$AE = 1{\cdot}64$
D	x	$+2{\cdot}5AD - 2{\cdot}5CD - 1{\cdot}0DE + 2 = 0$	$DE = -1{\cdot}64$
	y	$+1{\cdot}0AD - 1{\cdot}0DC - 1{\cdot}5DE - 1 = 0$	$CD = -2{\cdot}44$
C	x	$+2{\cdot}5CD - 2{\cdot}5BC + 1{\cdot}5CE + 2 = 0$	$CD = -2{\cdot}45$
	y	$+1{\cdot}0CD - 1{\cdot}0BC - 0{\cdot}5CE - 1 = 0$	$CE = -1{\cdot}63$
B	x	$2{\cdot}5BC + 4{\cdot}0BE = 0$	$BC = -2{\cdot}63$
	y	$1{\cdot}0BC + 0{\cdot}5BE + 1{\cdot}8 = 0$	$BE = 1{\cdot}64$
E	x	$3{\cdot}5AE + 1{\cdot}0DE - 1{\cdot}5CE - 4{\cdot}0BE = 0$	
	y	$2{\cdot}5AE + 1{\cdot}5DE + 0{\cdot}5CE - 0{\cdot}5BE = 0$	

The tension coefficients thus found are multiplied by the lengths of the members to give the load in each of them as in Table 1.3.

TABLE 1.3

Member	Length	Tension Coefficient	Force
AD	$\sqrt{7{\cdot}25} = 2{\cdot}7$	$-3{\cdot}90$	10·51 kN compression
DC	2·7	$-2{\cdot}44$	6·61 kN compression
CB	2·7	$-2{\cdot}63$	7·10 kN compression
AE	$\sqrt{18{\cdot}5} = 4{\cdot}3$	$+1{\cdot}64$	7·04 kN tension
BE	$\sqrt{16{\cdot}25} = 4{\cdot}05$	$+1{\cdot}64$	6·60 kN tension
CE	$\sqrt{2{\cdot}5} = 1{\cdot}58$	$-1{\cdot}63$	2·57 kN compression
DE	$\sqrt{3{\cdot}25} = 1{\cdot}8$	$-1{\cdot}64$	2·95 kN compression

Example 1.4 The space frame shown in plan in Fig. 1.17 has the pinned supports A, B and C at the same level. DE is horizontal and at a height of 2 m above the plane of the supports. Calculate the forces in the members when the frame carries loads of 8 kN and 4 kN acting in a horizontal plane at E and D respectively.

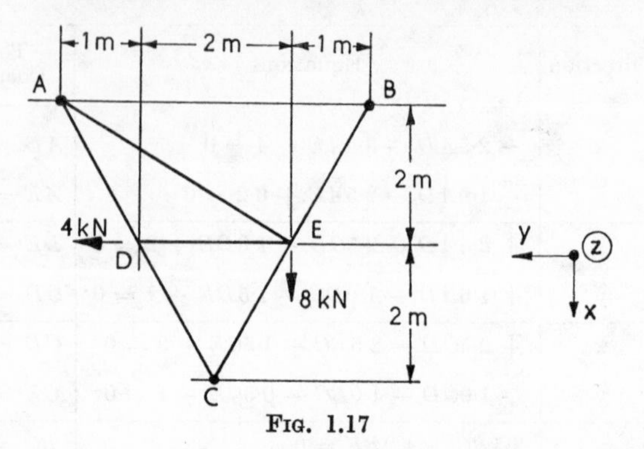

Fig. 1.17

Solution. In the case of a space frame having pinned supports the number of bars for a statically determinate frame is $3j_f$ where j_f = number of free joints. The number of equilibrium equations will be $3j_f$, hence sufficient equations are obtained to solve for the unknown tension coefficients.

The tension coefficient equations are built up as shown in Table 1.4 (*a*) using the positive directions indicated in the figure. The forces in the members are given in Table 1.4 (*b*).

TABLE 1.4 (*a*)
Equilibrium Equations

Joint	Direction	Equations	Tension Coefficients
D	x	$-2AD + 2CD = 0$	$AD = CD = 0$
	y	$1AD - 2DE - 1DC + 4 = 0$	$DE = 2{\cdot}0$
	z	$2AD + 2CD = 0$	
E	x	$2CE - 2AE - 2BE + 8 = 0$	$CE = -2{\cdot}0$
	y	$2DE + 3AE + 1CE - 1BE = 0$	$AE = 0$
	z	$2CE + 2BE + 2AE = 0$	$BE = 2{\cdot}0$

TABLE 1.4 (*b*)
Forces in Members

Member	Length (m)	Tension Coefficients	Load
AD	3·0	0	0
DC	3·0	0	0
DE	2·0	2·0	4 tons tension
AE	4·12	0	0
CE	3·0	− 2·0	6 tons compression
BE	3·0	+ 2·0	6 tons tension

Examples on Tension Coefficients

Using the method of tension coefficients determine the forces in the various plane frames shown in Figs. 1.18–1.24.

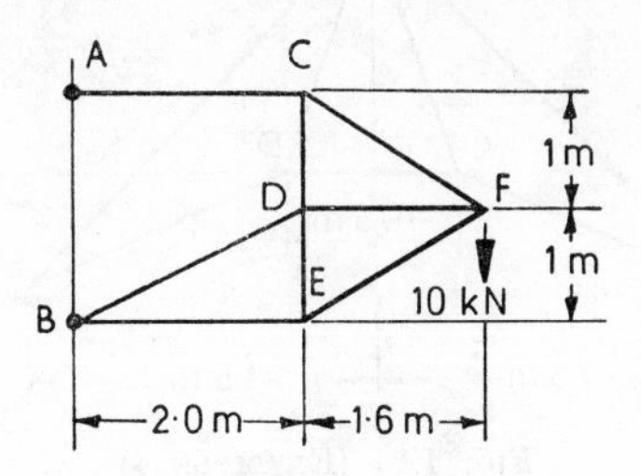

Fig. 1.18 (Example 1)

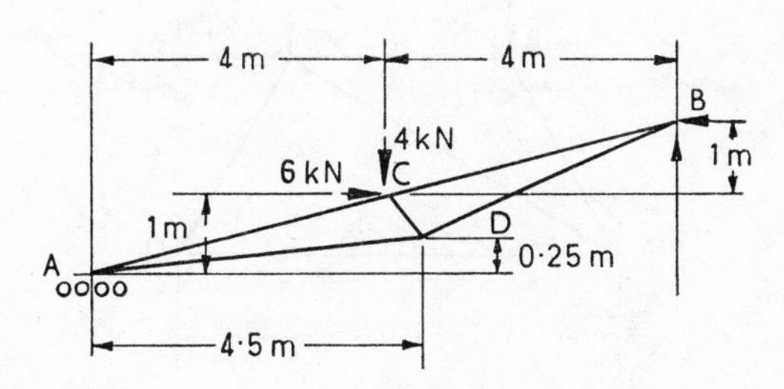

Fig. 1.19 (Example 2)

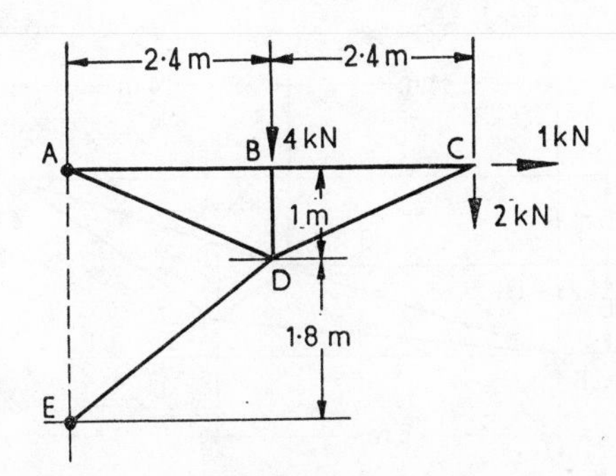

Fig. 1.20 (Example 3)

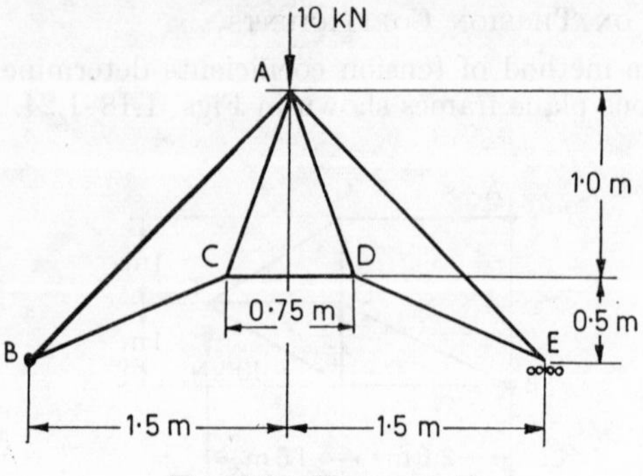

FIG. 1.21 (EXAMPLE 4)

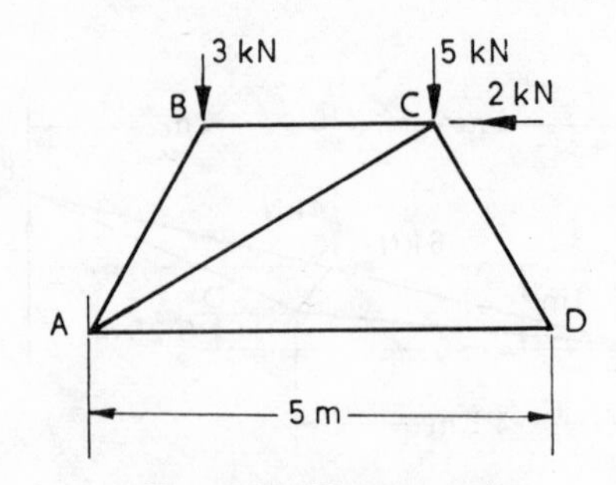

FIG. 1.22 (EXAMPLE 5)

Horizontal reaction shared equally between each support. Members AB BC and CD all equal in length

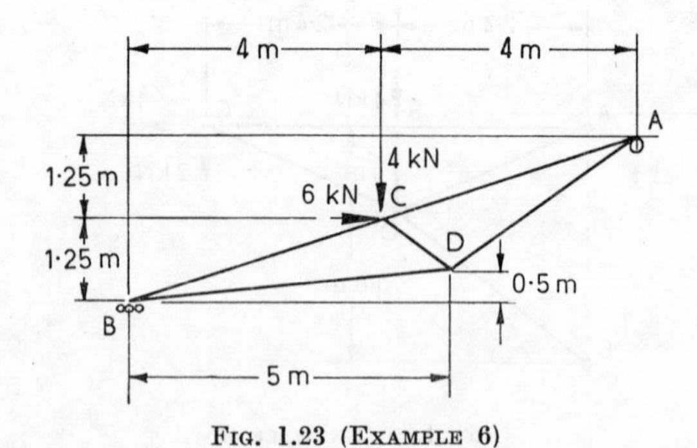

FIG. 1.23 (EXAMPLE 6)

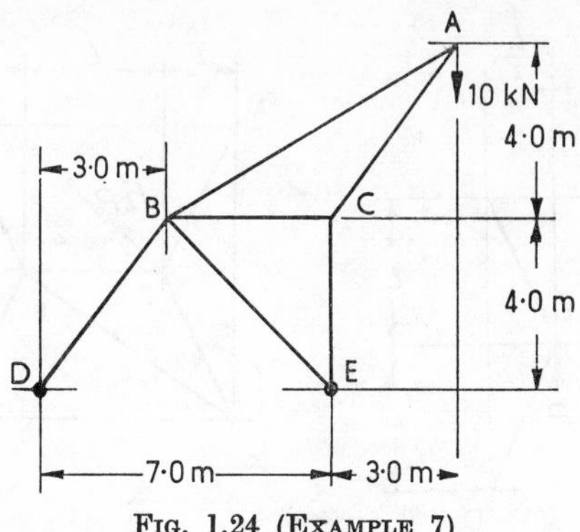

FIG. 1.24 (EXAMPLE 7)

Answers

1. $AC = +18$, $CF = +21$, $CD = -11{\cdot}25$, $DF = -20$, $EF = +2{\cdot}36$, $DE = -1{\cdot}25$, $BD = -22{\cdot}3$, $BE = +2{\cdot}0$.

2. $BC = -17{\cdot}5$, $BD = +12{\cdot}30$, $CD = -5{\cdot}58$, $AD = 14{\cdot}18$, $AC = -14{\cdot}55$.

3. $AB = +5{\cdot}8$, $BC = +5{\cdot}8$, $BD = -4{\cdot}0$, $AD = +2{\cdot}22$, $CD = -5{\cdot}2$, $DE = -8{\cdot}56$.

4. $AB = -12{\cdot}7 = AE$, $BC = DE = +9{\cdot}87$, $CD = +7{\cdot}5$, $AC = AD = +4{\cdot}28$.

5. $AB = -3{\cdot}45$, $BC = -1{\cdot}72$, $CD = -4{\cdot}20$, $AC = -2{\cdot}74$, $AD = +3{\cdot}10$.

6. $AC = -15{\cdot}00$, $AD = +9{\cdot}94$, $BD = +13{\cdot}80$, $CD = -6{\cdot}98$, $BC = -14{\cdot}4$.

7. $BD = 5{\cdot}35$, $BE = 4{\cdot}55$, $BC = -13{\cdot}1$, $AB = 15{\cdot}0$, $AC = -21{\cdot}8$, $CE = -17{\cdot}4$.

The space frames shown in Figs. 1.25–1.26 are pinned to the supports A, B, C and D. The member EF is horizontal and distant L above the supports. The loads indicated act in a horizontal plane. Determine the forces by the method of tension coefficients.

Answers

8. $EF = -H$, $FC = +\dfrac{2}{\sqrt{3}}H$, $FD = \frac{3}{4}P - H$, $FB = -\frac{3}{4}P$.

9. $AE = -\frac{3}{4}P$, $CE = +\frac{3}{4}P$, $EF = -P$, $FC = 1{\cdot}03P$, $FD = -\frac{3}{4}P$, $FB = 0$.

10. A pin-jointed space frame is illustrated pictorially in Fig. 1.27. A, B, C and D are hinged points of attachment to a rigid vertical wall. The plane ADGE is horizontal and the plane EFG is vertical. A vertical load of 5·0 kN is applied at joint F.

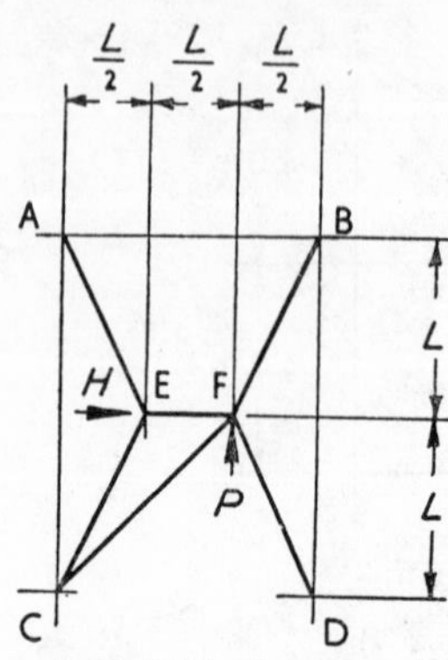

FIG. 1.25 (EXAMPLE 8)

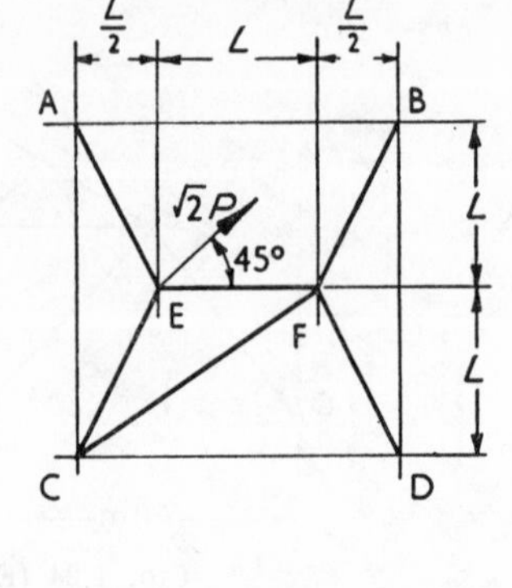

FIG. 1.26 (EXAMPLE 9)

Determine the magnitude and nature of the forces set up in the members. (*Glasgow*)

Answer. $AE = -15{\cdot}0$, $BE = +11{\cdot}2$, $BF = +5{\cdot}34$, $CF = -5{\cdot}34$, $DE = +6{\cdot}25$, $EF = -6{\cdot}25$.

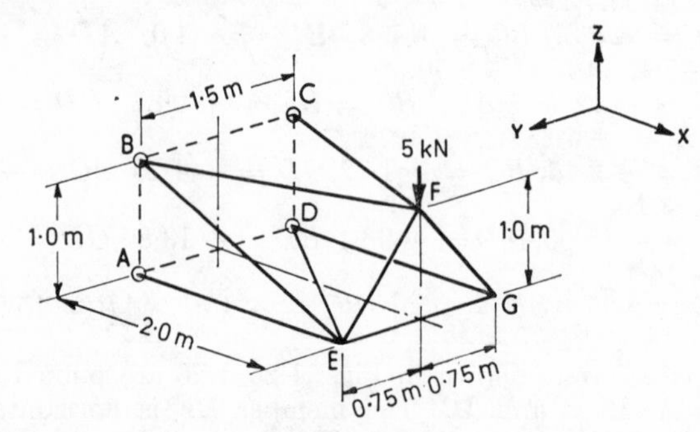

Fig. 1.27

11. A hexagon ABCDEF formed from six bars is supported in a horizontal plane by struts from six points, a, b. c, d, e, f, which are fixed to pinned supports. A vertical load of 10 kN is carried at each of the points A, B, C, D, E and F. Calculate the load in each member of the frame. (See Fig. 1.28.) (*Glasgow*)

Answer. AB (and other horizontal members) $= 5/\sqrt{3}$, Aa (and other inclined members) $= -10/\sqrt{3}$.

12. A derrick crane is attached to a prismatic braced structure at D, E and F as shown in Fig. 1.29. Find the forces in the members of the space frame so formed when the crane supports a vertical load of 20 kN at G. (*Glasgow*)

Answer: $AD = 38{\cdot}5$, $CF = BE = -29{\cdot}3$, $FE = 51{\cdot}3$, $DF = DE = -67{\cdot}5$, $GD = 123$, $GE = GF = -67{\cdot}7$. Others zero.

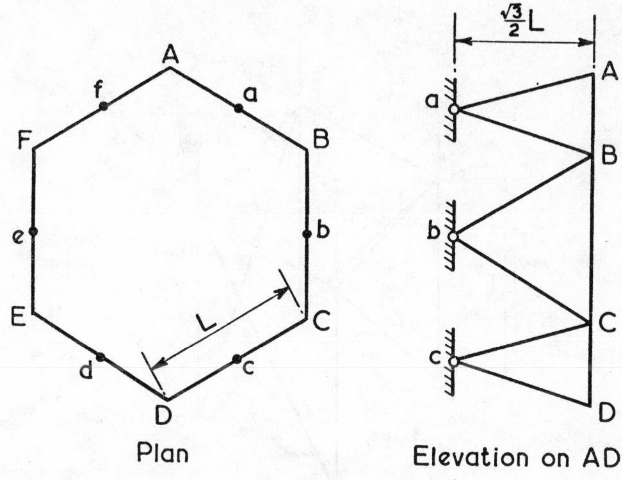

FIG. 1.28

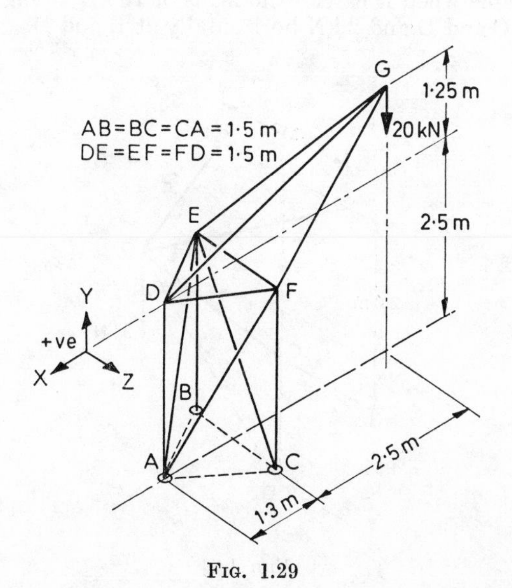

FIG. 1.29

13. The square frame ABCD shown in Fig. 1.30 is connected to pinned supports a, b, c and d by the members Aa, Bb, Cc and Dd, and aB, bD, dC and cA. Calculate the forces in the members

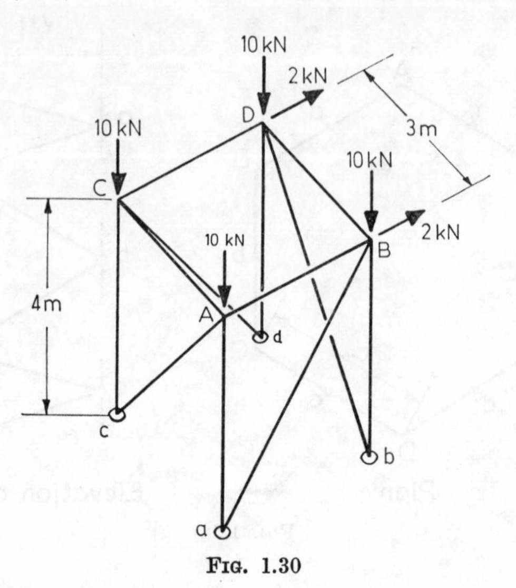

FIG. 1.30

of the frame when it carries the loads of 10 kN acting vertically at A, B, C and D and 2 kN horizontally at B and D. (*Glasgow*)

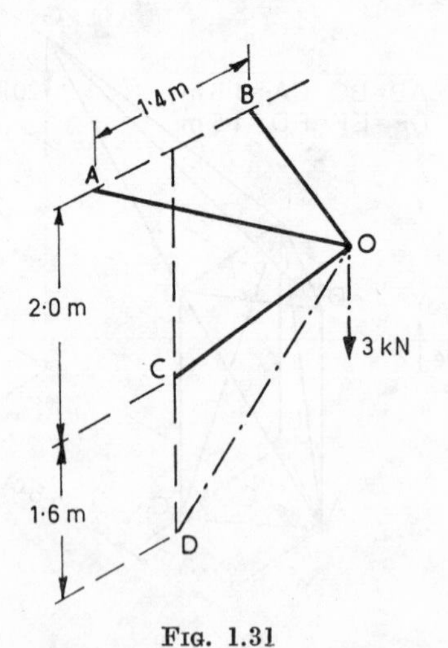

FIG. 1.31

Answer.

$$Aa = Dd = -10\text{ kN}; \quad Bb = -\frac{38\text{ kN}}{3}; \quad Ce = -\frac{22\text{ kN}}{3};$$

$$CD = +2\text{ kN}; \; Cd = -\frac{10\text{ kN}}{3}; \; Ba = +\frac{10\text{ kN}}{3}.$$

14. Fig. 1.31 represents a simple space frame which has three members OA, OB and OC. A rope carrying a load of 3 kN passes over a pulley at O to winding gear at D. The points A, B, C and D are in the same vertical plane, A and B are at the same level and C and D are in a vertical line which bisects AB. The lengths of the members are: AO = OB = 2·4 m, OC = 2·6 m.

Determine the forces in the three members OA, OB and OC, stating for each whether it is in tension or compression.

(*London*)

Answer. $AO = BO =$ 2·7 kN tension; $CO =$ 8·1 kN compn.

Chapter 2

Influence Lines and Rolling Loads

THE INITIAL DIFFICULTY with influence lines is in the understanding of what an influence line really is. Bending moment and similar diagrams are familiar and generally easily understood. These diagrams give the values of a function at all points in the span when the load occupies a certain fixed position. For example,

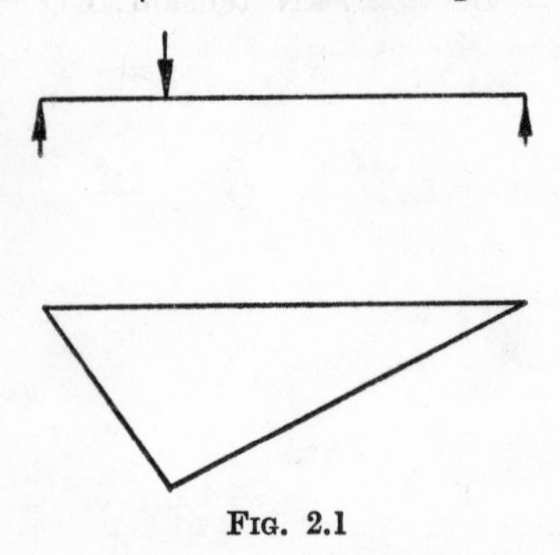

FIG. 2.1

the bending moment diagram shown in Fig. 2.1 is a diagram showing how the bending moment varies at all points on the span when the loading occupies the position shown in Fig. 2.1.

In many problems, however, the engineer has to deal with loads which move and can therefore occupy any position on the span. Influence lines are used for this.

Influence Line for Bending Moment

To draw an influence line a *given point* on the span is selected and a diagram drawn to show how the value of a function, e.g. bending moment, varies at this point as a single unit load occupies different positions on the span.

This diagram is called the influence line for bending moment at the given point. The difference between an influence line and a bending moment diagram is that the latter shows how the bending moment at all points in the span varies when the load remains fixed in a given position. The influence line for bending moment at a given point shows how the bending moment *at that point* varies as the load occupies different positions on the span. The diagram shown in Fig. 2.2 is the influence line for bending moment at X. The ordinate m_y at a point Y gives the bending moment at X when unit load is at Y.

If loads W_1, W_2 and W_3 occupy positions on the span such that the ordinates to the influence line diagram are m_1, m_2 and m_3

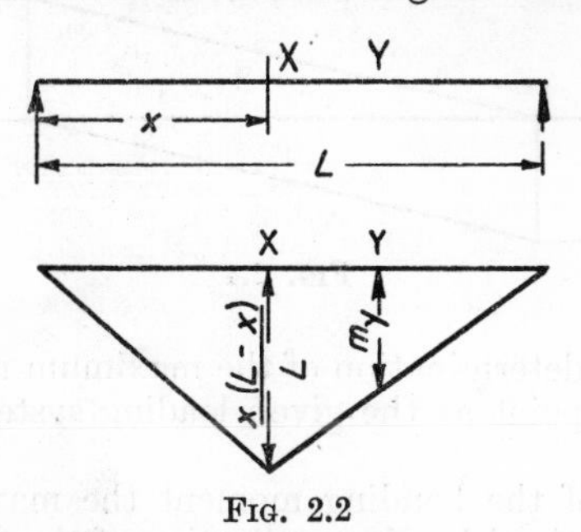

FIG. 2.2

then the bending moment at X due to the three loads when they are in that particular position is given by $W_1m_1 + W_2m_2 + W_3m_3$.

The effect of a uniformly distributed load is obtained by taking the area under the influence line covered by the load and multiplying by the intensity of loading.

Maximum Moment and Shear Force Diagrams

The engineer is in general not troubled about the variation of bending moment at a given point but with the maximum value

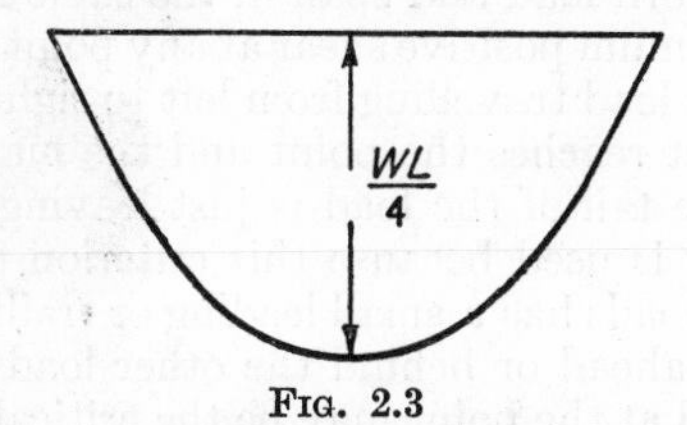

FIG. 2.3

of bending moment which can occur at that point. Thus influence lines lead on to maximum moment, shear force, etc., diagrams. A maximum moment diagram for a single load, W, is a diagram showing the maximum moment at all points on the span as a load W moves across it. Such a diagram is shown in Fig. 2.3. It is a parabola, the envelope of the apices of the triangles forming the influence lines for bending moment at the different points.

The diagram for maximum shearing force due to a single load, W, consists of two straight lines as shown in Fig. 2.4. At any given point the maximum shear can be either positive or negative in sign according as the load is immediately to the left or right of the section under consideration.

The maximum moment diagram for a single rolling load leads on naturally to the cases of a number of rolling loads and a uniformly distributed load. The problem with such loading systems

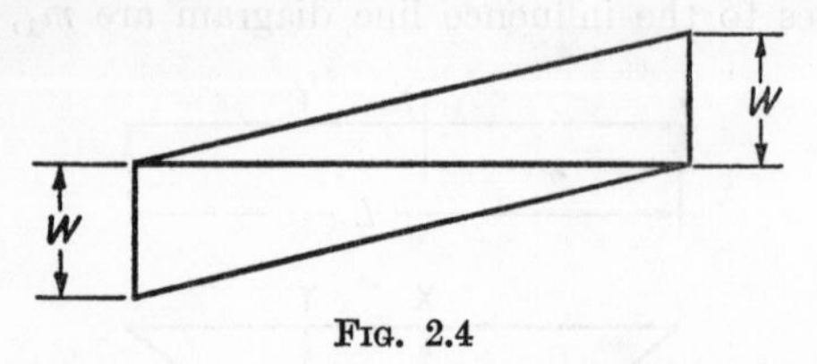

FIG. 2.4

is generally the determination of the maximum moment or shear at a particular point as the given loading system traverses the span.

In the case of the bending moment the maximum value at any point is obtained by the application of the following rules—

1. The maximum bending moment at any point occurs when a wheel is at that point.
2. The maximum bending moment under any wheel occurs when that wheel and the centre of gravity of the system are equidistant from the centre of the span.
3. In the case of a uniformly distributed load shorter in length than the span of the beam the maximum bending moment at any point occurs when the load is so placed that the point divides both load and span in the same ratio.
4. The maximum positive shear at any point occurs normally, in the case of a load travelling from left to right, when the head of the load just reaches the point and the maximum negative shear when the tail of the load is just leaving the point. The word normally is used because this criterion may not be true if the train of loads has a small leading or trailing load which is some distance ahead or behind the other loads. In such cases the second load at the point may be the critical one for positive shear and the penultimate one for negative shear.

Examples on rolling loads and influence lines can be divided into two main classes, (i) those dealing with beams and (ii) those dealing with braced structures. It is intended to follow this division in this chapter.

Rolling Loads on Beams

Example 2.1. A simply supported beam of 10 m span is subjected to a uniform dead load covering the span of 60 kN/m and a uniform live load (longer than the span) of 100 kN/m. Determine (*a*) the maximum and minimum shear force at the left-hand quarter point; (*b*) the maximum bending moment at the same point; (*c*) the range over which the shearing force may have positive or negative values.

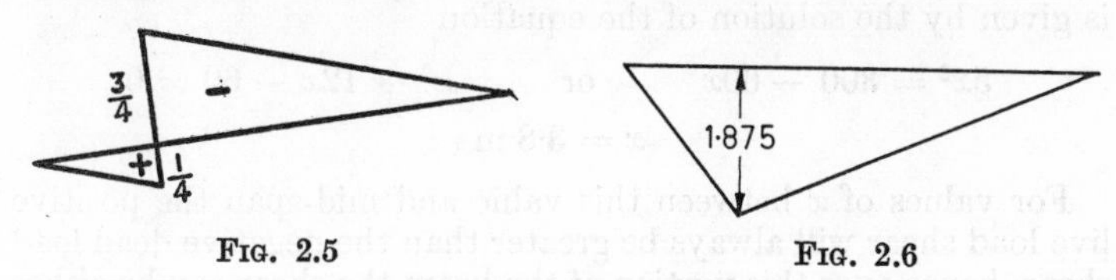

FIG. 2.5 FIG. 2.6

Solution. In dealing with examples where both dead and live loads are mentioned the student must remember that the dead load is always present. The live load may or may not be present and when it is it can occupy any position on the span. In any given example it can be placed so as to give the maximum effect.

This example is concerned with the point distant 2·5 m from the left-hand support.

The influence line for shear at this point is as given in Fig. 2.5.

For distributed loads the area under the influence line has to be taken.

The positive area $= \frac{1}{2} \times 2{\cdot}5 \times \frac{1}{4} = 0{\cdot}313$ m units

The negative area $= \frac{1}{2} \times 7{\cdot}5 \times \frac{3}{4} = 2{\cdot}813$ m units

Shear due to dead load $= 60 \times (0{\cdot}313 - 2{\cdot}813) = -150$ kN

Positive shear due to live load $= 100 \times 0{\cdot}313 = 31{\cdot}3$ kN

Negative shear due to live load $= 2{\cdot}813 \times 100 = 281{\cdot}3$ kN

Due to combined loads the maximum and minimum values of negative shear are 431·3 and 118·7 kN respectively.

The influence line for bending moment at the quarter point is as shown in Fig. 2.6. The maximum bending moment at this point occurs when the live load covers the whole span.

$$\text{The area under the influence line} = \tfrac{1}{2} \times 1{\cdot}875 \times 10 = 9{\cdot}375 \text{ m units}$$

$$\therefore \quad \text{Maximum bending moment} = 160 \times 9{\cdot}375 = 150 \text{ kN/m}$$

To solve the third part of this problem the student must determine the point in the left-hand portion of the span where the positive shear due to live load is equal to the negative shear due to dead load.

At a point distance x from the left-hand support the positive shear due to a live load of 100 kN/m run is equal to

$$100 \times \tfrac{1}{2} \times x \times \frac{x}{10} = 5x^2$$

The negative shear due to a dead load of 60 kN/m

$$= 300 - 60x$$

Hence the distance x from the left-hand support to the point at which positive live load shear equals negative dead load shear is given by the solution of the equation

$$5x^2 = 300 - 60x \qquad \text{or} \qquad x^2 + 12x - 60 = 0$$

$$x = 3{\cdot}8 \text{ m}$$

For values of x between this value and mid-span the positive live load shear will always be greater than the negative dead load shear, hence over this portion of the beam the shear can be either positive or negative depending on the position of the load.

In a similar way it can be shown that at the sections between a point distant 3·8 m from the right-hand support and the centre of the span the negative shear due to live load will always be greater than the positive shear due to dead load.

Therefore reversals of stress can take place over the central 2·4 m of the beam.

Example 2.2. Determine the maximum bending moment which can occur when a train of loads of 50, 100, 150 and 100 kN separated by distances of 2·5 m, 2 m and 2 m respectively crosses a span of 80 m.

Also find the maximum positive and negative shearing forces at a point 10 m from the left-hand support.

Solution. The train of loads is shown in Fig. 2.7.

The position of the centre of gravity of the load system must first be obtained. Let it be at a distance $\bar{x}$ from the trailing

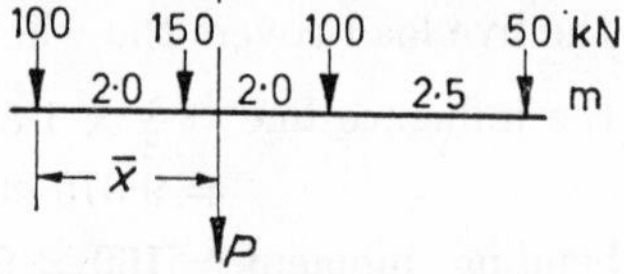

Fig. 2.7

100 kN load. Taking moments about the line of action of this load gives

$$400\bar{x} = 150 \times 2 + 100 \times 4 + 50 \times 6{\cdot}5 = 1025$$

Hence $\quad \bar{x} = 2{\cdot}563$ m or 0·563 m from the 150 kN load.

The maximum bending moment at any point occurs when a wheel is at that point and the maximum bending moment under any wheel when that wheel and the c.g. of the system are equidistant from mid-span. Hence a possible position of the loads for maximum bending moment on the span is as shown in Fig. 2.8

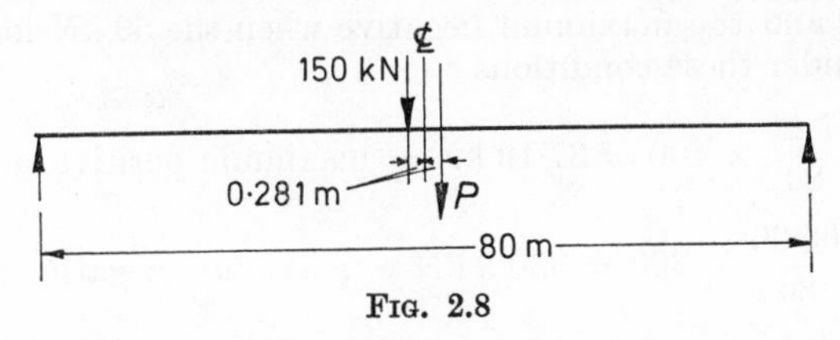

FIG. 2.8

when the 150 kN load is at a distance of $\frac{0{\cdot}563}{2}$, i.e. 0·281 m from the centre of the span. With the load in this position the reaction, R_A, at the left-hand support is given by

$$R_A = \frac{39{\cdot}72}{80} \times 400$$

The moment under the 150 kN load, M_{150}, is given by

$$M_{150} = R_A \times 39{\cdot}72 - 100 \times 2 = 7670 \text{ kN/m}$$

The alternative position for the maximum bending moment would be under the 100 kN load when it was distant $\frac{1{\cdot}437}{2}$, i.e. 0·718 m from the centre. It should be obvious that the moment in this case is less than M_{150}, but its value will be calculated to check this.

The reaction, R_B, at the right-hand support is given by

$$R_B = \frac{39{\cdot}28}{80} \times 400$$

The moment, M_{100}, under the 100 kN load is then given by

$$M_{100} = R_B \times 39{\cdot}28 - 50 \times 2{\cdot}5 = 7575 \text{ kN/m}$$

With the train of loads as shown in Fig. 2.7 the maximum positive shear at any point occurs when the 50 kN leading load is at the point and the maximum negative shear when the 100 kN trailing load is at the point, the maximum positive shear being R_B and the maximum negative shear being R_A. For a point distant 10 m from the left-hand support and with the previously mentioned conditions—

$$R_B = \frac{6{\cdot}06}{80} \times 400 = 30{\cdot}3 \text{ kN} = \text{maximum positive shear}$$

$$R_A = \frac{67{\cdot}44}{80} \times 400 = 337{\cdot}3 = \text{maximum negative shear}$$

If the load system is reversed and crosses from left to right with the 100 kN load leading and the 50 kN trailing then the maximum positive shear will occur when the 100 kN load is at the point and the maximum negative when the 50 kN load is at it, and under these conditions

$$R_B = \frac{7{\cdot}44}{80} \times 400 = 37{\cdot}19 \text{ kN} = \text{maximum positive shear}$$

$$R_A = \frac{66{\cdot}06}{80} \times 400 = 330{\cdot}3 \text{ kN} = \text{maximum negative shear}$$

Example 2.3. A three-pinned structure ABC comprises two beams AB and BC. The span AC is 40 m and the rise to the central hinge B is 12 m. Vertical columns spaced at 6·67 m centres transfer the force actions to the rib from simply supported beams. Draw the influence line diagram for the bending moment and normal thrust at section X, 10 m from the left hand support.

Determine the maximum bending moment and thrust at this section due to a live load system, comprising four loads of 20 kN each at 5-m centres crossing the span. *(Glasgow)*

Solution. The three-pinned structure is shown in Fig. 2.9.

If H_A and V_A are the horizontal and vertical reactions respectively at A, then the normal thrust, T, at X is given by

$$T = H_A \cos\theta + V_A \sin\theta - [W_1 \sin\theta]$$

where W_1 = load carried by column 1 and only applies when the load is to the left of 2. θ is the inclination to the horizontal of the member AB.

The moment, M, at X is given by

$$M = -(V_A \times 10) + (H_A \times 6) + [W_1 \times \tfrac{10}{3}]$$

The square bracket term applies only when the load is to the left of 2.

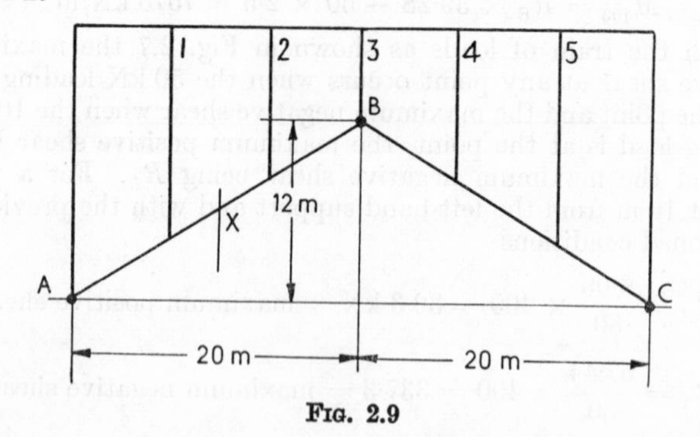

Fig. 2.9

The problem consists in the first instance in the determination of the values of H and V when unit load reaches the structure through columns 1, 2, 3, 4 and 5 respectively.

For unit load at 1, $V_A = \frac{5}{6}$

Moments about B give $H_A = \dfrac{\frac{5}{6} \times 20 - 1 \times 13\frac{1}{3}}{12} = 0{\cdot}278$

When unit load is at 5, H_A is also equal to 0·278 but $V_A = \frac{1}{6}$

For unit load at 2, $V_A = \frac{4}{6}$

Moments about B give $H_A = 0{\cdot}555$

When unit load is at 4, $H_A = 0{\cdot}555$

and $V_A = \frac{2}{6}$

With unit load at 3, $V_A = \frac{3}{6}$

and moments about B give $H_A = 0{\cdot}833$

To obtain the influence line for normal thrust the values of T are tabulated as shown in Table 2.3 (*a*).

TABLE 2.3 (*a*)

Load pt.	1	2	3	4	5
$H \cos \theta$	0·238	0·472	0·717	0·472	0·238
$V \sin \theta$	0·422	0·344	0·258	0·172	0·086
$W \sin \theta$	0·515				
T	0·145	0·816	0·975	0·644	0·324

The influence line for M is obtained from the values given in Table 2.3 (*b*).

TABLE 2.3 (*b*)

Load pt.	1	2	3	4	5
$-10V_A$	− 8·33	− 6·67	− 5·00	− 3·33	− 1·67
$+6H_A$	+ 1·67	+ 3·33	+ 5·00	+ 3·33	+ 1·67
$+\frac{10}{3} W_1$	+ 3·33				
M	− 3·33	− 3·33	0	0	0

The influence lines are as shown in Figs. 2.10 and 2.11.

The train of loads is as shown in Fig. 2.12. Placing wheel 2 at mid-span and determining the ordinates under the other

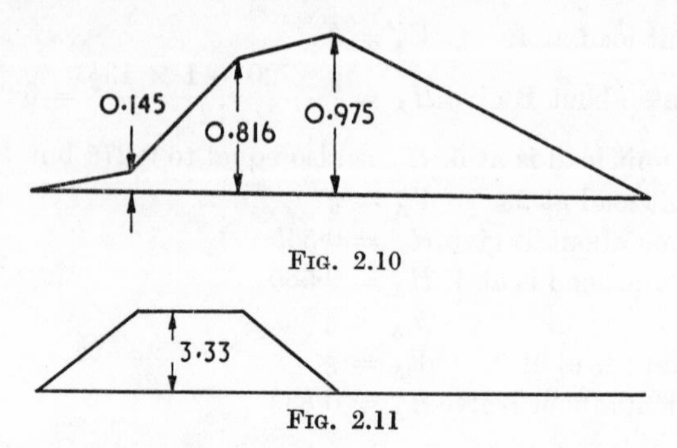

Fig. 2.10

Fig. 2.11

wheels from the influence line for T then gives the thrust when the load is in this position as

$$T = 20(0{\cdot}85 + 0{\cdot}975 + 0{\cdot}73 + 0{\cdot}485) = 60{\cdot}8 \text{ kN}$$

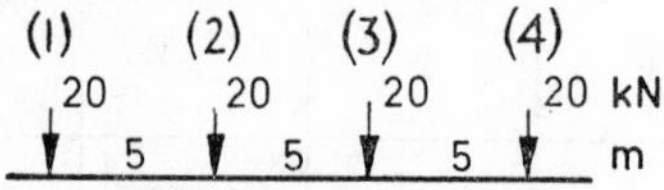

(1) (2) (3) (4)
20 20 20 20 kN
5 5 5 m

Fig. 2.12

When wheel 3 is at mid-span the sum of the ordinates is slightly less hence maximum thrust = 60·8 kN.

The maximum moment occurs when the wheels are placed symmetrically about X, giving

$$M_{\max} = 20 \times 2 \times (3{\cdot}33 + 1{\cdot}25) = 183{\cdot}2 \text{ kN m}$$

Examples on Rolling Loads on Beams

1. A uniformly distributed load of length 40 m and intensity 3 kN/m run crosses a bridge of 100 m span. Determine the maximum bending moment on the bridge and the maximum bending moment and shearing force at a point distant 25 m from the left-hand support.

Answer. 2,400 kN m; 1,800 kN m; 66 kN.

2. Two loads of 4 kN and 12 kN respectively separated by a distance of 6 m cross a beam of span 50 m from left to right, the smaller load leading. Calculate

(i) The position and magnitude of the maximum bending moment on the girder.

(ii) The maximum shearing force at a point distant 15 m from the left-hand support.

Answer. 188·2 kN m at 0·75 m to left of centre; 10·7 kN.

3. A bridge spanning 30 m is composed of two 15 m joists supporting a 10 m joist as shown in Fig. 2.13. The live load to

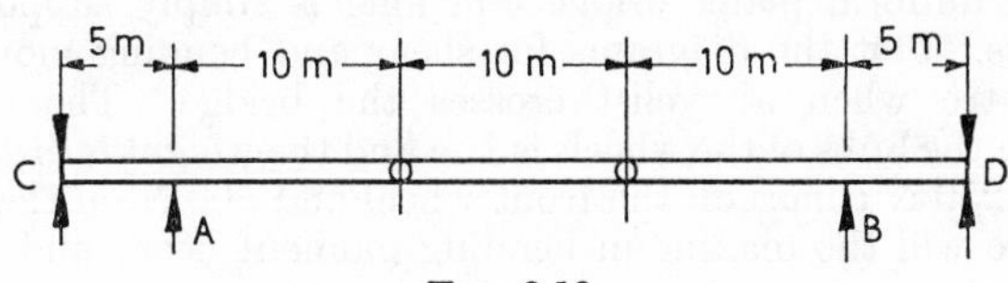

Fig. 2.13

be carried is a knife-edge load of 150 kN and a uniformly distributed load of 10 kN/m run. Calculate the tail weight which should be provided at the ends C and D and the maximum bending moment which can occur on the bridge. (*Glasgow*)

Answer. 500 kN; 2500 kN m.

4. A simply supported beam, span 15 m is traversed by the load system shown in Fig. 2.14. Compare the equivalent uniform load intensities for the maximum possible bending moment in

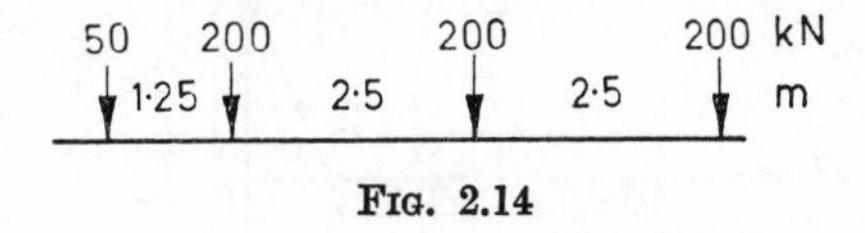

Fig. 2.14

the span, the maximum bending moment at the left-hand quarter point and the maximum shear force at this point. The load system cannot be reversed. (*Glasgow*)

Answer. 65·8 kN m; 72·8 kN m; 142 kN m.

5. A road bridge consists of two cantilevers AB, CD each 15 m long and a suspended span BC whose length is 20 m. A concentrated load of 10 kN moves across the span from A to D. By means of influence lines show the effect of his load on—

(*a*) Bending moment at A.
(*b*) Shear at B.
(*c*) Bending moment at mid-point of BC.
(*d*) Shear at D.

Answer. Maximum values (*a*) 150 kN m; (*b*) 10 kN; (*c*) 50 kN m; (*d*) 10 kN.

6. A beam of 30-m span is crossed by two loads of 50 kN and 100 kN separated by a distance of 6 m. Draw to scale a diagram showing the maximum bending moment that occurs at any point on the beam. At what point does the greatest bending moment occur and what is its value?

Answer. 980 kN m at 1 m from centre.

7. A uniform plank bridge 5 m long is simply supported at the ends. Plot the diagrams for shear and bending moment at the centre when a cyclist crosses the bridge. The distance between the hubs of the wheels is 1 m and the weight is distributed so that 250 N comes on the front wheel and 500 N on the rear.

Where will the maximum bending moment occur and what is its value?

Answer. 1·35 kN m at 0·167 m from left of centre.

8. A bridge of 100 m span weighs 10 kN/m run and is traversed by a uniformly distributed live load of 30 kN/m run. Determine the length of truss which will require to be counterbraced (*see* p. 38).

Answer. Central $33\frac{1}{3}$ m.

9. A girder is launched across a gap in the manner shown in Fig. 2.15. The axis of the roller at A is fixed in position and the

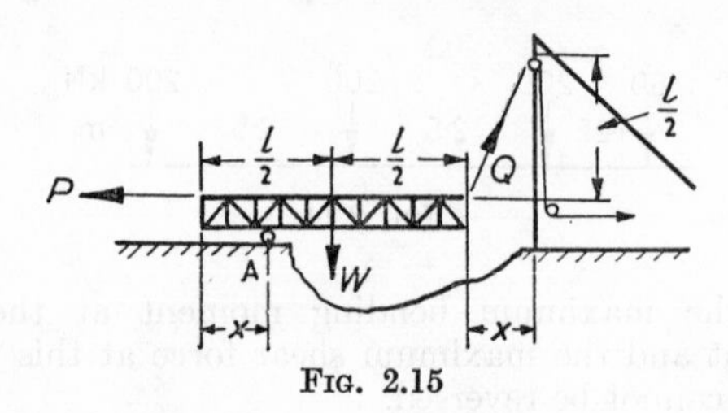

Fig. 2.15

horizontal pull P in the preventer tackle is adjusted to keep the girder horizontal. Find expressions for the values of the pulls, Q and P, for values of x less than $\frac{l}{2}$ and show that the maximum value of P is about $0{\cdot}17\ W$. (*Cambridge*)

Answer. $P = \dfrac{Wx\,(l - 2x)}{l\,(l - x)}$

10. Three axle loads of magnitudes 30, 60 and 60 kN respectively pass over a bridge of 40 m span. The horizontal distance between the axles taken in order are 5 m and 10 m. Find the greatest bending moment produced by the loads. (*Cambridge*)

Answer. 1130 kN m.

11. The given load system, Fig. 2.16, crosses a beam simply supported over a span of 60 m.

Determine (*a*) the maximum B.M. and the maximum shear force at a section 20 m from the left-hand end; (*b*) the maximum bending moment on the span. (*Aberdeen*)

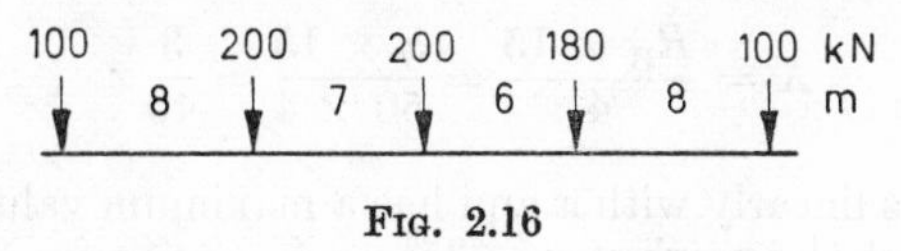

Fig. 2.16

Answer. (*a*) 7900 kN m, 333 kN; (*b*) 9000 kN m.

12. A beam is simply supported over a span of 60 m and subjected to a uniform dead load of 10 kN/m covering the span and a live load consisting of a uniform load longer than the span of 20 kN/m plus a knife edge load of 100 kN which may take up any position on the span. Determine the maximum and minimum shearing force at the left-hand quarter point and the range over which the shearing force may have positive or negative values due to the combined dead and live load. (*Aberdeen*)

Answer. 562·5 kN, 87·5 kN; central 20 m.

13. Wheel loads of 10, 20, 20, 20 and 20 kN at spacings of 9, 6, 6 and 6 m cross a girder of 50 m span. What is the maximum bending moment at a point 20 m from the left-hand end? (*St. Andrews*)

Answer. 775 kN m.

Influence Lines for Braced Structures

Example 2.4. A Pratt (N) truss has a span of 50 m, a depth of 4 m and is divided into ten equal panels. Draw the influence lines for the loads in the top boom of the eighth panel from the left-hand support and for the diagonal in the sixth panel from the same support when the bottom chord is loaded. Determine the loads in these members when a uniform load of 40 kN/m run 10 m long crosses the bridge.

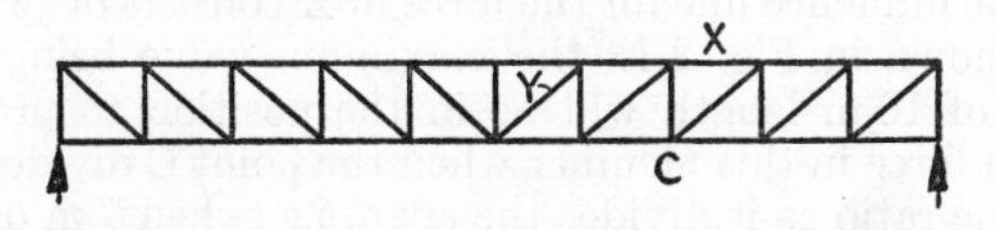

Fig. 2.17

Solution. The truss is shown in Fig. 2.17, the members in question being X and Y.

To get the force in X cut a vertical section through this member and take moments about C. When the load is to the left of C and distance x from the left-hand support then consideration of the equilibrium of the portion to the right of the section under consideration gives

$$X = \frac{R_B \times 15}{4} = \frac{x \times 15}{50 \times 4} = \frac{3}{40}x$$

This varies linearly with x and has a maximum value when the unit load is at C, i.e. when $x = 35$.

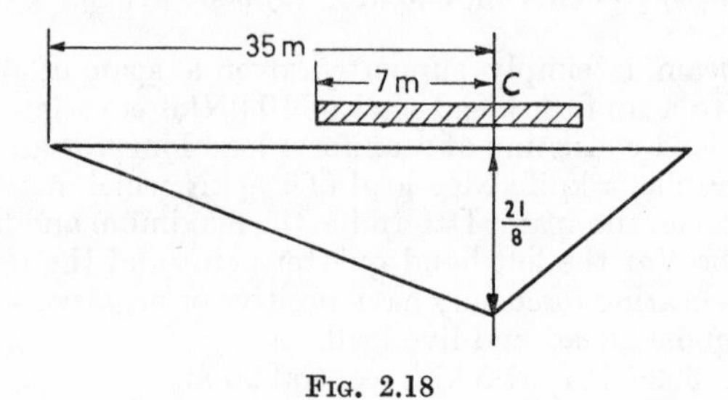

FIG. 2.18

When the load is to the right of C consideration of the equilibrium of the portion of the beam to the left of the section gives

$$X = \frac{R_A \times 35}{4} = \frac{(50 - x) \times 35}{50 \times 4}$$

This also varies linearly with x and has a maximum value when

$$x = 35$$

Thus the influence line for the force in X consists of two straight lines as shown in Fig. 2.18 the maximum value being $\frac{21}{8}$. The live load of 10 m length will be in the position to produce the maximum force in this member when the point C divides the load in the same ratio as it divides the span, i.e. when 7 m of the load is to the left of C and 3 m to the right. The force in X is then equal to the area of the two trapezia multiplied by the intensity of loading, i.e.

$$X = 40 \times 2{\cdot}625 \left\{\frac{7}{2}\left(1 + \frac{28}{35}\right) + \frac{3}{2}\left(1 + \frac{12}{15}\right)\right\} = 945 \text{ kN}$$

To get the force in Y cut a vertical section through this member and consider the equilibrium of the vertical forces. When the load is to the left of the panel cut by the section and distance x from the left-hand support then

$$Y \sin \theta = R_B = \frac{x}{50}$$

i.e.

$$Y = \frac{x}{50} \operatorname{cosec} \theta$$

where θ = inclination of sloping member to horizontal.

This varies linearly with x and has a maximum value when

$$x = 25$$

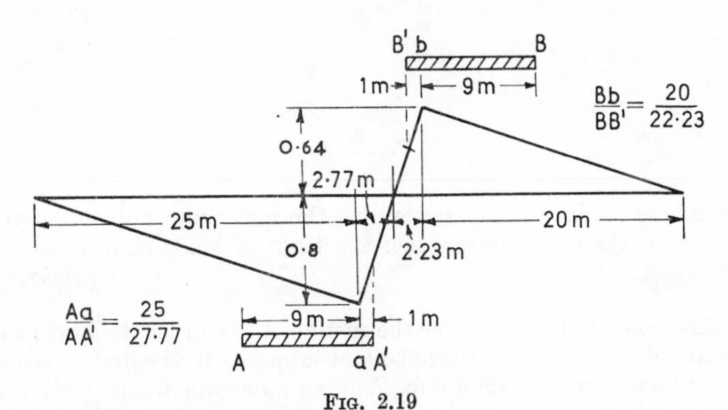

FIG. 2.19

When the load is to the right of the panel under consideration the equilibrium of the vertical forces gives

$$Y \sin \theta = R_B - 1$$

hence

$$Y = -\left(\frac{50 - x}{50}\right) \operatorname{cosec} \theta$$

This again varies linearly with x and has a maximum negative value when $x = 30$.

The completed influence line consists therefore of these two straight lines joined by a third as shown in Fig. 2.19.

The maximum positive force in Y will occur when the load is in the position AA′ and the maximum negative when it covers the length BB′. In each case the force is equal to the area under the influence line multiplied by the intensity of loading.

$$\text{Maximum tension in Y} = 40 \times \tfrac{1}{2} \times \{10(0{\cdot}80 + 0{\cdot}512)\}$$
$$= 262{\cdot}4 \text{ kN}$$

$$\text{Maximum compression in Y} = 40 \times \tfrac{1}{2} \times \{10(0{\cdot}64 + 0{\cdot}352)\}$$
$$= 198{\cdot}5 \text{ kN}$$

Example 2.5. A framed girder of triangular shape spans 30 m and has a maximum depth of 5 m (see Fig. 2.20). Draw influence line diagrams to show the variations of the forces in the

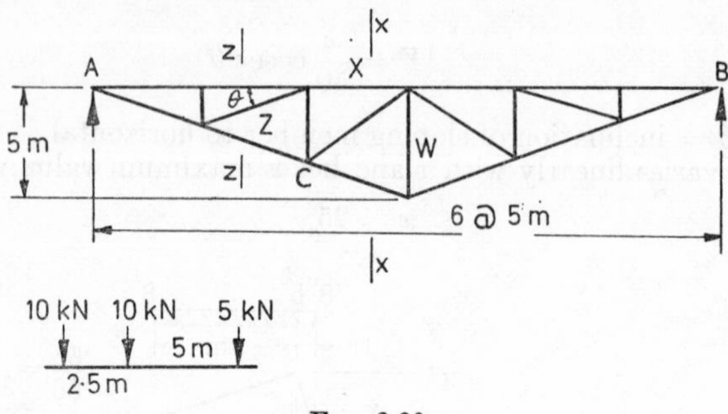

FIG. 2.20

members X, W and Z. Determine the maximum and minimum forces in these members when the train of loads shown crosses the span. (*Glasgow*)

Solution. For Z, consider the section Z–Z and let the unit load be at a distance x from the left-hand support. In the first instance let x be not greater than 5 m. Taking moments about A gives

$$1 \times x = Z \times 10 \sin \theta$$

Thus
$$Z = \frac{\sqrt{10}}{10} \times x$$

This varies linearly with x and has a maximum value when $x = 5$. In the second instance consider values of x greater than 10 m. Moments about A give $Z = 0$ for all positions of the load where x is greater than 10 m. When x varies between 5 m and 10 m then a proportion equal to $\frac{10 - x}{5}$ goes to the panel point immediately to the left of section Z–Z. Thus the force Z varies linearly as x varies between 5 m and 10 m, the final influence line being as shown in Fig. 2.21.

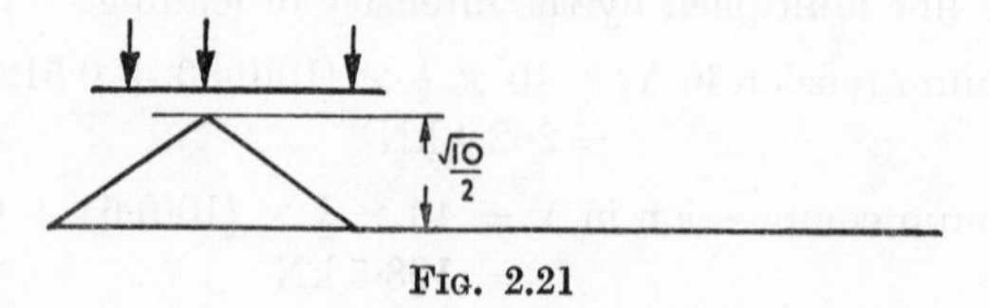

FIG. 2.21

The position of the load for maximum force in this member is as shown in Fig. 2.21. The value of the force is

$$10\left(\frac{\sqrt{10}}{2}+\frac{\sqrt{10}}{4}\right)=7\cdot5\sqrt{10}=23\cdot7\text{ kN}$$

For X, consider the section X–X and take moments about C. This gives, when the load is distant x from A,

$$X\times\frac{10}{3}=\left(\frac{30-x}{30}\right)\times10-[(10-x)]$$

the term in square brackets only occurring when it is positive in sign.

This gives a linear variation for X, as shown in Fig. 2.22, the maximum value occurring when $x=10$, in which case $X=2$.

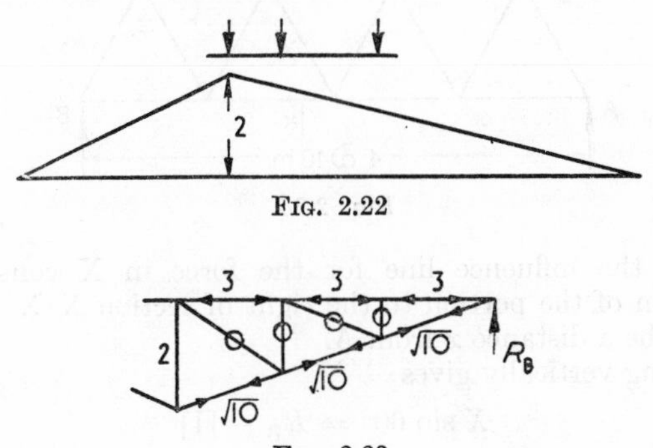

FIG. 2.22

FIG. 2.23

The position of the load train for the maximum force in X is also shown in Fig. 2.22. The value of the force in X is given by

$$X=10\times2+10\times1\tfrac{3}{4}+5\times1\tfrac{1}{4}=43\cdot75\text{ kN}$$

For W, consider the equilibrium of the portion to the right of X–X. Stressing by inspection, the forces in the members, in terms of R_B, when the load is to the left of X–X are given in Fig. 2.23. Since R_B varies linearly with x, the distance of the load from the left-hand support, the influence line for W consists of two straight lines as shown in Fig. 2.24, the maximum value

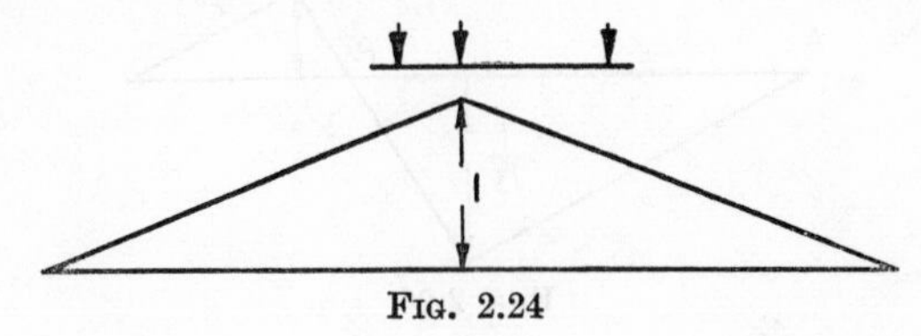

FIG. 2.24

being unity. The position of the load train to give the maximum value is also shown in the figure. The force in W is then given by

$$W = 10(1 + \tfrac{5}{6}) + 5 \times \tfrac{2}{3} = 21\tfrac{2}{3}\ \text{kN}$$

Example 2.6. A four-panel Warren girder whose span is 40 m is formed from equilateral triangles. Determine the maximum load in the diagonal member just to the right of mid-span when a live load whose length is greater than the span, of intensity 10 kN/m run crosses the girder, the dead weight of which is 2 kN/m run.

Solution. The girder is as shown in Fig. 2.25, the member in question being marked X.

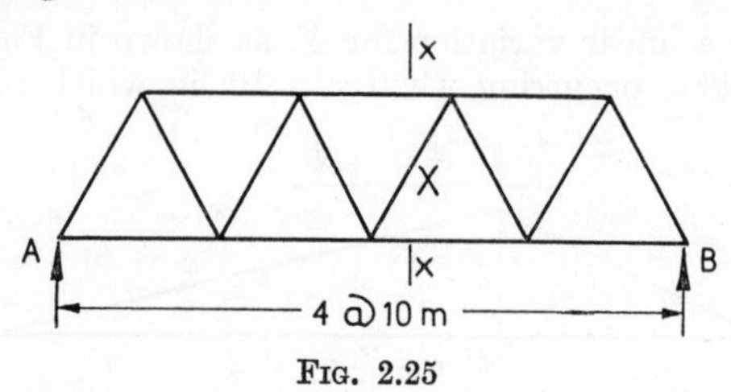

FIG. 2.25

To get the influence line for the force in X consider the equilibrium of the portion to the right of section X–X. Let the unit load be a distance x from A.

Resolving vertically gives

$$X \sin 60° = R_B - [1]$$

The term in brackets applies when $x > 30$.

For all values of x, $R_B = \dfrac{x}{40}$, hence the relationship between X and x is linear, the influence line being as shown in Fig. 2.26.

$$\text{The positive area} = \frac{80}{3} \times \frac{1}{\sqrt{3}} \times \frac{1}{2} = \frac{40}{3\sqrt{3}}\ \text{m units}$$

$$\text{The negative area} = \frac{40}{3} \times \frac{1}{2\sqrt{3}} \times \frac{1}{2} = \frac{10}{3\sqrt{3}}\ \text{m units}$$

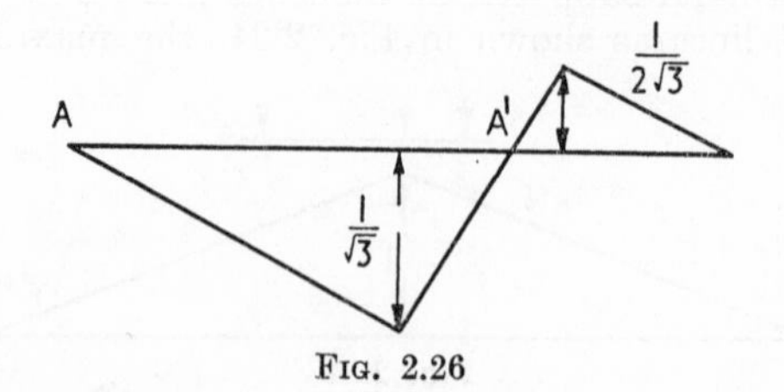

FIG. 2.26

The live load will occupy the position A–A′ on the span to give the maximum force but the dead load must cover the whole span. Hence the maximum force, X_{max}, in X is given by

$$X_{max} = 10 \times \frac{40}{3\sqrt{3}} + 2\left(\frac{40}{3\sqrt{3}} - \frac{10}{3\sqrt{3}}\right) = 88{\cdot}7 \text{ kN}$$

Counterbracing

If the truss shown in Example 2.4 has to carry a dead load of 20 kN per metre run then the force in the member Y due to this dead load is $20 \times \frac{1}{2} \times (0{\cdot}80 \times 27{\cdot}77 - 0{\cdot}64 \times 22{\cdot}23) = 79{\cdot}3$ kN tension.

The resultant force in Y due to the combined dead and live loads is therefore 341·7 kN tension or 119·2 kN compression.

The sloping members in a Pratt Truss, however, are generally long and slender and whilst such a member can be designed to take a tension of 341·7 kN a much heavier section would be required to take a compression of 119·2 kN. If, however, in the panel in question there was a member crossing the diagonal opposite to Y the forces in such a member would be equal and opposite to Y, i.e. a compression of 341·7 kN and a tension of 119·2 kN. If, therefore, the bracing system consists of two diagonals one of which, Y, is designed for a tension of 341·7 kN and the other designed for a tension of 119·2 kN, the structure would be adequately designed despite the fact that neither of the two diagonals could resist compression. Such a panel is said to be counterbraced. Problems of counterbracing therefore virtually mean determining the length of bridge in which the shear can be both positive and negative in value (*see* Example 2.1).

Examples on Braced Girders

1. A five-panel Warren truss is formed of equilateral triangles and is loaded on the bottom chord. Draw the influence lines for the diagonal member and the top chord member immediately to the left of the centre point of the truss.

Answer. + 0·46; − 0·46; 1·39.

2. Plot the influence lines for the forces in the members JK and KD of the frame shown in Fig. 2.27, when a load rolls across the bottom chord. Calculate the maximum forces in these members when the train of loads shown crosses the bridge.

Answer. $JK = 40{\cdot}8$ kN; $KD = 32{\cdot}4$ kN.

3. Calculate the maximum loads in the members CD and EF of the Pratt truss shown in Fig. 2.28, which has a span of 36 m and is 8 m high, when a point load of 20 kN crosses the lower chord. Neglect the weight of the truss.

Answer. + 12·5 kN; − 8·33 kN; + 12·5 kN.

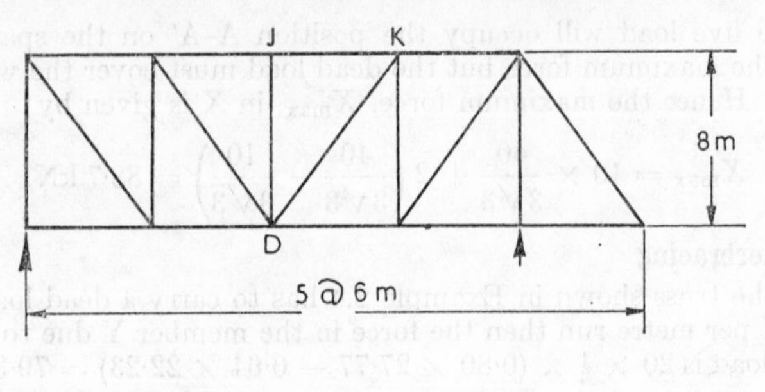

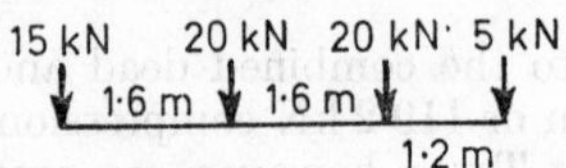

FIG. 2.27

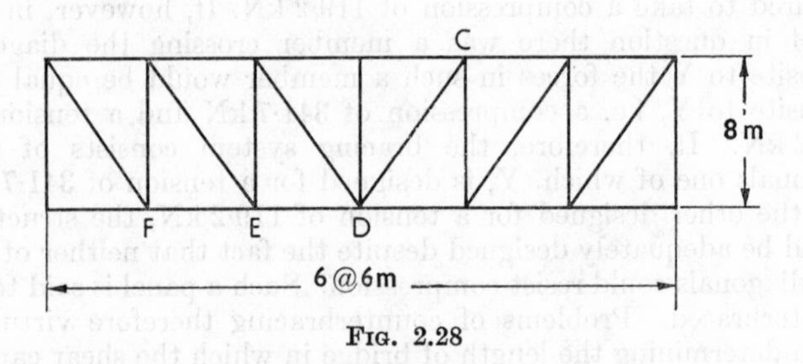

FIG. 2.28

4. Determine the maximum loads which can occur in the member CD of the frame shown in Fig. 2.29 when a uniformly distributed load of 2 kN/m crosses the span.

Answer. + 6·75 kN; − 3·01 kN.

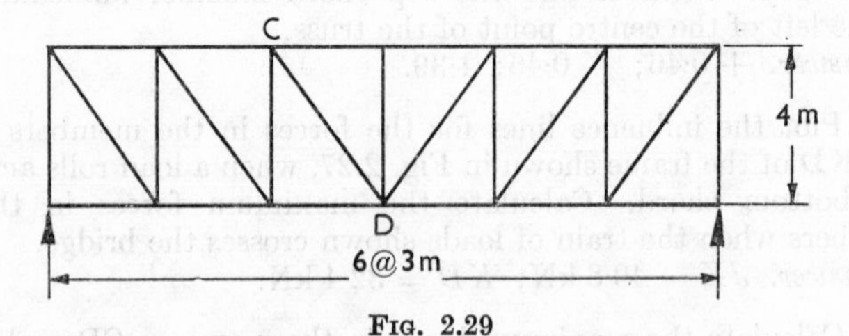

FIG. 2.29

5. Draw the influence line for the force in the member CD of the Warren girder shown in Fig. 2.30 and determine from it the maximum load which occurs in the member when a uniformly distributed load of intensity w/unit length crosses the bridge. The length of the load is greater than the span.

Answer. $+ 0{\cdot}156\,wL$; $- 1{\cdot}406\,wL$.

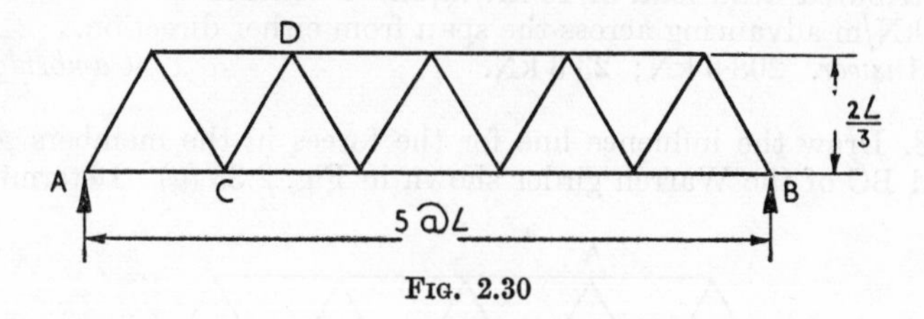

FIG. 2.30

6. Draw the influence lines for the forces in the members AC and AB of the truss shown in Fig. 2.31 for a load on the bottom boom. From these lines determine the maximum forces in these

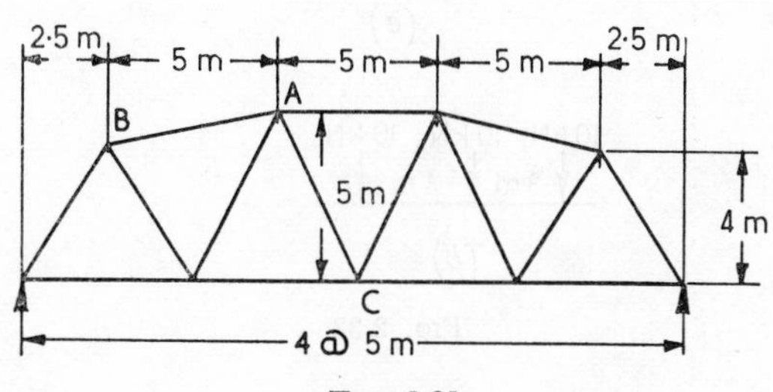

FIG. 2.31

members when a uniformly distributed load of 20 kN/m run (longer than the span) crosses the bridge the dead load of which is 10 kN/m run.

Answer. 102·5 kN; 250 kN.

7. The bridge shown in Fig. 2.32 has a span of 44 m in 11 equal bays and is freely supported at its ends. PS = 8 m; SQ = 4 m and the angle RPQ is a right angle.

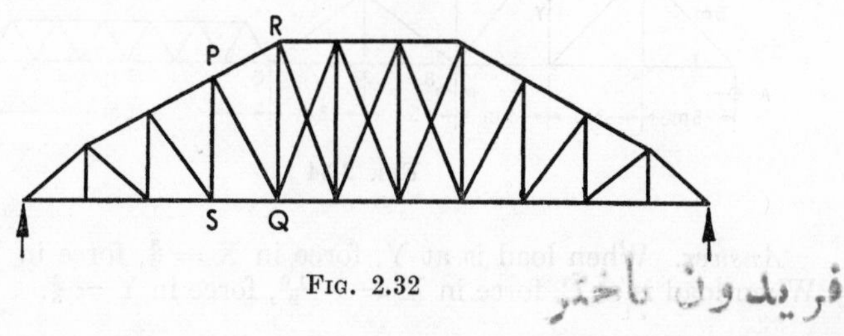

FIG. 2.32

Draw the influence line diagram for the force in the member PQ. Hence determine the greatest tension and compression in this member when the loads on the frame consist of a uniformly distributed dead load of 10 kN/m and a continuous live load of 30 kN/m advancing across the span from either direction.

Answer. 208·5 kN; 23·3 kN. (*Cambridge*)

8. Draw the influence line for the forces in the members AB and BC of the Warren girder shown in Fig. 2.33 (*a*). Determine

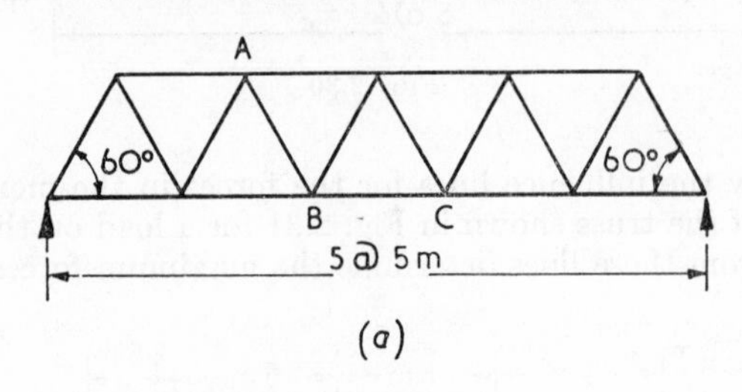

FIG. 2.33

the maximum loads in these members when the train shown in Fig. 2.33 (*b*) crosses the span.

Answer. 14·3 kN; – 1·62 kN; 27·7 kN.

9. The anchor, cantilever and suspended spans for a bridge are shown in the Fig. 2.34. Draw the influence line for the force in members X and Y. (*Aberdeen*)

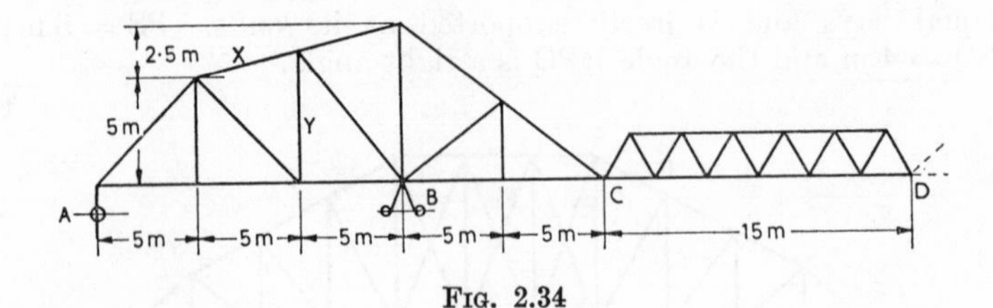

FIG. 2.34

Answer. When load is at Y, force in X = $\frac{5}{9}$, force in Y = $\frac{4}{5}$. When load is at C, force in X = $-\frac{10}{9}$, force in Y = $\frac{2}{5}$.

10. A deck bridge of the outline shown in Fig. 2.35 is designed to span a river running in a wide and shallow valley, maximum headroom being required for navigation.

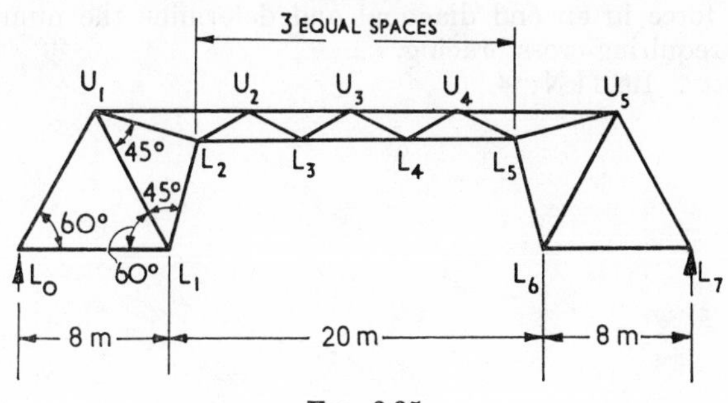

FIG. 2.35

The live load is 12 m long and has an intensity of 30 kN/m. The self weight may be taken as 15 kN/m.

Calculate the maximum forces in the members L_1L_2, U_1U_2, U_1L_2. (*Durham*)

Answer. 331 kN; 2440 kN; 2360 kN.

11. A Pratt truss of 60 m span is formed from 9 equal square panels and carries a live load of 150 kN/m run. Calaulate the maximum forces in the diagonal member in the third panel from the left-hand support.

Answer. 3180; 354.

12. A Warren girder having a span of 40 m consists of four equal panels as shown in Fig. 2.36. Draw the influence lines for

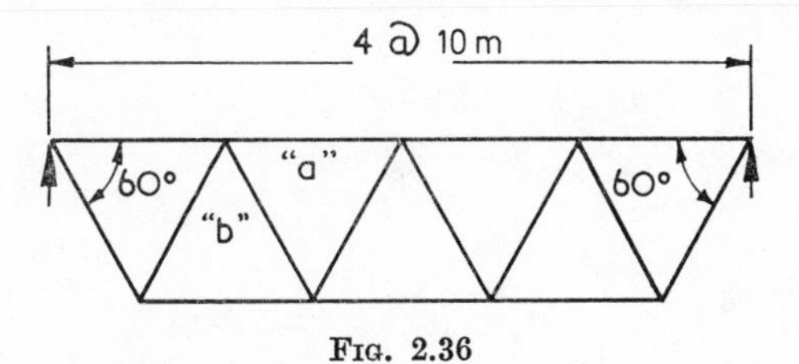

FIG. 2.36

members "a" and "b" and find the maximum force in "a" due to a distributed live load of 20 kN/m and 15 m long. (*Glasgow*)

Answer. 234·5 kN.

13. An N-girder spans 40 m and comprises 10 panels having a depth of 4 m. It supports a dead load of 15 kN/m run on the bottom boom panel points and a live load of 60 kN/m run having a length greater than the span. Evaluate the maximum tensile force in an end diagonal and determine the number of panels requiring cross bracing.

Answer. 1910 kN; 4.

Chapter 3

Theorems of Elasticity

THE ONLY FUNDAMENTAL PRINCIPLE involved in the solution of the examples given in the previous chapters was the application of the equations of static equilibrium. In order to deal with the remainder of the work in the average syllabus on structures it is necessary to know and to be able to apply some of the theorems of elasticity. It is proposed in this chapter to state some of these, and to show a few applications.

The Principle of Superposition

This principle states that the effect of a number of loads applied simultaneously is the sum of the effects of the loads applied singly. Its main use in problems on structures is when they have an asymmetrical load applied. Such a load, as shown in Fig. 3.1,

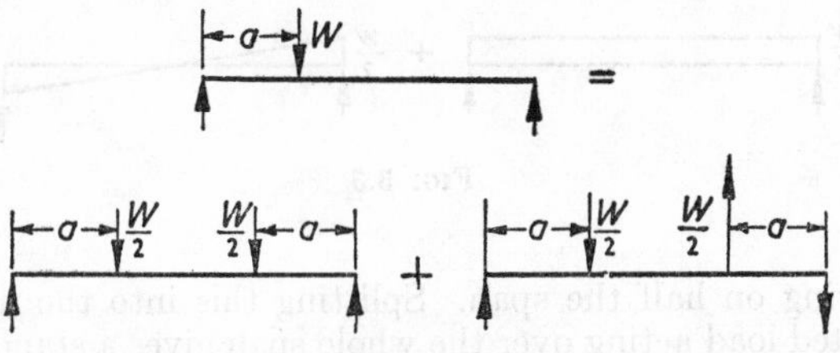

FIG. 3.1

can be split into a symmetrical load plus a skew symmetrical one. The symmetrical load is generally a standard case and the average student will generally know the answer.

The skew symmetrical load has the following properties—

1. The total load on the beam is zero. This is important in dealing with the arch problems described in Chapter VIII and the stiff-jointed frames given in Chapter IX, for if the resultant load on the structure is zero there is no horizontal thrust at the supports.

2. There is no bending moment at mid-span. This means that the mid-span point can virtually be taken as a pin and moments taken about it to determine reactions, etc.

3. There is no deflexion at mid-span.

4. The value of a function in the right-hand half of the span is equal and opposite to the value of the same function at the corresponding point in the left-hand portion of the span.

In addition to the point load shown in Fig. 3.1 problems involving two common cases of distributed load can be solved more easily if superposition is used. They are shown in Figs. 3.2 and 3.3. The first illustrates the case of a uniformly distributed

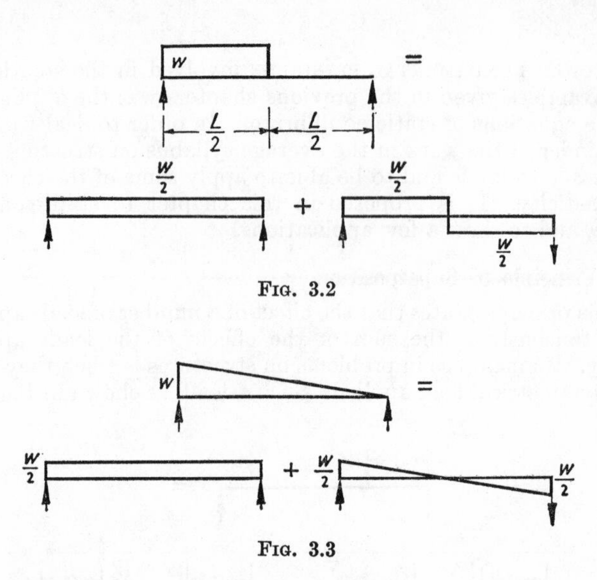

FIG. 3.2

FIG. 3.3

load acting on half the span. Splitting this into the uniformly distributed load acting over the whole span gives a standard case plus the skew-symmetrical case with no central deflexion or bending moment. The second shows the common fluid-pressure loading. This again can be split into a uniformly distributed load plus a skew-symmetrical one. In many problems the load shown in Fig. 3.3 is more easily dealt with than the standard fluid-pressure loading.

As an example of the application consider the following.

Example 3.1. A beam of span L carries a downward load of intensity w/unit length on the left-hand half of the span and an upward load of $\dfrac{wL}{4}$ at mid-span. Calculate the deflexion at the centre of the span.

Solution. The loading in the question shown in Fig. 3.4 splits up into three loads—

(i) A uniformly distributed load over the whole span of intensity $\frac{w}{2}$,

(ii) A skew symmetrical load of $\frac{w}{2}$ downwards on the left-hand half and $\frac{w}{2}$ upwards on the right-hand half,

(iii) An upward load of $\frac{wL}{4}$ at mid-span.

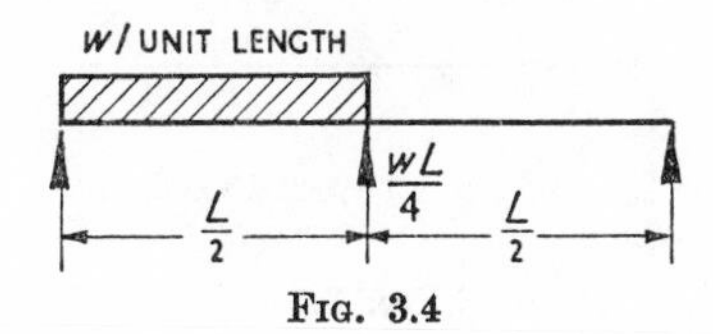

FIG. 3.4

The downward deflexion due to load 1 $= \frac{5}{384} \cdot \frac{w}{2} \cdot \frac{L^4}{EI}$

The upward deflexion due to load 3 $= \frac{1}{48} \cdot \frac{wL}{4} \cdot \frac{L^3}{EI}$

There is no deflexion due to load 2.

Hence the resultant deflexion $= \frac{5wL^4}{768EI} - \frac{4wL^4}{768EI} = \frac{wL^4}{768EI}$

Clerk–Maxwell's Reciprocal Theorem

The form in which this theorem is usually known and applied is that which states that the deflexion in an elastic structure at a point A in the direction of W_1 when unit load acts at a point B in a direction W_2 is the same as the deflexion at B in the direction W_2 when unit load acts at A in the direction of W_1.

The theorem forms the basis of the Muller–Breslau Theorem for determining influence lines for redundant structures and is the fundamental on which model analysis of structures is based. Problems dealing with model analysis are dealt with in Chapter 11.

It can, however, be applied directly to the solution of problems of the type given below.

Example 3.2 A steel section cantilever 7 m long carries loads of 5, 10, 3, 6 and 2 kN at distances 1, 2, 3, 4 and 6 m respectively from the fixed end. If $E = 200$ kN/mm² and $I = 135 \times 10^6$ mm⁴ units, what is the deflexion at the free end?

Solution. The deflexion at the free end due to a load, W, at a distance a from the fixed end is, by Clerk–Maxwell's theorem, the same as the deflexion at a distance a from the fixed end when a load, W, acts at the free end.

Instead therefore of determining the deflexions at the free end when loads act at points distant 1, 2, 3, 4 and 6 m from the fixed end the deflexion at these points will be determined when a load acts at the free end.

Let a load, W, act at the free end and measure x from the fixed end (see Fig. 3.5).

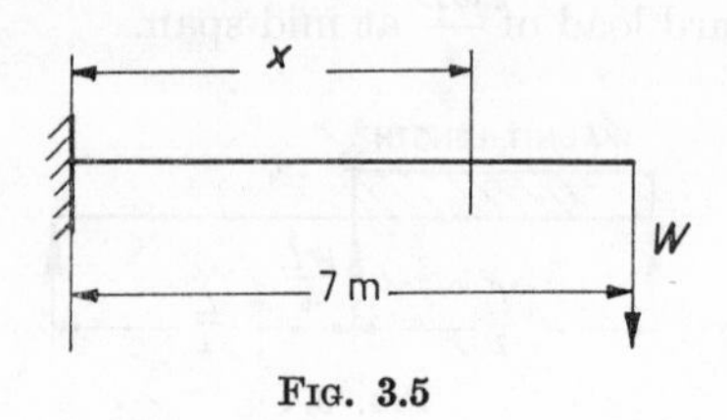

FIG. 3.5

Then $$M_x = EI\frac{d^2y}{dx^2} = W(7 - x)$$

$$EI\frac{dy}{dx} = W\left(7x - \frac{x^2}{2}\right) + A$$

But when $x = 0$, $\frac{dy}{dx} = 0$, hence $A = 0$.

Integrating again gives

$$EIy = W\left(3{\cdot}5x^2 - \frac{x^3}{6}\right) + B$$

But when $x = 0$, $y = 0$, hence $B = 0$.

When $x = 1$ $\quad y = 3{\cdot}33\frac{W}{EI}$

$x = 2$ $\quad y = 12{\cdot}67\frac{W}{EI}$

$x = 3$ $\quad y = 27\frac{W}{EI}$

$x = 4$ $\quad y = 45{\cdot}33\frac{W}{EI}$

$x = 6$ $\quad y = 90\frac{W}{EI}$

Hence the deflexion at the free end due to the loading system given in the question is given by

$$\Delta = \frac{1}{EI}(5 \times 3{\cdot}33 + 10 \times 12{\cdot}67 + 3 \times 27 + 6 \times 45{\cdot}33 + 2 \times 90)$$

$$= \frac{677}{EI}\text{ m} = 25{\cdot}1\text{ mm.}$$

The Reciprocal Theorem is sometimes given in the general form due to Betti. In this form it states: in a stable structure having a linear load-displacement diagram the work done by one system of forces acting through the corresponding displacements due to a second system of forces equals the work done by the second system of forces when acting through the corresponding displacements due to the first system of forces.

This generalized form of the theorem is used to solve the following example.

Example 3.3. A beam ABCDE is supported at A and E. A forced vertical deflexion δ_B at B causes vertical deflexions of δ_C at C and δ_D at D. A forced vertical deflexion δ_d at D causes vertical deflexions of δ_b and δ_c at B and C. Show that if supports are placed at B and D to the same level as A and E a vertical load W at C induces a reaction of

$$\frac{\delta_B \delta_c - \delta_b \delta_C}{\delta_B \delta_d - \delta_b \delta_D} \,.\, W$$

at D. (*London*)

Solution. There are here three systems of forces, shown in Figs. 3.6(a), (b) and (c) respectively and tabulated, with their corresponding deflexions, in Table 3.1.

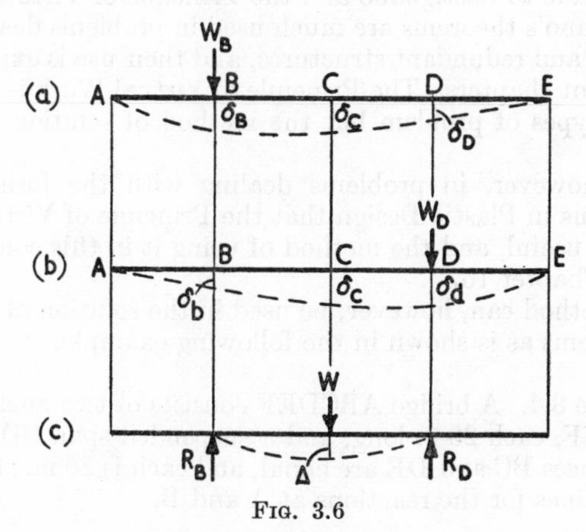

FIG. 3.6

TABLE 3.1

System	Forces (B)	(C)	(D)	Deflexions (B)	(C)	(D)
I	W_B	O	O	δ_B	δ_C	δ_D
II	O	O	W_D	δ_b	δ_c	δ_d
III	$-R_B$	W	$-R_D$	O	Δ	O

Applying the Maxwell-Betti theorem to systems I and III gives—

$$W_B \times O + O \times \Delta + O \times O = -R_B\delta_B + W\delta_C - R_D\delta_D$$

whence $R_B = \dfrac{1}{\delta_B}(W\delta_C - R_D\delta_D)$.

Applying the theorem to systems II and III gives—

$$-R_B\delta_b + W\delta_c - R_D\delta_d = O$$

Substituting for R_B gives—

$$-\frac{\delta_b}{\delta_B}(W\delta_C - R_D\delta_D) + W\delta_c - R_D\delta_d = O$$

which simplifies to

$$R_D = \frac{\delta_B\delta_c - \delta_b\delta_C}{\delta_B\delta_d - \delta_b\delta_D} \cdot W$$

The other theorems of elasticity commonly used by the student are those due to Castigliano and the Principle of Virtual Work.

Castigliano's theorems are much used in problems dealing with deflexions and redundant structures, and their use is explained in the relevant chapters. The Principle of Virtual Work is also used in these types of problem but the method of solution is almost identical.

It is, however, in problems dealing with the formation of mechanisms in Plastic Design that the Principle of Virtual Work is so very useful, and the method of using it in this connexion is given in Chapter 16.

This method can, however, be used in the solution of influence line problems as is shown in the following example.

Example 3.4. A bridge ABCDEF consists of two anchor spans AB and EF, each 20 m long, and a suspended span CD = 40 m. The distances BC and DE are equal, and each is 20 m. Draw the influence lines for the reactions at A and B.

Solution. The line diagram for the bridge is shown in Fig. 3.7. In order to obtain the reaction at A a virtual displacement of Δ is given at that point. Then since B, C, D, E and F are pins, the deflected form of the structure is as shown in Fig. 3.8(*a*).

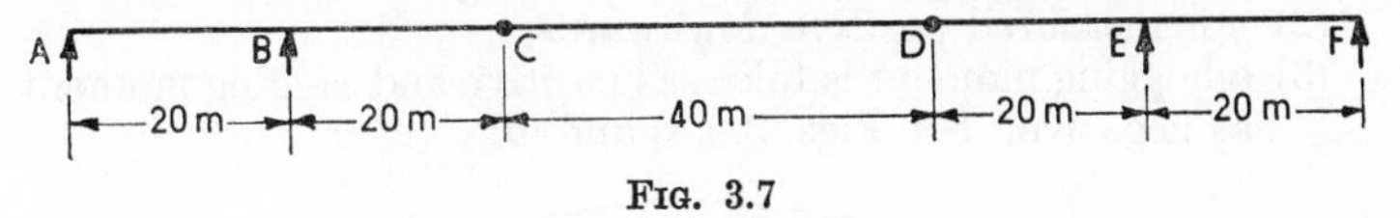

FIG. 3.7

The displacement at a point on AB distance x from B is $\frac{x}{20} \times \Delta$.

If a load W is acting at this point then by the principle of virtual work

$$R_A \times \Delta = W \frac{x}{20} \times \Delta$$

or

$$R_A = \frac{x}{20} \cdot W$$

Hence, if the displacement at A in Fig. 3.8(*a*) is made equal to unity, this diagram becomes the influence line for the reaction at A, the positive sign denoting upward reaction.

In a similar way, if a vertical displacement of Δ is given at B the deflected form allowing for the other pinned supports is as shown in Fig. 3.8(*b*).

From A to C the deflexion at any point distant x from A is given by $\frac{x}{20} \times \Delta$. If a load W acts at this point then—

$$R_B \times \Delta = W \frac{x}{20} \times \Delta$$

Hence, if Δ is made equal to unity Fig. 3.8(*b*) becomes the influence line for the reaction at B.

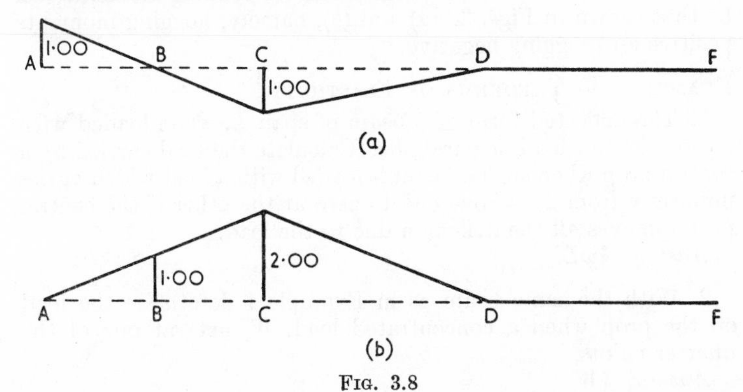

FIG. 3.8

In the solution of example 3.2 the relationship $EI\,\dfrac{d^2y}{dx^2} = M$ has been used. This expression holds in this form when—

(1) x is measured positive from left to right,
(2) y is measured positive downwards,
(3) a hogging moment is taken as positive and sagging moment as negative. See Figs. 3.9(a) and (b).

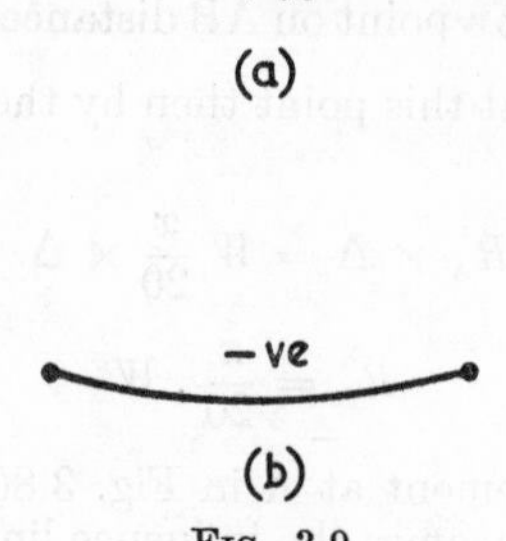

Fig. 3.9

One difficulty with this convention is that the commonest beam is the simply supported one, and with gravity loading on such a beam the bending moment according to the above convention is negative. This disadvantage, however, is in the authors opinion outweighed by that of writing down the relationship $EI\,\dfrac{d^2y}{dx^2} = \mathrm{M}$.

If the convention that sagging moments are positive is adopted, then with the positive direction of y as downwards the relationship is $EI\,\dfrac{d^2y}{dx^2} = -M$.

In this book, however, the sign taken for bending moments will be that shown in Figs. 3.9(a) and (b), namely, hogging moments positive and sagging negative.

Examples on Theorems of Elasticity

1. The deflected form of a beam of span L, when loaded with a central point load, is parabolic. Calculate the load carried by a central prop when such a beam is loaded with a load which varies uniformly from $2w$ at one end to zero at the other if the central prop removes all the deflexion due to the load.

Answer. $\frac{2}{3}wL$.

2. With the same beam as in Example 1 determine the load on the prop when a concentrated load, W, acts at one of the quarter points.

Answer. $\frac{3}{4}W$.

3. Determine the expression for the influence line for deflexion at a point distant $L/4$ from the left-hand support in a simply supported beam of uniform section whose span is L.

Answer. $y = \dfrac{x}{8EI}\left(\dfrac{7l^2}{16} - x^2\right) + \left[\dfrac{1}{6EI}\left(x - \dfrac{L}{4}\right)^3\right].$

4. Determine the central bending moment and deflexion for the beam loaded as shown in Fig. 3.10.

Answer. $\frac{7}{24}WL$; $0{\cdot}0310\,\dfrac{WL^3}{EI}$.

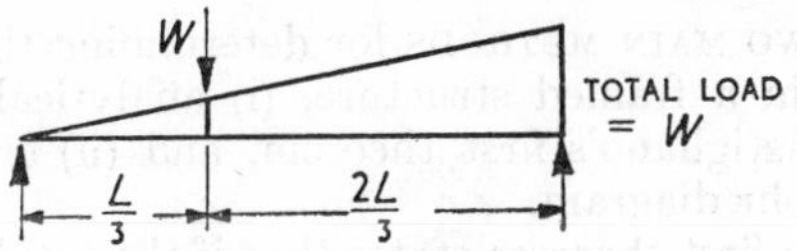

FIG. 3.10

5. State Clerk-Maxwell's Reciprocal Theorem.

An elastic member is pinned to a drawing board at its ends A and B. When a moment M is applied at A, A rotates θ_A, B rotates θ_B, and the centre deflects δ_1. The same moment M applied to B rotates B θ_C and deflects the centre through δ_2. Find the moment induced at A when a load W is applied to the centre in the direction of the measured deflexions, both A and B being restrained against rotation.

Answer. $\left(\dfrac{\delta_1\theta_C - \delta_2\theta_B}{\theta_A\theta_C - \theta_B{}^2}\right) W.$

6. State Clerk-Maxwell's Reciprocal Theorem.

A point load of 1 kN is placed at the centre of a beam simply supported over a span of 5 m causing a maximum deflexion of 100 mm. The beam is of variable cross-section so that between the supports its deflected form is parabolic. If one end of the beam extends 1 m beyond the suport, what load must be applied at this end to lift the centre 50 mm assuming the beam cannot leave its supports? *(London)*

Answer. 0·625 kN.

Chapter 4

Deflexions of Framed and Other Structures

THERE ARE TWO MAIN METHODS for determining the deflexion at given points in a framed structure, (i) analytical, using strain energy and Castigliano's first theorem, and (ii) graphical, using the Williot-Mohr diagram.

Castigliano's first theorem states that if the total strain energy of a structure be partially differentiated with respect to an applied load the result gives the displacement of the point of application of the load in its line of action. Expressed symbolically—

$$\frac{\partial U}{\partial W} = \Delta$$

where U = total strain energy of system
W = an applied load
Δ = displacement of point of application of load in its line of action

In general the energy in a framed structure is direct energy, thus giving

$$\Delta = \sum \frac{PL}{AE} \cdot \frac{\partial P}{\partial W}$$

where P = load in any member
L = length of that member
AE = extensibility of that member

or

$$\Delta = \sum \frac{fL}{E} \cdot \frac{\partial P}{\partial W}$$

where f = stress in the member.

The main disadvantage of the analytical method is that it only gives displacements of a loaded point in the direction of the load. If the displacement at an unloaded point is required then an imaginary load, which is later equated to zero, must be placed at this point to enable the analysis to be carried out. A similar method must be employed if the displacement at a loaded point but not in the direction of the load is required.

If the energy is due to bending then the displacement of a loaded point is given by

$$\Delta = \int \frac{M_X}{EI_X} \cdot \frac{\partial M_X}{\partial W} \, dx$$

where M_X = bending moment at a point X
EI_X = flexural rigidity of the beam at X

It is seen that these expressions for deflexion contain the partial derivatives $\partial P/\partial W$ and $\partial M_X/\partial W$. These partial derivatives can be expressed by p and m_x respectively,

where $p = \partial P/\partial W$ = load in any member due to unit load acting at point at which deflexion is required,
$m_x = \partial M_x/\partial W$ = bending moment at any point X, due to unit load at the point where the deflexion is required.

The fact that unit loads can be applied at the points is the basis of the "influence co-efficient" method of solution referred to in some textbooks. In general, the arithmetic and algebra is similar to that if Castigliano's theorems are applied, but there are some examples where the influence co-efficient method is easier. It is particularly so in the case of bending energy where the loading is one of the standard cases, since the expression for deflexion in the case of beams of uniform section becomes $\Delta = \frac{1}{EI}\int M_x m_x dx$

where M_x = bending moment at X due to the external loads,
m_x = bending moment at X due to unit load at the point where the deflexion is required.

If the diagram for m_x is that for a function which varies uniformly from $x = 0$ to $x = l$, then the expression for the integral is equal to the area of the M_x diagram multiplied by the ordinate of the m_x diagram at the position of the centre of area of the M_x diagram.

For example, if the M_x diagram is that for a uniformly distributed load of intensity w on a span, l, the M_x diagram is a parabola, the area of which is $\frac{1}{12}wL^3$. If the m_x diagram is that due to moment m_0 applied at the point $x = l$, then the m_x diagram is a triangle, and the height of the ordinate at the centre of gravity of the parabola is $\frac{m_0}{2}$. Thus the $\int M_x m_x dx = \frac{1}{12} wL^3 \times \frac{m_0}{2}$.

This method is demonstrated in Example 4.2, also in Example 5.5 (see page 99). In the latter example a straightforward graphical integration has been carried out and in many cases this is as simple as the application of the rule given above.

The graphical method has the advantage of giving the absolute displacement of every point on the frame. It cannot be applied to structures which are not framed. A difficulty which a student will find with the Williot diagram is in anticipating how far it will spread and thus selecting a suitable scale and starting point. It is, however, very unlikely that more than two attempts are required before deciding on these two points. For tutorial purposes also it is necessary to have a frame with angles generally 60° or 45°.

Many examples can be solved by either of the methods, but the worked examples have been subdivided into each class. This division has not, however, been made in the examples for the student to work out. He should either use his judgment as to which method is better or alternatively use both so as to give himself practice.

Examples Worked by the Analytical Method

Example 4.1. The frame ABCD shown in Fig. 4.1 is supported at A and B and carries a load of 120 kN at C; all the

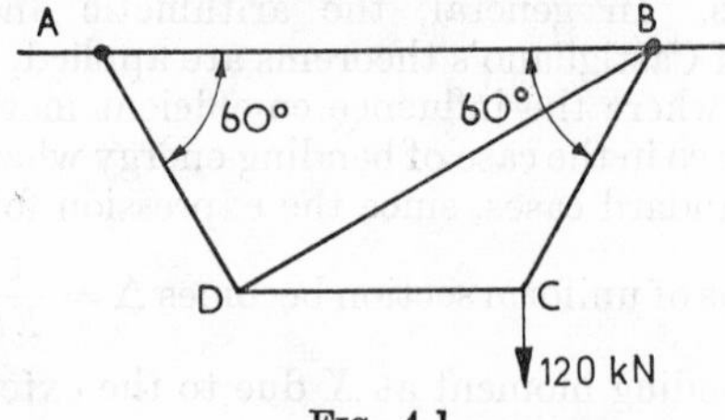

FIG. 4.1

tension members of the frame have an area of $1{\cdot}2 \times 10^3$ mm² and the compression members $1{\cdot}8 \times 10^3$ mm².

Calculate the deflexion of the point C if AD = DC = CB = 2 m and $E = 200$ kN/mm².

Solution. This example will first be solved by the use of Castigliano I, but for problems requiring the deflexion of a single load acting on a structure the alternative solution given, which is a straightforward application of strain energy, is easier.

It is a framed structure. Consequently the energy is direct energy and the deflection, Δ_C, is given by

$$\Delta_C = \sum \frac{PL}{AE} \cdot \frac{\partial P}{\partial W}$$

where P = load in member
L = its length
AE = its extensibility
W = load at C

The loads in the members must first be obtained. These are required not only to determine P but also $\frac{\partial P}{\partial W}$. The latter is more easily determined if the load at C is taken as W and not 120 kN. The factor of 120 kN can be introduced at the end of the calculations.

Taking a load of W at C the loads in the members are obtained by inspection and are as shown in Fig. 4.2, the only compression member being DB.

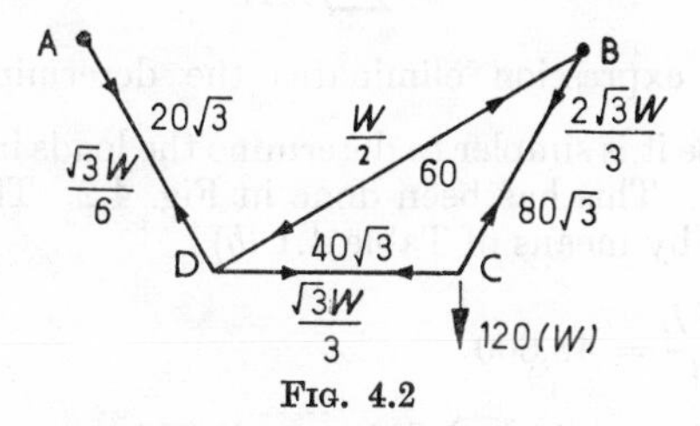

FIG. 4.2

The work is tabulated (this is recommended in all cases where a summation is involved) in Table 4.1 (*a*). The negative sign for P for member DB indicates a compressive force.

$$\sum \frac{PL}{AE} \cdot \frac{\partial P}{\partial W} = \frac{24{\cdot}46W}{7{\cdot}2}$$

The lengths of the members have been given in millimetres hence the deflexion given by the above summation will also be in millimetres. Since the area has been given in square millimetres and the load in kiloNewtons care must be taken to see that E is given in kiloNewtons per square millimetre.

TABLE 4.1 (*a*)

Member	Length (mm)	Area (mm²)	P	$\frac{\partial P}{\partial W}$	$\frac{PL}{A} \cdot \frac{\partial P}{\partial W}$
AD	2×10^3	$1{\cdot}2 \times 10^3$	$\frac{W}{2\sqrt{3}}$	$\frac{1}{6}\sqrt{3}$	$\frac{W}{7{\cdot}2}$
DB	$2\sqrt{3} \times 10^3$	$1{\cdot}8 \times 10^3$	$-\frac{W}{2}$	$-\frac{1}{2}$	$\frac{2\sqrt{3}W}{7{\cdot}2}$
DC	2×10^3	$1{\cdot}2 \times 10^3$	$\frac{W}{\sqrt{3}}$	$\frac{1}{\sqrt{3}}$	$\frac{4W}{7{\cdot}2}$
CB	2×10^3	$1{\cdot}2 \times 10^3$	$\frac{2W}{\sqrt{3}}$	$\frac{2}{\sqrt{3}}$	$\frac{16W}{7{\cdot}2}$

Substituting the values for E and W gives the deflexion in millimetres as

$$\Delta_C = \frac{24{\cdot}46 \times 120}{200 \times 7{\cdot}2} = 2{\cdot}04 \text{ mm}$$

Using the straightforward relationship between increase of strain energy and work done gives

$$\tfrac{1}{2}W\Delta_C = \sum \frac{P^2L}{2AE}$$

Using this expression eliminates the determination of $\dfrac{\partial P}{\partial W}$ and in this case it is simpler to determine the loads in the members straight away. This has been done in Fig. 4.2. The summation is determined by means of Table 4.1 (*b*).

$$\sum \frac{P^2L}{A} = 48{,}800$$

$$\Delta_C = \frac{1}{WE}\sum \frac{P^2L}{A} = \frac{48\,800}{120 \times 200} = 2{\cdot}04 \text{ mm.}$$

TABLE 4.1 (*b*)

Member	Length (mm)	Area (mm²)	P	$\dfrac{P^2L}{A}$
AD	2×10^3	$1{\cdot}2 \times 10^3$	$20\sqrt{3}$	20×10^2
DB	$2\sqrt{3} \times 10^3$	$1{\cdot}8 \times 10^3$	60	$40\sqrt{3} \times 10^2$
DC	2×10^3	$1{\cdot}2 \times 10^3$	$40\sqrt{3}$	80×10^2
CB	2×10^3	$1{\cdot}2 \times 10^3$	$80\sqrt{3}$	320×10^2

Example 4.2. The structure shown in Fig. 4.3(*a*) consists of an upright cantilever AB of length $3R$ and a semicircular portion BC of radius R. The flexural rigidity is constant throughout. It

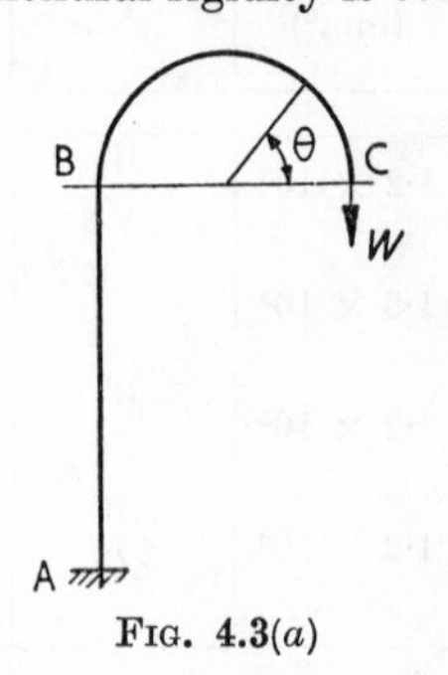

FIG. 4.3(*a*)

carries a load, W, acting vertically at C. Determine the vertical deflexion of C and the horizontal deflexion of B. (*St. Andrews*)

Solution. The point B is unloaded but since its deflexion is required an imaginary load, H, which will afterwards be equated to zero must be placed there.

The energy is due to bending only, consequently

$$\Delta_C = \frac{\partial U}{\partial W} = \int \frac{M_X}{EI} \cdot \frac{\partial M_X}{\partial W} \cdot dx$$

and
$$\Delta_B = \frac{\partial U}{\partial H} = \int \frac{M_X}{EI} \cdot \frac{\partial M_X}{\partial H} \cdot dx$$

Taking firstly the semicircular portion BC and measuring θ as shown, then the moment M_{xC} at a point X_C is given by

$$M_{xC} = WR(1 - \cos\theta)$$

$$\frac{\partial M_{xC}}{\partial W} = R(1 - \cos\theta)$$

$$dx = R d\theta$$

Hence
$$\frac{\partial U}{\partial W} = \frac{WR^3}{EI}\int_0^{\pi}(1 - \cos\theta)^2 \, d\theta$$

$$= \frac{WR^3}{EI}\left[\theta - 2\sin\theta + \frac{\theta}{2} + \sin\frac{2\theta}{4}\right]_0^{\pi}$$

$$= \frac{WR^3}{EI}\left(\frac{3\pi}{2}\right)$$

For the straight portion AB, the moment, M_{xB}, at a point distance x from B is given by

$$M_{xB} = 2WR + Hx$$

$$\frac{\partial M_{xB}}{\partial W} = 2R$$

$$\frac{\partial M_{xB}}{\partial H} = x$$

$$\frac{\partial U}{\partial W} = \frac{1}{EI}\int_0^{3R}(4WR^2 + 2HRx)dx = \frac{12WR^3}{EI} + \frac{9HR^3}{EI}$$

$$\frac{\partial U}{\partial H} = \frac{1}{EI}\int_0^{3R}(2WRx + Hx^2)\, dx = \frac{1}{EI}\left[WRx^2 + \frac{Hx^3}{3}\right]_0^{3R}$$

$$= \frac{9WR^3 + 9HR^3}{EI}$$

But $H = 0$ and for the whole structure

$$\frac{\partial U}{\partial W} = \frac{WR^3}{EI}\left(\frac{3\pi}{2} + 12\right)$$

$$\frac{\partial U}{\partial H} = \frac{9WR^3}{EI}$$

Hence the vertical deflexion of C is $\dfrac{16{\cdot}71WR^3}{EI}$ and the horizontal deflexion of B, $\dfrac{9WR^3}{EI}$.

The second part of this example will be worked out by the influence coefficient method integrating by the rule which was given on page 57.

The bending moment diagram for the portion AB due to the applied loading is shown in Fig. 4.3(*b*). It is a rectangular diagram of uniform bending moment intensity $2WR$, and the area of the diagram is $2WR \times 3R = 6WR.^2$

The bending moment diagram due to unit load applied horizontally at B is a triangle, as shown in Fig. 4.3(*c*).

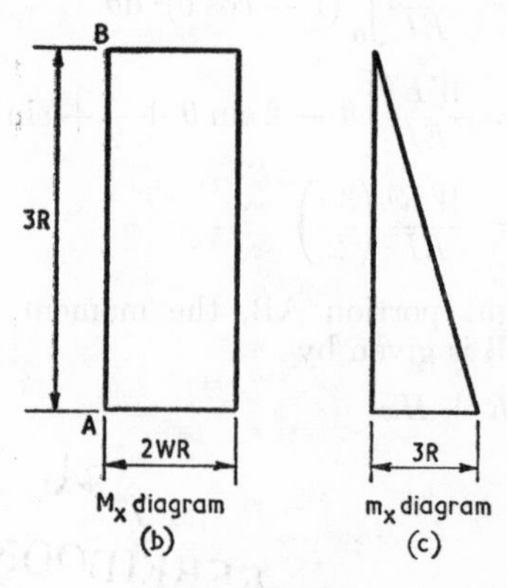

FIG. 4.3(*b*) (*c*)

The centre of area of the M_x diagram is at mid-height, at which point the ordinate of the m_x diagram has the value $\frac{3}{2}R$.

Hence $\dfrac{1}{EI}\displaystyle\int m_x M_x . dx = \dfrac{1}{EI} . 6WR^2 . \dfrac{3}{2}R = \dfrac{9WR^3}{EI}$

Hence the deflexion at B measured in a horizontal direction is $\dfrac{9WR^3}{EI}$.

Example 4.3. A frame consists of a vertical member DB 1 metre long and two equilateral triangles BAD and BCD on each side of it. At A there is a hinge to a rigid support and at D, the lower end of the member BD, there is a roller support. A load of 10 kN is carried at C. All tension members are 125 mm² and all compression members are 250 mm² in area.

Find the vertical deflexion of C and the movement of H on the rollers. Take $E = 200$ kN/mm.²

Solution. The frame is shown in Fig. 4.4, force H being applied at D in order to determine the movement on the rollers.

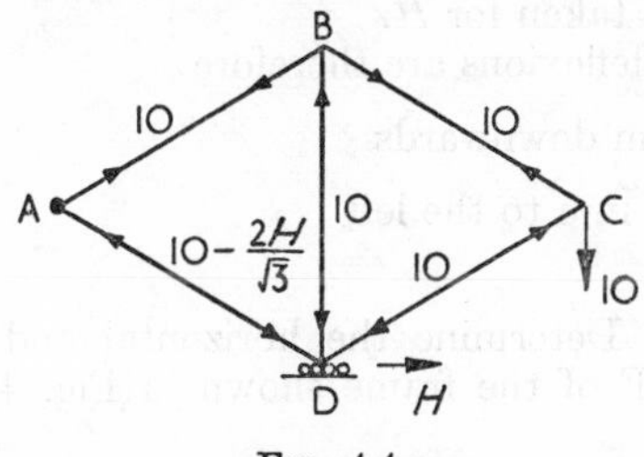

FIG. 4.4

This force is of course zero. Using the first theorem of Castigliano gives

$$\Delta_C = \frac{\partial U}{\partial W_{W=10}} = \sum \frac{PL}{AE} \cdot \frac{\partial P}{\partial W}$$

and

$$\Delta_D = \frac{\partial U}{\partial H_{H=0}} = \sum \frac{PL}{AE} \cdot \frac{\partial P}{\partial H}$$

The forces in the members are obtained by inspection, the results being shown in Fig. 4.4. The summations to obtain the deflexions are carried out in Table 4.3. The signs are important in this example in order to check on the assumed movement of D.

TABLE 4.3

Member	Length (mm)	Area (mm²)	P (kN)	$\frac{\partial P}{\partial W}$	$\frac{\partial P}{\partial H}$	$\frac{PL}{A} \cdot \frac{\partial P}{\partial W}$
AD		250	− 10	− 1	$+\frac{2}{\sqrt{3}}$	40
AB		125	+ 10	+ 1		80
BD	1000	250	− 10	− 1		40
BC		125	+ 10	+ 1		80
CD		250	− 10	− 1		40

$$\Delta_C = \frac{1}{E}\sum \frac{PL}{A}\cdot\frac{\partial P}{\partial W} = \frac{280}{200} = 1{\cdot}4 \text{ mm}$$

$$\Delta_D = \frac{1}{E}\sum \frac{PL}{A}\cdot\frac{\partial P}{\partial H}$$

$$= -\,40 \times \frac{2}{\sqrt{3}} \times \frac{1}{200} = -\,0{\cdot}231.$$

The negative value for Δ_D indicates that it is in the opposite direction to that taken for H.

The required deflexions are therefore

At C, 1·4 mm downwards

At H, 0·231 mm to the left

Example 4.4. Determine the horizontal and vertical movements of point F of the frame shown in Fig. 4.5. The load is

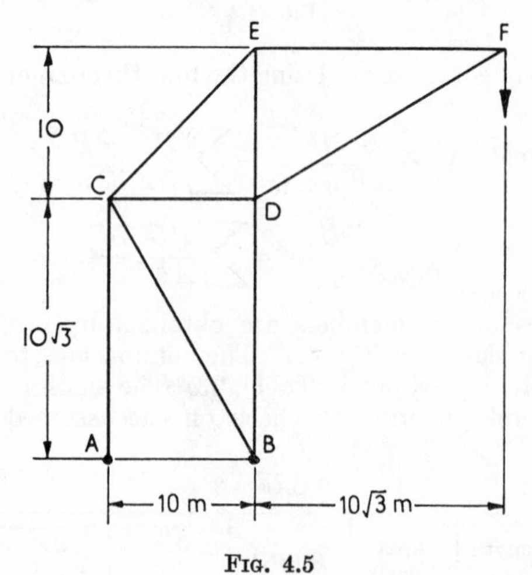

FIG. 4.5

such that the tension members of the frame are stressed to 120 N/mm² and the compression members to 60 N/mm.² Take $E = 200$ kN/mm.² *(Glasgow)*

Solution. Let W = vertical load at F and H = imaginary horizontal load.

Then the vertical movement of F, Δ_{FV}, is given by

$$\Delta_{FV} = \frac{\partial U}{\partial W} = \sum \frac{PL}{AE} \cdot \frac{\partial P}{\partial W}$$

and the horizontal movement of F, Δ_{FH}, is given by

$$\Delta_{FH} = \frac{\partial U}{\partial H} = \sum \frac{PL}{AE} \cdot \frac{\partial P}{\partial H}$$

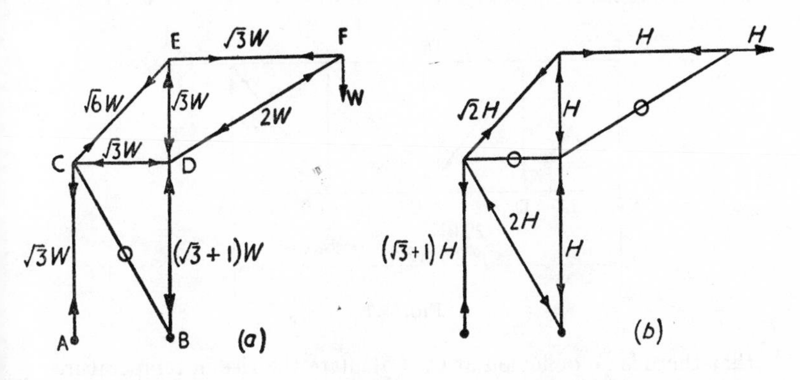

FIG. 4.6

The loads in the members are determined by inspection. Fig. 4.6 (*a*) shows the loads produced by *W* and Fig. 4.6 (*b*) the loads produced by *H*.

The summation is carried out by means of the table shown below.

TABLE 4.4

Member	Length (m)	P/A (N/m²)	$\frac{\partial P}{\partial W}$	$\frac{\partial P}{\partial H}$	$\frac{PL}{A} \cdot \frac{\partial P}{\partial W}$	$\frac{PL}{A} \cdot \frac{\partial P}{\partial H}$
AC	$10\sqrt{3}$	120	$+\sqrt{3}$	$+(\sqrt{3} \pm 1)$	3600	$3600 + 1200\sqrt{3}$
CE	$10\sqrt{2}$	120	$+\sqrt{6}$	$+\sqrt{2}$	$2400\sqrt{3}$	2400
EF	$10\sqrt{3}$	120	$+\sqrt{3}$	$+1$	3600	$1200\sqrt{3}$
BD	$10\sqrt{3}$	-60	$-(\sqrt{3} + 1)$	-1	$1800 + 600\sqrt{3}$	$600\sqrt{3}$
BC	20	—	—	—	—	—
CD	10	-60	$-\sqrt{3}$	0	$600\sqrt{3}$	—
DE	10	-60	$-\sqrt{3}$	-1	$600\sqrt{3}$	600
DF	20	-60	-2	0	2400	—

Summing up the various terms gives

$$\Delta_{FV} = \frac{18\,680}{200} = 93{\cdot}4 \text{ mm}$$

$$\Delta_{FH} = \frac{11\,800}{200} = 59{\cdot}0 \text{ mm}$$

Example 4.5. The top chord and vertical members in the truss shown in Fig. 4.7 have an area of 2000 mm.² The lower chord and diagonals are 1200 mm² in area. A load of 20 kN is carried at C and at the same time the top chord is heated with the result

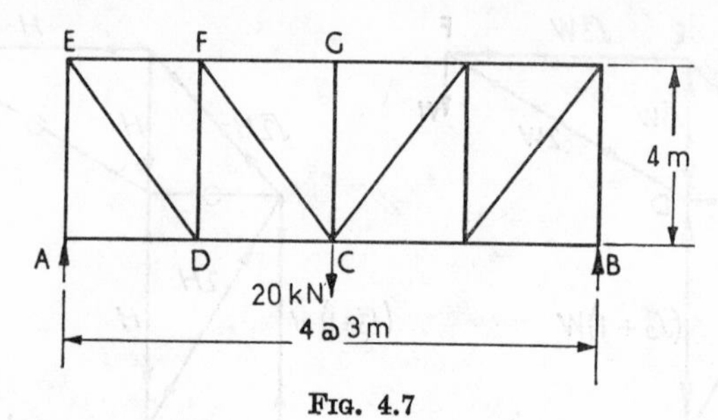

Fig. 4.7

that there is no deflexion at C. Calculate the rise in temperature of the top chord.

Take $E = 200$ kN/mm² and $\alpha = 1{\cdot}1 \times 10^{-5}$ per °C.

(*Glasgow*)

Solution. Before solving this problem it will be advisable to study the deflexion produced by a change of temperature.

The deflexion at a given point in the line of action of a load, W, acting at that point is given by

$$\Delta = \frac{\partial U}{\partial W} = \sum \frac{PL}{AE} \cdot \frac{\partial P}{\partial W}$$

this expression also holding when W is put equal to zero.

The term $\dfrac{PL}{AE}$ is the change in length of the member under load: if this is denoted by ε, then the expression for deflexion can be rewritten

$$\Delta = \Sigma\varepsilon \,.\, \frac{\partial P}{\partial W}$$

The effect of a change of temperature is to cause a change in length equal to $L\alpha t$ where L = length of the member, α = coefficient of linear expansion and t = change of temperature.

The deflexion Δ_T due to change of temperature may therefore be obtained from the expression

$$\Delta_T = \Sigma L\alpha t \frac{\partial P}{\partial W}$$

where P is load in the member due to load W acting at the point where the deflexion is required.

The solution of Example 4.5 is as follows.

Since the frame is symmetrical only the left-hand half will be considered. The loads in the members when a load, W, acts vertically at C are shown in Fig. 4.8.

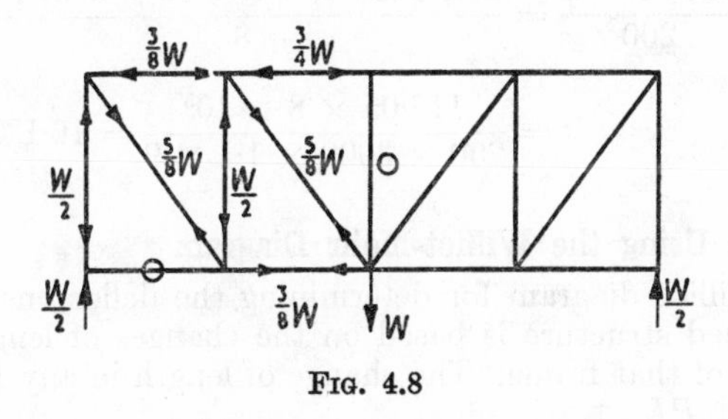

FIG. 4.8

The vertical deflexion of C due to a 20 kN load is calculated by means of the table below.

TABLE 4.5

Member	Length (mm)	Area (mm²)	P	$\frac{\partial P}{\partial W}$	$\frac{PL}{A} \cdot \frac{\partial P}{\partial W}$
AD	3000	1200	0	0	0
DC	3000	1200	+ 7·5	0·375	7·05
EF	3000	2000	− 7·5	− 0·375	4·75
FG	3000	2000	− 15·0	− 0·75	18·75
AE	4000	2000	− 10·0	− 0·50	11·25
FD	4000	2000	− 10·0	− 0·50	11·25
ED	5000	1200	+ 12·5	+ 0·625	32·5
FC	5000	1200	+ 12·5	+ 0·625	32·5

Summing up the last column gives the deflexion, Δ_{C2} due to the 2-ton load as

$$\Delta_{C2} = \frac{118{\cdot}05 \times 2}{200} \text{ mm}$$

The deflexion at C due to the heating of the top chord members Δ_{CT} is given by

$$\Delta_{CT} = 2 \times \Sigma L\alpha t \frac{\partial P}{\partial W} \quad \text{(for the members EF and FG)}$$

$$= 2 \times 3000 \times 1{\cdot}1 \times 10^{-5} \times t \times (\tfrac{3}{8} + \tfrac{3}{4})$$

(The negative signs for $\frac{\partial P}{\partial W}$ show that C moves upward due to an expansion of the top chord members.)

The deflexions Δ_{CT} and Δ_{C2} are equal hence

$$\frac{118{\cdot}05 \times 2}{200} = \frac{2 \times 3000 \times 1{\cdot}1 \times 10^{-5} \times 9 \times t}{8}$$

giving
$$t = \frac{118{\cdot}05 \times 8 \times 10^5}{200 \times 3000 \times 1{\cdot}1 \times 9} = 16{\cdot}1^\circ\text{C}$$

Examples Using the Williot-Mohr Diagram

The Williot diagram for determining the deflexions at points in a framed structure is based on the changes of length in the members of that frame. The change of length in any member is given by $\frac{PL}{AE}$ (the symbols having the meaning previously defined). The preliminary calculations, namely the determination of the load, P, in each member, are therefore the same as in the strain energy analysis. Having calculated the change in length the Williot diagram is drawn assuming that one point remains fixed and that one member does not change direction. If these conditions are satisfied in the problem then the displacement can be obtained from the Williot diagram alone. If they are not then a Mohr diagram must be superimposed on the Williot diagram to allow for the rotation of the member assumed fixed. By measuring from the Mohr diagram to the corresponding point on the Williot diagram the deflexion of that point is obtained.

Example 4.6. Use the Williot diagram to solve the problem given in Example 4.4.

TABLE 4.6

Member	AC	CE	EF	BC	BC	CD	DE	DF
$\frac{fL}{E}$	$+6\sqrt{3}$	$+6\sqrt{2}$	$+6\sqrt{3}$	$-3\sqrt{3}$	0	-3	-3	-6

Solution. The change in length of any member $= \dfrac{PL}{AE} = \dfrac{fL}{E}$.
The changes in lengths of the members are given in table 4.6.

The negative signs indicate reductions and the positive signs extensions

Since the supports A and B are pinned both remain fixed in position and the direction AB remains unchanged. Consequently the deflexions can be determined directly from a Williot diagram without a Mohr diagram. (This applies in all cases where the frame is pinned to both supports.) The diagram is given in Fig. 4.9.

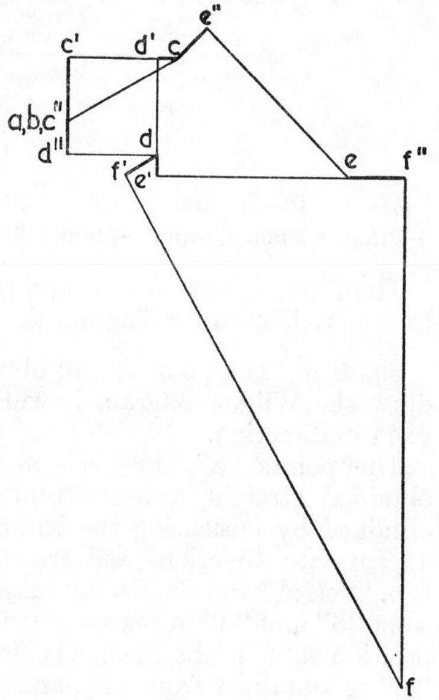

Using small letters to denote the positions of the panel points on the Williot diagram "a, b" is the starting point and "c" is the next point to fix. This is done by drawing ac′ $= 6\sqrt{3}$ units in the direction of the movement of C relative to A, i.e. in the direction AC: the member BC is unstressed hence c″ coincides with b. Lines are now drawn through c′ perpendicular to AC and through c″ perpendicular to BC the intersection of the perpendiculars giving the point "c."

Fig. 4.9

The point "d" is now fixed: cd′ is drawn 3 units long in the direction DC and bd″ is drawn $3\sqrt{3}$ units long in the direction DB. Perpendiculars to DC and DB respectively are drawn through d′ and d″, the intersection giving the position of "d."

Point "e" on the diagram is fixed from d and c by drawing de′ 3 units long in the direction ED and ce″ $6\sqrt{2}$ units long in the direction CE. Perpendiculars to DE and CE are drawn through e′ and e″ respectively their intersection giving "e."

A similar construction is used to determine "f." Then the horizontal distance "af" gives the horizontal displacement of F and the vertical distance of "f" from "a" gives the vertical displacement.

Scaling these values gives results approximately equal to the answers obtained using the analytical method (see page 64).

Example 4.7. Under a given load system the changes in length of the bars of the frame shown in Fig. 4.10 are as follows

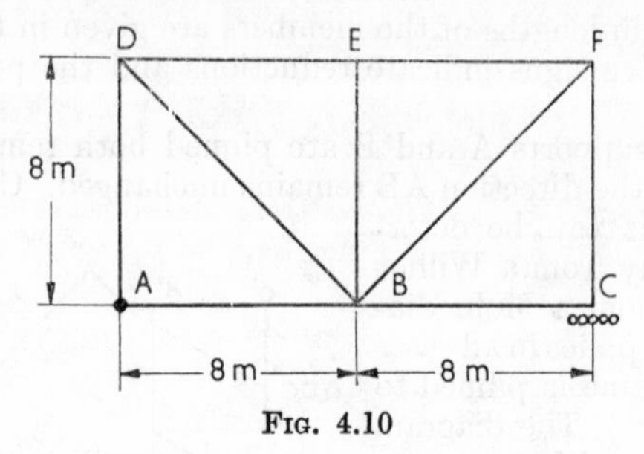

Fig. 4.10

AB	BC	DE	EF	AD	BE	CF	DB	BF
+ 3mm	+ 6mm	−3mm	−3mm	−3mm	0	−1·5mm	+ 4·5mm	+ 6mm

The frame is pinned at A and rests on rollers at C. Determine from a Williot–Mohr diagram the deflexion of E.

Solution. The point A will obviously be fixed but in order to draw the Williot diagram it will be assumed that AD remains fixed in direction.

The points "a" and "d" on the Williot diagram are thus obtained straight away. From them the position of "b" is obtained by measuring the strained lengths ab′ and db″ in the appropriate directions and erecting the perpendiculars b′b and b″b. (*Note*: b″ and b virtually coincide.) The point "e" is obtained from "b" and "d" in the standard way and from "e" and "b" the position of "f" is obtained. The final point on the Williot diagram, "c," is obtained from "b" and "f." The diagram is given in Fig. 4.11.

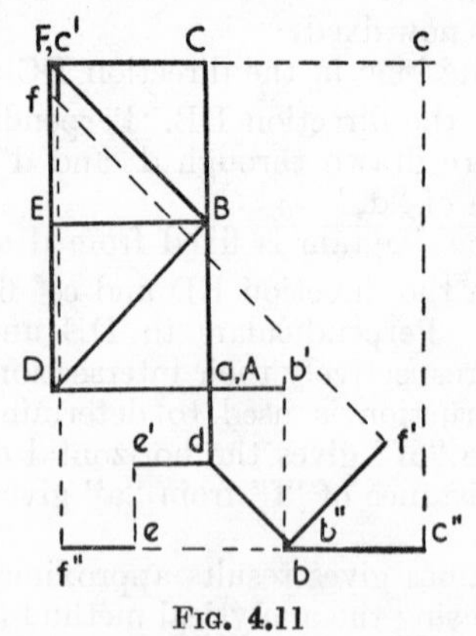

Fig. 4.11

The support, C, is a roller bearing therefore it can only move in a horizontal direction. The final deflexion of C, however, is obtained by measuring from C on the Mohr diagram to "c" on the Williot diagram. Hence the point C on the Mohr diagram must lie on a horizontal line through "c." The point A does not move and to construct the Mohr diagram a vertical is drawn through A to cut the horizontal through "c" in C. If capital letters are used for the Mohr diagram and small letters for the Williot then "a" and A coincide and the Mohr diagram is the original truss configuration drawn to scale on AC as base. This is denoted by ABCFED.

The deflexion of E in the frame is given by the distance Ee and on scaling is found to be 13 mm vertically and 3·5 mm horizontally.

Example 4.8. The bottom chord of a Warren girder slopes at 30° to the horizontal. Each member is 3 m long. When loaded as shown in Fig. 4.12 the stress in the tension members is

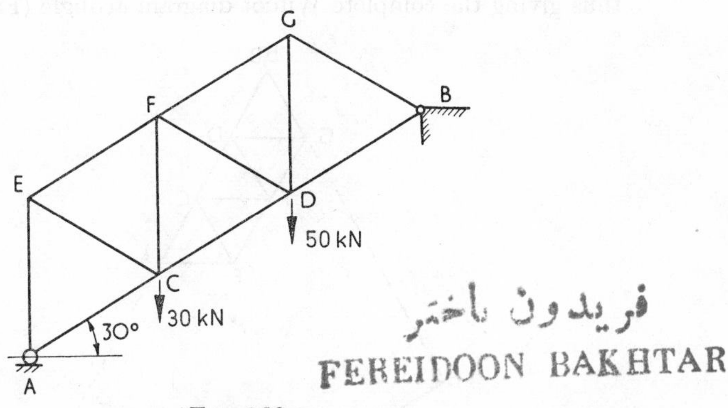

FIG. 4.12

80 N/mm² and that in the compression members 60 N/mm². The support A yields 75 × 10⁻³ mm/kN reaction. Taking CD as reference find the deflexions of the joints E, C and D.

Take $E = 200$ kN/mm².

Solution. Taking moments about B gives $R_A = \frac{110}{3}$ kN.

Resolving vertically gives $R_B = \frac{130}{3}$ kN.

Using the method of inspection to determine the forces in the frame gives the following results—

Tension members: CD, DB, EC, FD and DG

Compression members: AE, EF, FG, GB and FC

Unstressed member: AC

Since the stresses in the compression members are 60 N/mm² and their length is 3 m the reduction in length of each compression member is

$$\frac{60 \times 3000}{200 \times 10^3} = 0{\cdot}9 \text{ mm}$$

The increase in length of the tension members is 1·2 mm.

The load on support A acts vertically consequently the settlement due to load is in a vertical direction having a magnitude of

$$\frac{110}{3} \times \frac{75}{10^3} = 2{\cdot}75 \text{ mm}$$

The question asks for CD to be taken as the reference member for the Williot diagram. Since CD is a tension member the line "cd" is 1·2 m long. On this as base the Williot diagram is built up in the following way; "f" is fixed from "c" and "d"; "e" is fixed from "c" and "f"; and "a" from "c" and "e"; "g" is obtained from "f" and "d," and finally "b" from "g" and "d" thus giving the complete Williot diagram acdbgfe (Fig. 4.13).

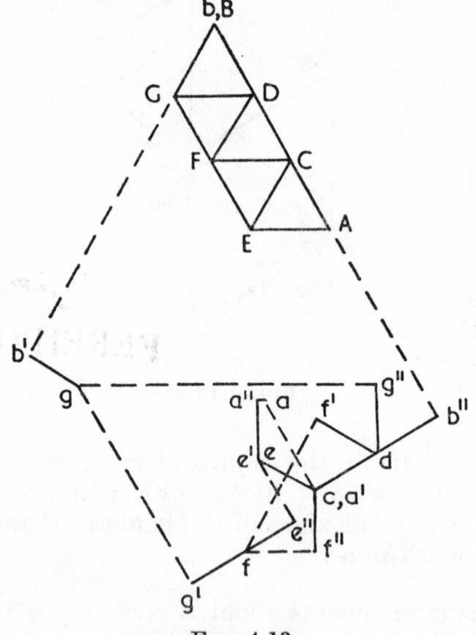

FIG. 4.13

In order to draw the Mohr diagram the following facts are used—

1. B is a pinned joint and therefore does not move; hence B on the Mohr diagram coincides with "b" on the Williot.

2. The vertical movement of A is 2·75 mm hence A on the Mohr diagram is 2·75 mm vertically above "a" on the Williot diagram.

3. The Mohr diagram is perpendicular to the original truss configuration hence its base is on a line inclined at 30° to the vertical through B.

The last two factors enable A on the Mohr diagram to be obtained. On AB as base the complete diagram ACDBGFE is drawn.

The deflexions of the joints E, C and D are obtained by measuring from these points on the Mohr diagram to the corresponding points on the Williot diagram.

Scaling the distances gives—

Movement of E: 3·9 mm downwards + 0·05 mm to the right
" " C: 5·5 mm " + 0·50 mm " " "
" " D: 5·9 mm " + 2·1 mm " " "

Examples on Deflexions

1. Fig. 4.14 shows a pin-jointed framework pinned to the wall at G and H and loaded as shown. If all the members have the

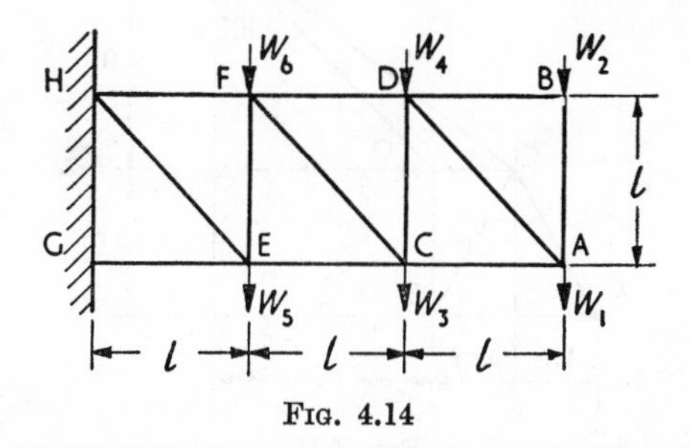

Fig. 4.14

same cross-sectional area and Young's Modulus, find the vertical displacements of A and F.

Answer.
$$\Delta_A = \frac{L}{AE}\{29{\cdot}48(W_1 + W_2) + 16{\cdot}65W_3 + 17{\cdot}65W_4 + 5{\cdot}83W_5 + 6{\cdot}83W_6\}$$

$$\Delta_F = \frac{L}{AE}\{6{\cdot}83(W_1 + W_2) + 5{\cdot}83(W_3 + W_4) + 3{\cdot}83W_5 + 4{\cdot}83W_6\}$$

2. The Warren girder shown in Fig. 4.15 is loaded so that the members indicated by heavy lines are in compression, the stress in them being 75 N/mm². Determine the vertical deflexion of C if the stress in the tensile members is 125 N/mm.² Take $E = 200$ kN/mm².

Answer. 8·4 mm.

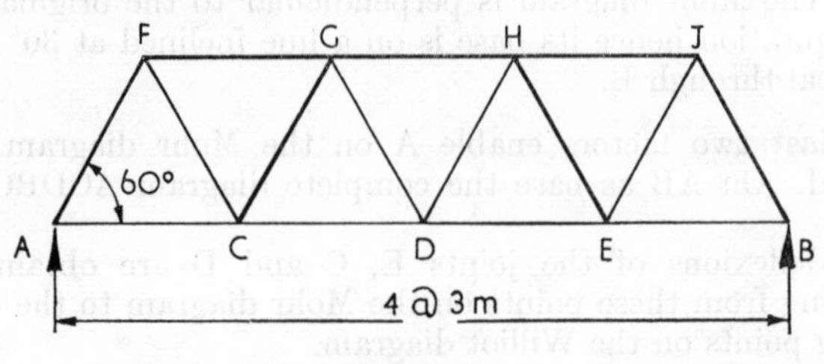

FIG. 4.15

3. Calculate the deflexion under the load for the frame shown in Fig. 4.16 which is pinned to the supports A and B. All the members are 1,300 mm² in area and $E = 200$ kN/mm².

Answer. 120 mm.

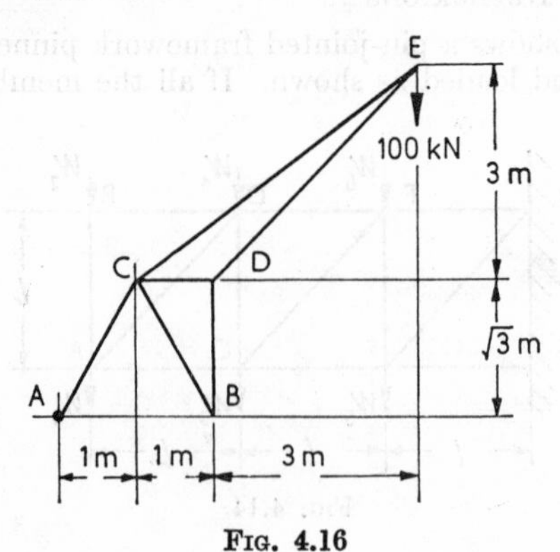

FIG. 4.16

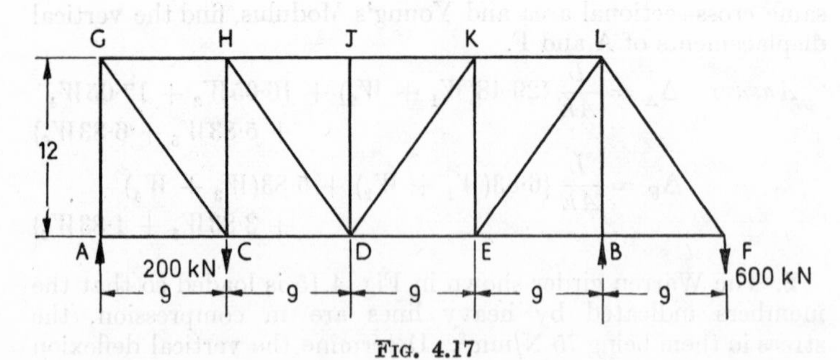

FIG. 4.17

4. If the loaded members of the frame shown in Fig. 4.17 are stressed to 125 N/mm² and $E = 200$ kN/mm² determine the vertical deflexion at the point F. Dimensions in metres.

Answer. 52·6 mm.

5. The frame shown in Fig. 4.18 has a stiff joint at C and is fully fixed at D. If the flexural rigidity for CD is twice that for AC and BC determine the vertical deflexion of B when the frame carries a load of W at A.

Answer. $\dfrac{WL^3}{4EI}$ upward.

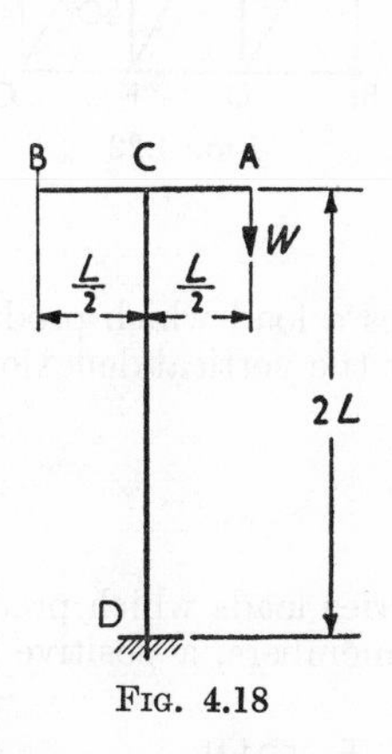

FIG. 4.18

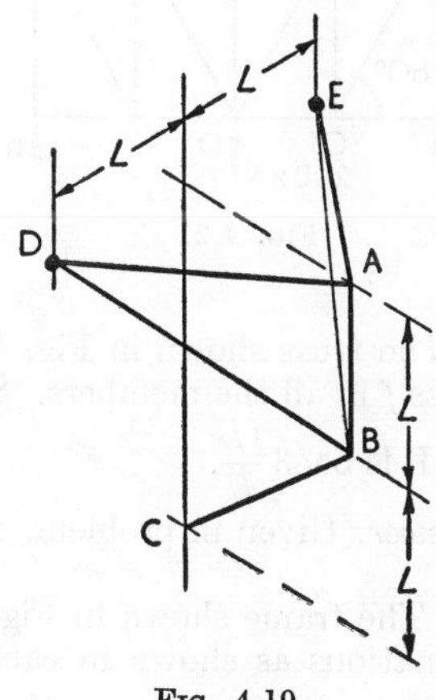

FIG. 4.19

6. A rigid bar AB carries a load, W, and is supported in space as shown in Fig. 4.19. If all the supporting members have the same extensibility aE, calculate the vertical deflexion of A.

(*St. Andrews*)

Answer. $1{\cdot}36\,\dfrac{WL}{aE}$.

7. Calculate the deflexion under the load for the frame shown in Fig. 4.20. The members are all 1300 mm² in cross-section and the supports A and B are pins. $E = 200$ kN/mm².

Answer. 19·8 mm

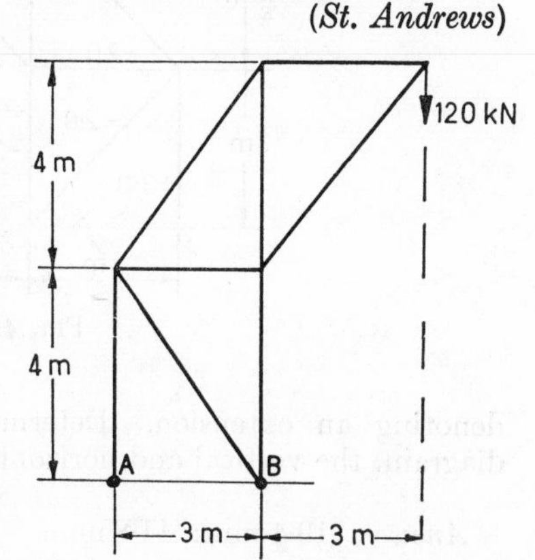

FIG. 4.20

8. Find the deflexion of point C in the frame shown in Fig. 4.21 when a vertical load of 200 kN acts at D.

The cross-sectional areas of the members are—

Upper and lower chords, 4000 mm²; Diagonals, 1350 mm²; Verticals, 2000 mm². Take $E = 200$ kN/mm².

Answer. 6·32 mm.

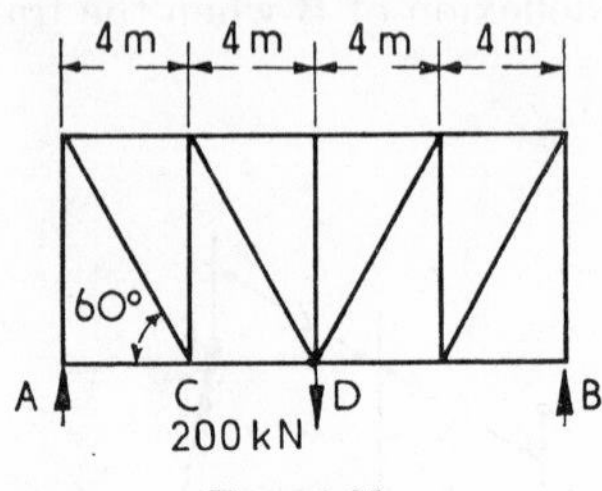

FIG. 4.21

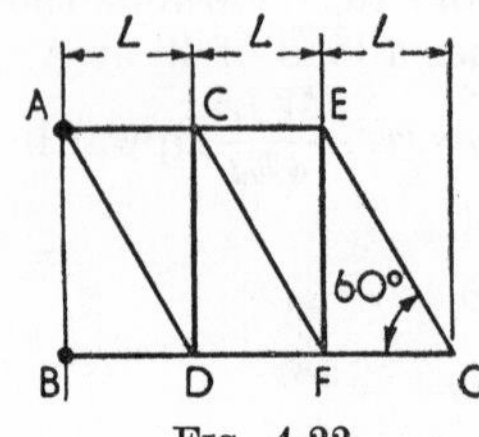

FIG. 4.22

9. The truss shown in Fig. 4.22 carries a load which produces a stress f in all the members. Show that the vertical deflexion at point F is $5\sqrt{3}\dfrac{fL}{E}$.

Answer. Given in problem.

10. The frame shown in Fig. 4.23 carries loads which produce deformations as shown in each of the members, a positive sign

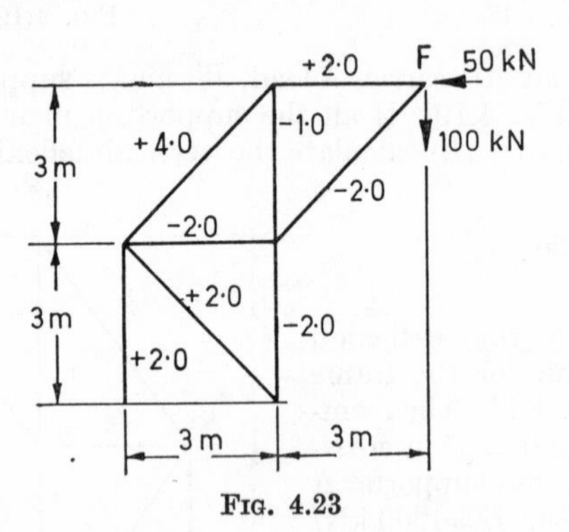

FIG. 4.23

denoting an extension. Determine, preferably by a Williot diagram, the vertical and horizontal displacements of F.

(*St. Andrews*)

Answer. 19·4 mm; 11·8 mm.

11. The frame shown in Fig. 4.24 carries a load at G which produces a stress of 30 N/mm² in each member. Determine

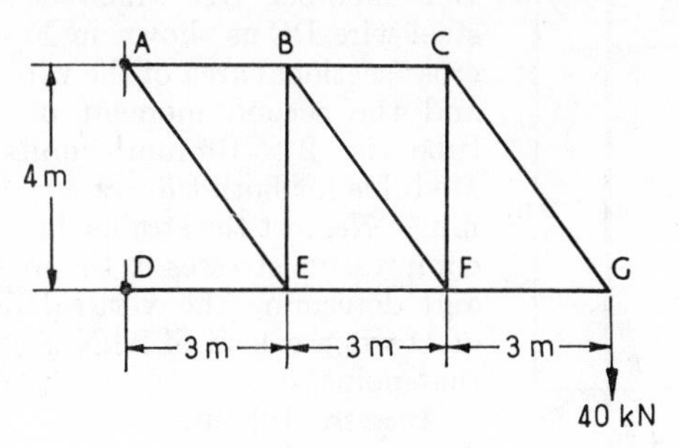

FIG. 4.24

by means of a Williot diagram the horizontal movement of F. $E = 200$ kN/mm².

Answer. 0·89 mm towards D.

12. Calculate in terms of P the magnitude of the horizontal load which must be applied at C in the frame shown in Fig. 4.25

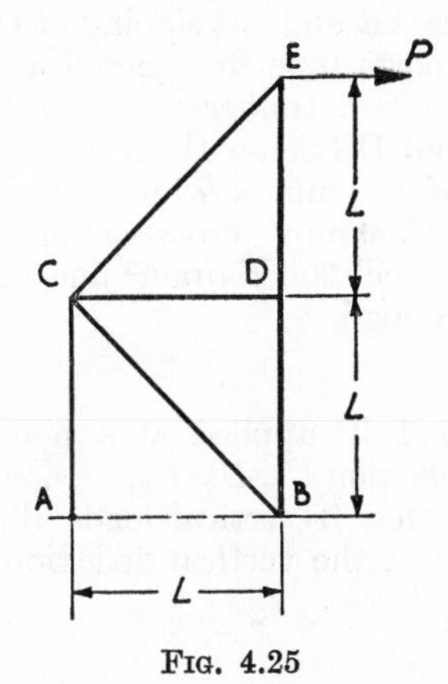

FIG. 4.25

in order that the horizontal deflexion of E is zero. State the vertical deflexion of C under the combined loading. The cross-sectional areas of the members are equal. (*Glasgow*)

Answer. $2{\cdot}41P$; $\dfrac{0{\cdot}41PL}{AE}$.

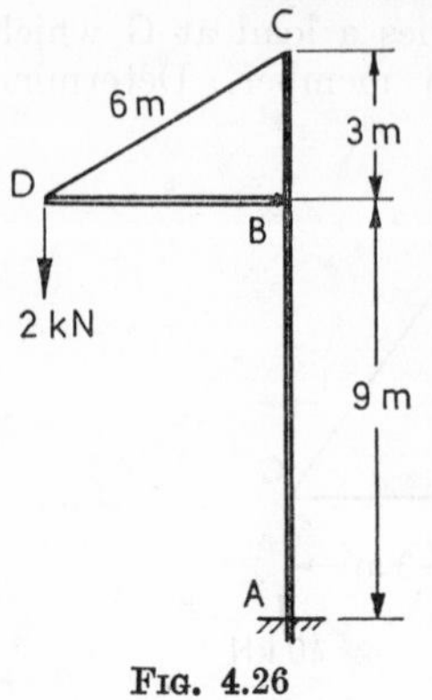

FIG. 4.26

13. A vertical steel tube ABC is encastred at the base A and has hinged to it at B a member BD which is stayed by a steel wire DC as shown in Fig. 4.26. The cross-sectional area of the wire is 32·0 mm² and the second moment of area of the tube is 2×10^6 mm⁴ units. Young's Modulus for both tube and wire is 200 kN/mm². Neglect the strains due to the direct compressive stresses in the mast and strut and determine the vertical displacement of D when a load of 2 kN is suspended at that point. (*London*)

Answer. 1·26 m.

14. A rectangular davit ABC of constant section is rigidly fixed at A. The vertical leg AB is 5 m long and the horizontal arm BC is 2 m long, the joint at B being rigid. Calculate the horizontal and vertical deflexions of C when a vertical load of 30 kN is suspended from it. The section has $I = 120 \times 10^6$ mm⁴ units and $E = 200$ kN/mm². (*London*)

Answer. 36·5 mm horizontal; 33·0 mm vertical.

15. A high pitched roof is supported by trusses made up as follows: Top rafters AB and AC sloping at 60° to the horizontal and resting on supports (one free to roll and one hinged) at B and C, consisting of two timber members 200 mm × 100 mm; two members DB and DC below them sloping at 30° to the horizontal and made of 150 mm × 75 mm timbers; and a vertical steel rod AD of 1300 mm² cross-section. If E for wood = 10 kN/mm² and for steel 200 kN/mm² find the vertical deflexion of A under a 50 kN load. (*London*)

Answer. 6·0 mm.

16. A vertical load W applied at a point P in a structure produces vertical deflexion $U_1, U_2, U_3, \ldots$ at points 1, 2, 3, ... respectively. Prove that if vertical loads $W_1, W_2, W_3, \ldots$ are applied at 1, 2, 3, ... the vertical deflexion of the point P will be $\sum \frac{W_1 U_1}{W}$.

Fig. 4.27 shows a cantilever frame attached to a rigid vertical support. All the members have the same cross-section a and the inclined members are at 60° to the horizontal. Show that with the given loading the deflexion at A is $\frac{418Wl}{3aE}$. (*Cambridge*)

Answer. Given in question.

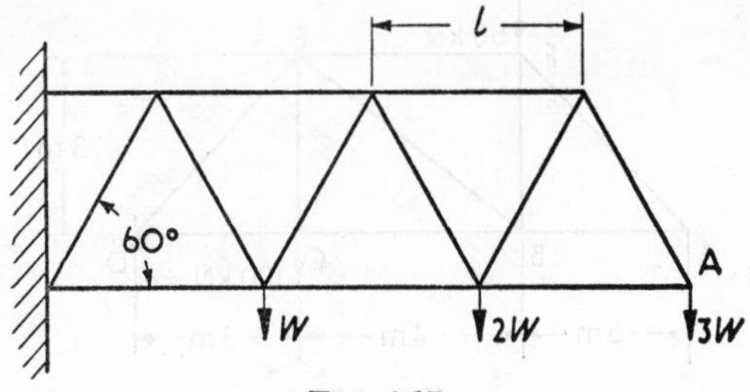

FIG. 4.27

17. A bar AB of uniform cross-section is bent into the form of a semicircle as shown in Fig. 4.28. It is encastred vertically at A to a fixed rigid support and at B to a rigid arm BC. The line BC is horizontal and C is the centre of the semicircle.

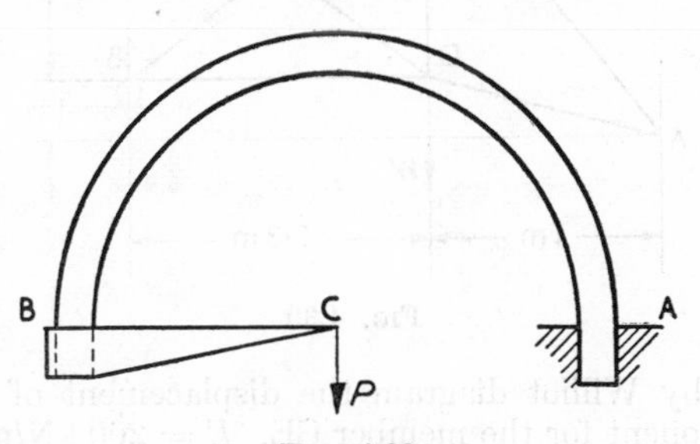

FIG. 4.28

If a vertical force, P, be applied at C show that, neglecting the strains due to forces normal to radial cross-sections and to shear, the arm BC remains horizontal and obtain an expression for the displacement of C. Also show that whatever be the direction of P the total displacement of C remains the same in amount and takes place in the direction of P. *(Cambridge)*

Answer. $\dfrac{\pi P R^3}{2EI}$.

18. The frame shown in Fig. 4.29 is supported on a pin at A and a roller bearing at D. The members have an area of 2000 mm² and $E = 200$ kN/mm². Determine the vertical displacements of F and C and the movement of D on the rollers.

Answer. D = 1·94 mm; F = 2·13 mm; C = 3·95 mm.

19. The truss shown in Fig. 4.30 is hinged at A and simply supported at B. Under a load, W, acting at D the ties and struts are stressed to 90 N/mm² and 60 N/mm² respectively.

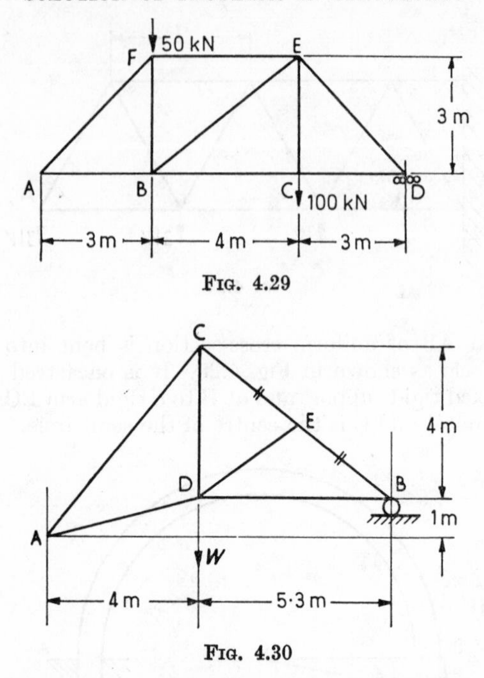

Fig. 4.29

Fig. 4.30

Determine by Williot diagram the displacement of B and the swing component for the member CE. $E = 200$ kN/mm^2.

(Aberdeen)

Answer. 6·75 mm; 0·00207 radians.

20. The framed structure shown in the Fig. 4.31 is hinged to supports at A and on a roller bearing at B. Under a vertical load, W, at C, the tension members are stressed to 90 N/mm^2 and the

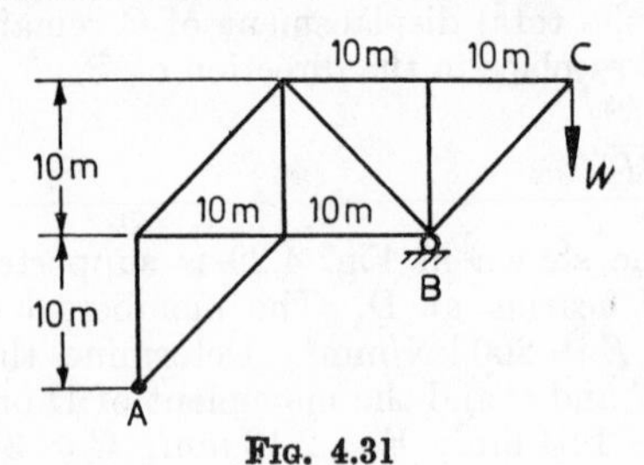

Fig. 4.31

compression members to 60 N/mm^2. Determine the horizontal movement at the roller bearing. $E = 200$ kN/mm^2.

Answer. 3·05 mm.

21. A rod of diameter d is bent as shown in Fig. 4.32, A, B, C and D being co-planar. The rod is rigidly fixed at B and a moment

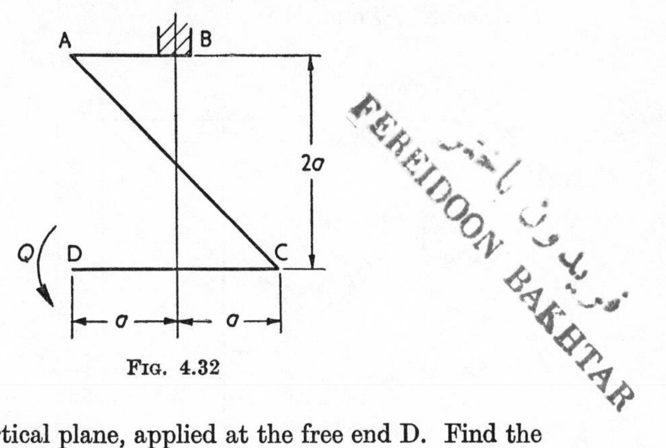

Fig. 4.32

Q, acting in a vertical plane, applied at the free end D. Find the vertical movement of D.

Answer. $\dfrac{341Qa^2}{\pi Ed^4}$.

22. The crane shown in Fig. 4.33 lifts a load of 50 kN. The rope lifting the load passes over pulleys at A and B and is attached to the ground at C. Find the vertical deflexion of the point A.

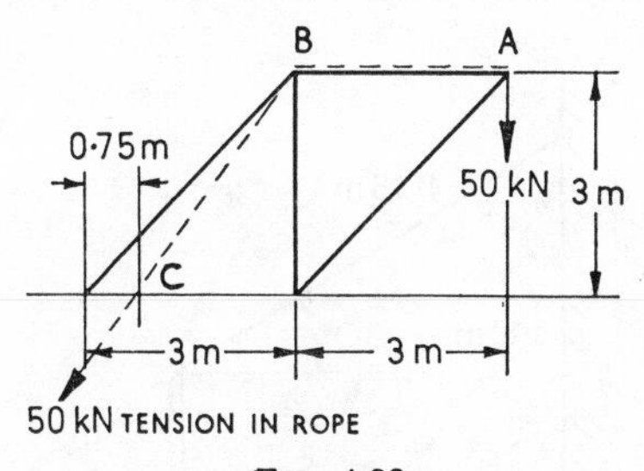

Fig. 4.33

The cross-sectional area of each member is 2000 mm² and the frame is pin-jointed throughout. $E = 200$ kN/mm². (*Glasgow*)

Answer. 1·97 mm.

23. A tower frame carries horizontal and vertical loads as shown in Fig. 4.34 which stress the ties to 120 N/mm² and the struts to 90 N/mm². Draw the Williot diagram displacement.

Determine the vertical movement of B, by jacking the column base there, required to counteract the horizontal movement of C due to the loading.

Answer. 7·7 mm.

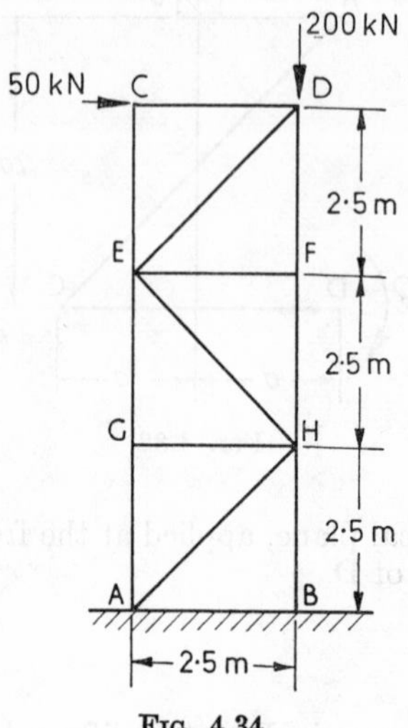

FIG. 4.34

24. A continuous member ACD is pinned to a wall at A and at D to the tie ED (Fig. 4.35). Find the vertical deflexion of C

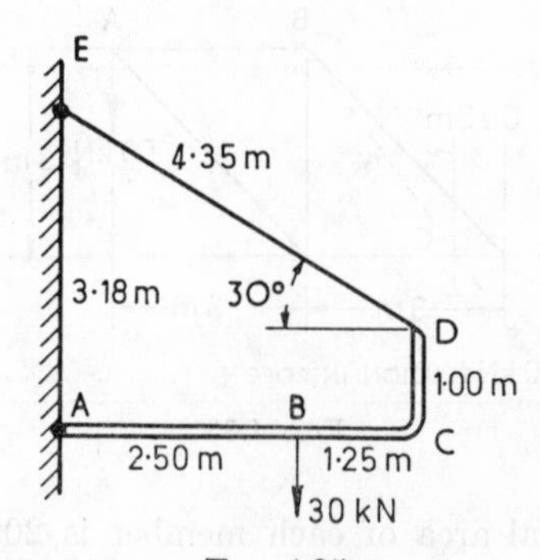

FIG. 4.35

due to 30 kN at B. I for ACD is 8×10^6 mm^4 and the area of the tie 322 mm^2. Neglect the direct deformations of ACD. $E = 200$ kN/mm^2. *(Glasgow)*

Answer. 48·5 mm.

Chapter 5

Redundant Pin-jointed Frameworks

THE FIRST CHAPTER in this book dealt with the analysis of statically determinate frames and gave the rules whereby it could be determined whether a frame satisfied this condition. A frame which cannot be analysed by the application of the laws of static equilibrium is known as statically indeterminate or redundant. The major part of the time in the average course on Theory of Structures is spent on dealing with frames which are redundant due to various reasons. The term "redundant frame" covers a wide field and it would be unwise to deal with the whole of the subject in one chapter.

Redundant frameworks can, however, be divided into three main groups.

1. Pin-jointed frames which are redundant because they have too many bars.
2. Frames which are redundant because the joints are stiff and not pinned.
3. Frames which are redundant because too many reactive forces have been provided.

The first class are described as redundant pin-jointed frameworks and are dealt with in this chapter. The second class are known generally as stiff-jointed frames and examples of this type are dealt with in Chapter 9. The third class covers such structures as arches which are dealt with in Chapter 8.

The main method for dealing with redundant frameworks is the application of the second theorem of Castigliano or the method of least work. If there are in a framework more bars than are required to satisfy the relationship, in the case of a plane frame, of $n = 2j - 3$, the additional bars constitute redundances. If R is the force in any redundant member then the second theorem of Castigliano states that

$$\frac{\partial U}{\partial R} = \lambda$$

when U = strain energy of the whole frame

λ = initial lack of fit of redundant member.

If there is no initial lack of fit in the redundant member then $\dfrac{\partial U}{\partial R} = 0$. This, however, is the condition when the strain energy is a minimum, i.e. the force in the redundant member is such that the least work is done when the structure moves into its deflected position.

If the member has an initial lack of fit then the assumed force in that member must be taken consistent with the lack of fit. If it is too short, a tension must be taken and if too long, a compression.

This chapter is concerned mainly with pin-jointed frames. In such a frame the loads are axial; in consequence the strain energy is direct energy or symbolically

$$U = \sum \frac{P^2 L}{2AE}$$

where P, L, A and E have the meanings previously given

and
$$\frac{\partial U}{\partial R} = \sum \frac{PL}{AE} \cdot \frac{\partial P}{\partial R}$$

There are, however, certain types of frame which combine bending energy with direct energy and these are dealt with in this chapter.

Example 5.1. Three members BA, CA and DA shown in Fig. 5.1 are pinned to a rigid member, BCD, and to each other at A.

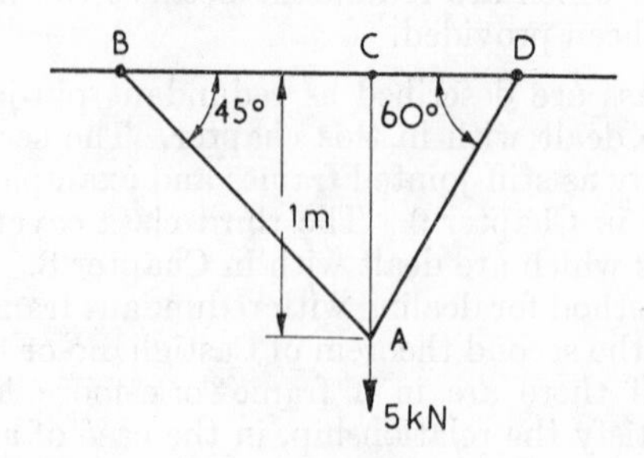

FIG. 5.1

The members AB and AC are of steel with areas 150 mm² and 100 mm² respectively. AD is a duralumin rod with area 200 mm². Determine the loads in all the members when a vertical load of 5 kN acts at A.

E for steel = 200 kN/mm². E for duralumin = 60 kN/mm².

Solution. In this frame, which has pinned supports, there is one free joint, A, consequently the number of bars required for a statically determinate frame is $2 \times 1 = 2$. There are three bars provided hence there is one redundancy.

Any of the three bars could be taken as redundant. Take AC as the redundant member and let the force in it $= R$ tons. Since the members have no initial lack of fit then

$$\sum \frac{PL}{AE} \cdot \frac{\partial P}{\partial R} = 0$$

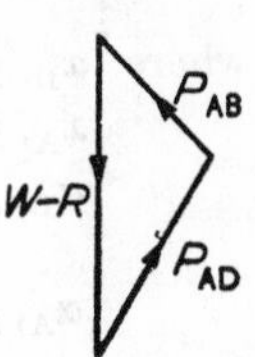

FIG. 5.2

The load, P, in each of the members must first be obtained. The triangle of forces for A shown in Fig. 5.2 can be used for this.

It gives

$$P_{AD} = \frac{2(5 - R)}{\sqrt{3} + 1} \qquad P_{AB} = \frac{\sqrt{2}(5 - R)}{\sqrt{3} + 1}$$

The summation is carried out as shown in the table below.

TABLE 5.1

Member	Length m	Area mm²	E (kN/mm²)	P	$\frac{\partial P}{\partial R}$	$\frac{PL}{AE} \cdot \frac{\partial P}{\partial R}$
AB	$1{\cdot}0\sqrt{2}$	150	200	$0{\cdot}519(5 - R)$	$-0{\cdot}519$	$-1{\cdot}375(5 - R) \times 10^{-2}$
AC	$1{\cdot}0$	100	200	R	$+1{\cdot}0$	$5{\cdot}00R \times 10^{-2}$
AD	$\frac{2{\cdot}0\sqrt{3}}{3}$	200	60	$0{\cdot}733(5 - R)$	$-0{\cdot}733$	$-5{\cdot}20(5 - R) \times 10^{-2}$

Summing gives $R(5{\cdot}20 + 5{\cdot}00 + 1{\cdot}38) = (1{\cdot}38 + 5{\cdot}20)$ leading to $R = 2{\cdot}85$ kN

Hence force in AC $= 2{\cdot}85$ kN

force in AB $= 0{\cdot}519(5 - 2{\cdot}85) = 1{\cdot}12$ kN

force in AD $= 0{\cdot}733(5 - 2{\cdot}85) = 1{\cdot}58$ kN

Example 5.2. Derive an expression for the force in a member of a framed structure in terms of the displacement of its ends under the applied loading.

Determine the vertical and horizontal movements of the point D in the frame shown in Fig. 5.3 when a load, W, acts vertically at that point. The bars are all made from the same material and the cross-sectional areas of AD, BD and CD are $4A$, A and $2A$ respectively. Determine also the forces in all the bars. (*Oxford*)

The answer to the bookwork in the first part of the question can be shown to be

$$T_{AB} = \left\{ \frac{(x_B - x_A)(\alpha_B - \alpha_A) + (y_B - y_A)(\beta_B - \beta_A)}{L^2} \right\} AE$$

where x_B, y_B are co-ordinates of B
x_A, y_A are co-ordinates of A
α_B, β_B are displacements of B in x and y directions respectively
α_A, β_A are displacements of A in x and y directions respectively
L = length of AB
AE = extensibility of bar AB

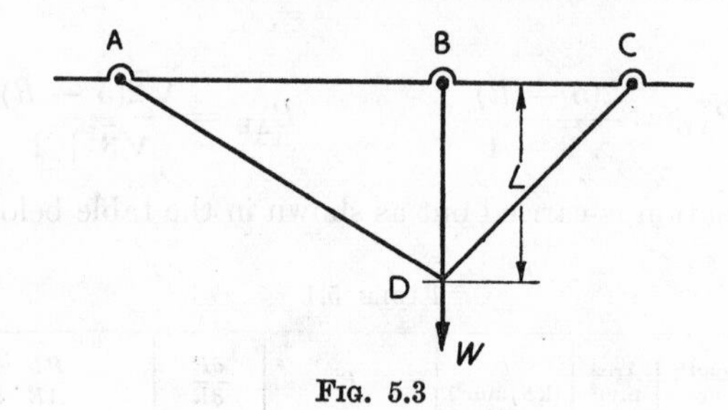

Fig. 5.3

This expression can be used to determine the forces and displacements of a redundant framework similar to that shown in the figure.

Let α, β be the displacements in the x and y directions of point D.

Then $$T_{AB} = \frac{(\sqrt{3}L \,.\, \alpha + L\beta)}{4L^2} 4AE = \frac{(\sqrt{3}\alpha + \beta)}{L} AE$$

$$T_{BD} = \frac{L\beta}{L^2} AE = \frac{\beta}{L} AE$$

$$T_{CD} = \frac{(-L\alpha + L\beta)}{2L^2} 2AE = \frac{-\alpha + \beta}{L} AE$$

Resolving horizontally at D gives

$$T_{AD} \cos 30° = T_{CD} \cos 45°$$

and substituting for T_{AD} and T_{CD} leads to the equation

$$\frac{\sqrt{3}}{2}(\sqrt{3}\alpha + \beta) = \frac{\sqrt{2}}{2}(\beta - \alpha)$$

Resolving vertically and substituting gives

$$T_{AD} \sin 30° + T_{CD} \sin 45° + T_{BD} = W$$

or
$$\frac{\sqrt{3}\alpha + \beta}{2} + \beta + \frac{\sqrt{2}}{2}(\beta - \alpha) = \frac{WL}{AE}$$

Solving the two simultaneous equations for α and β gives

$$\beta = 0{\cdot}455\frac{WL}{AE} \quad \text{and} \quad \alpha = -0{\cdot}033\frac{WL}{AE}$$

and the loads in the members are

$$T_{AD} = 0{\cdot}398W$$
$$T_{BD} = 0{\cdot}455W$$
$$T_{CD} = 0{\cdot}488W$$

Example 5.3. The bars of the pin-jointed frame shown in Fig. 5.4 are of the same material, and have the same cross-sectional

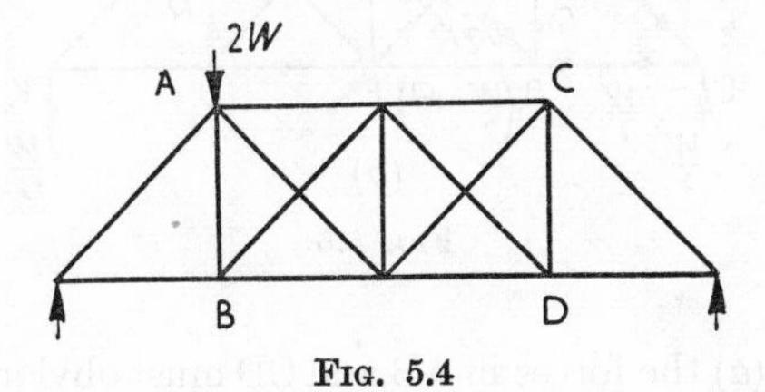

FIG. 5.4

area. Show that the forces in AB and CD are compressive and tensile respectively of magnitude

$$\frac{W}{2}\left(\frac{1 + 2\sqrt{2}}{3 + 4\sqrt{2}}\right) \qquad \textit{(London)}$$

Solution. The frame has 8 joints, consequently the number of bars required for a statically determinate frame is 13. There are, however, 15 bars so that the frame has two redundancies. One approach to the solution would be to take the frame as it stands, to let AB and CD be the redundant members with forces in them of R_1 and R_2 respectively and solve for R_1 and R_2 from the two simultaneous equations

$$\frac{\partial U}{\partial R_1} = \frac{\partial U}{\partial R_2} = 0$$

An alternative approach, which may be slightly longer but is certainly simpler, is to use the principle of superposition and replace the load $2W$ at A by loads of W acting downwards at A and C giving the symmetrical system shown in Fig. 5.5 (*a*) and a load of W downwards and upwards at A and C respectively giving the skew-symmetrical system shown in Fig. 5.5 (*b*).

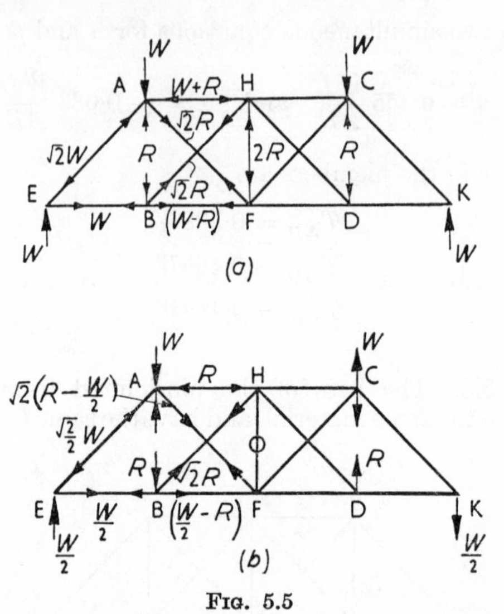

FIG. 5.5

In Fig. 5.5 (*a*) the forces in AB and CD must obviously be equal, and symmetry means that only one half the frame need be stressed. The skew symmetry of Fig. 5.5 (*b*) means that the force in any member is equal but opposite in sign to its corresponding member in the other half of the truss.

The forces in the members are indicated in Figs. 5.5 (*a*) and 5.5 (*b*).

Dealing first with the symmetrical case

$$\frac{\partial U}{\partial R} = \sum \frac{PL}{AE} \cdot \frac{\partial P}{\partial R} = 0$$

but since AE is constant

$$\Sigma PL \frac{\partial P}{\partial R} = 0$$

The summation (for half the truss taking half the energy for member HF) is carried out as shown in the table 5.3 (*a*).

TABLE 5.3 (a)

Member	Length	P	$\frac{\partial P}{\partial R}$	$PL\frac{\partial P}{\partial R}$
EB	L	$+W$	0	0
BF	L	$+(W-R)$	-1	$L(R-W)$
EA	$2L$	$-2W$	0	0
BH	$2L$	$+2R$	$+2$	$2\sqrt{2}RL$
AH	L	$-(W+R)$	-1	$+L(R+W)$
AB	L	$-R$	-1	RL
AF	$2L$	$+2R$	2	$2\sqrt{2}RL$
HF	L	$-2R$	-2	$\frac{1}{2}\times 4RL$

Summing up gives $\quad (5+4\sqrt{2})RL = 0$

i.e. $\quad R = 0$

The summation for the skew-symmetrical case is as shown in the table below.

TABLE 5.3 (b)

Member	Length	P	$\frac{\partial P}{\partial R}$	$PL\frac{\partial P}{\partial R}$
EB	L	$\frac{1}{2}W$	0	0
BF	L	$\left(\frac{W}{2}-R\right)$	-1	$\left(R-\frac{W}{2}\right)L$
EA	$\sqrt{2}L$	$-\frac{\sqrt{2}}{2}W$	0	0
BH	$\sqrt{2}L$	$2R$	2	$2\sqrt{2}RL$
AH	L	$-R$	-1	RL
AB	L	$-R$	-1	RL
AF	$\sqrt{2}L$	$\sqrt{2}\left(R-\frac{W}{2}\right)$	$\sqrt{2}$	$2\sqrt{2}\left(R-\frac{W}{2}\right)L$

Summing up the last column and equating to zero gives

$$R = \frac{W}{2}\left(\frac{1+2\sqrt{2}}{3+4\sqrt{2}}\right)$$

Since the forces in AB and CD under the symmetrical loading were zero the forces in these members under the combined loading are those from the skew-symmetrical loading given above, i.e. the forces in AB and CD, and compressive and tensile respectively of magnitude

$$\frac{W}{2}\cdot\frac{1+2\sqrt{2}}{3+4\sqrt{2}}$$

Example 5.4. A vertical mast of height $2h$, is subjected to a horizontal pull, P, at the top. It is held in position by two guys fixed to the mid-point and arranged in plan as shown in Fig. 5.6, each guy being anchored at a distance h from the base of the mast. The extensibility, AE, of each guy $= \frac{50EI}{h^2}$, where EI is the flexural rigidity of the mast in the plane of P. Determine the loads in the guys if the base of the mast is encastré in the plane of P but free to rotate in a plane perpendicular to this. (*St. Andrews*)

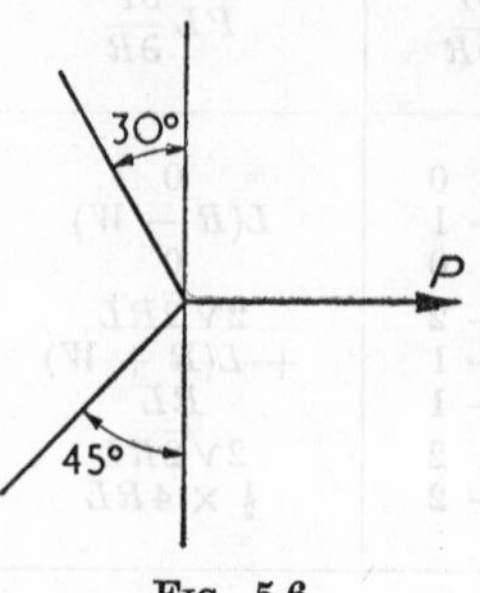

FIG. 5.6

Solution. Let the loads in the guys be $\sqrt{2}T_1$ and $\sqrt{2}T_2$. Then since there is no bending at the base in a plane perpendicular to P. moments about the base in this plane give

$$\frac{T_2}{\sqrt{2}} = \frac{\sqrt{3}}{2} T_1$$

hence
$$T_2 = \sqrt{\frac{3}{2}}\, T_1$$

The forces in the plane of the mast are therefore as shown in Fig. 5.7. The direct energy in the mast itself can be neglected but that in the ties must be included. Expressing T_2 in terms of T_1 leaves the latter as the only redundancy. Since there is no initial lack of fit then

$$\frac{\partial U}{\partial T_1} = 0$$

Consider first the mast

From A to B, measuring x from A

$$M_x = Px$$

$$\frac{\partial M_x}{\partial T_1} = 0$$

$$\therefore \qquad \frac{\partial U}{\partial T_1} = 0$$

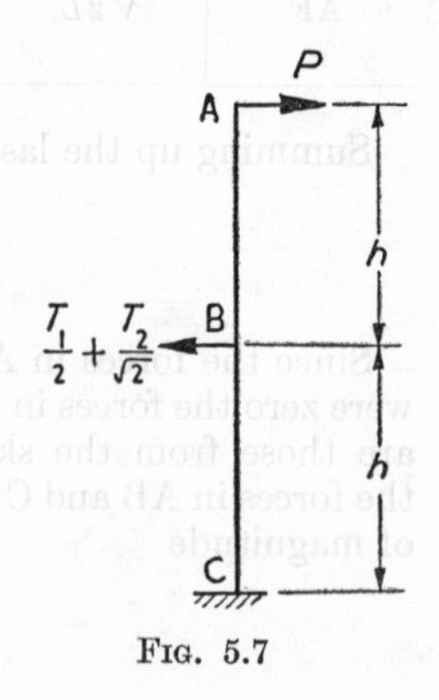

FIG. 5.7

From B to C measuring x from B

$$M_x = P(x + h) - \left(\frac{T_1}{2} + \frac{T_2}{\sqrt{2}}\right) x$$

$$= P(x + h) - \frac{T_1 x}{2}(1 + \sqrt{3})$$

$$\frac{\partial M_x}{\partial T_1} = -\frac{(\sqrt{3} + 1)}{2} x = -1{\cdot}366x$$

$$\therefore \qquad \left[\frac{\partial U}{\partial T_1}\right]_{BC} = \frac{1}{EI}\int_0^h \{1{\cdot}86Tx^2 - 1{\cdot}366Px(x + h)\}\,dx$$

$$= \frac{1}{EI}\left[\frac{1{\cdot}86T_1x^3}{3} - 1{\cdot}366P\left(\frac{x^3}{3} + \frac{hx^2}{2}\right)\right]_0^h$$

$$= \frac{h^3}{EI}(0{\cdot}62T_1 - 1{\cdot}142P)$$

For Tie 1,

$$P = \sqrt{2}T_1 \qquad L = \sqrt{2}h$$

$$\therefore \qquad \left[\frac{\partial U}{\partial T_1}\right]_1 = \frac{PL}{AE}\cdot\frac{\partial P}{\partial T_1} = \frac{2\sqrt{2}T_1h}{AE} = \frac{2\sqrt{2}T_1h^3}{50EI}$$

For Tie 2,

$$P = \sqrt{2}T_2 = \sqrt{3}T_1 \qquad L = \sqrt{2}h$$

$$\therefore \qquad \left[\frac{\partial U}{\partial T_1}\right]_2 = \frac{PL}{EA}\cdot\frac{\partial P}{\partial T_1} = \frac{3\sqrt{2}T_1h}{AE} = \frac{3\sqrt{2}T_1h^3}{50EI}$$

For the whole structure,

$$\frac{\partial U}{\partial T_1} = \left[\frac{\partial U}{\partial T_1}\right]_{BC} + \left[\frac{\partial U}{\partial T_1}\right]_1 + \left[\frac{\partial U}{\partial T_1}\right]_2 = 0$$

giving the equation

$$\frac{T_1h^3}{EI}\left(0{\cdot}62 + \frac{5\sqrt{2}}{50}\right) = \frac{1{\cdot}142Ph^3}{EI}$$

Hence $$T_1 = 1{\cdot}50P$$

The loads in the guys are then $2{\cdot}11P$ and $2{\cdot}59P$.

Example 5.5 A braced span, AB, is shown in Fig. 5.8 and supports two equal vertical loads of 100kN at the quarter points. Assuming all the joints to be pin-jointed determine the force in

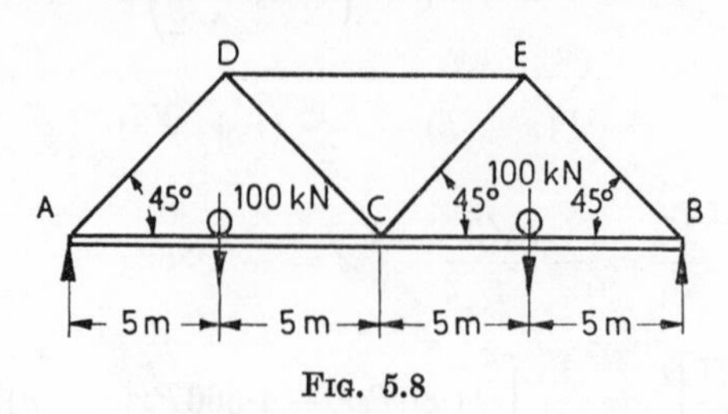

FIG. 5.8

the member DE and draw the bending moment diagram for the beam ACB. Neglect direct effects.

I (Beam) $= 1 \times 10^9\ \text{mm}^4$ units. Area (all members) $= 2 \times 10^3\ \text{mm}^2$. *(Glasgow)*

Solution. The propped beam is a common type of redundant structure the redundancy actually arising from the continuity of the beam over the prop. If in the case in point there were no continuity at C there would be no redundancy.

Since the question asks for the force in the member DE it is obviously simplest to take this as the redundant member. If the force in it is R then the forces in the other members are as shown in Fig. 5.9. In this figure the dimensions and loads shown in the

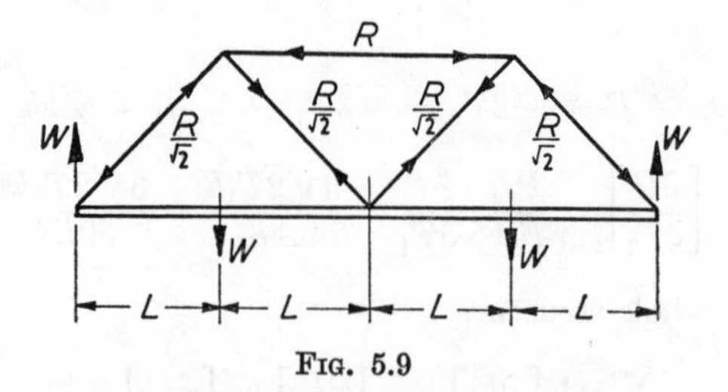

FIG. 5.9

example have been replaced by symbols so that algebra can be used instead of arithmetic for the earlier work in the analysis. (There is some debate about the advisability of this but the student can please himself whether he prefers to work with numbers or symbols; the author prefers the latter.)

Since there is no initial lack of fit then for the whole structure $\dfrac{\partial U}{\partial R} = 0$. The strain energy to be considered is the bending energy in the beam and the direct energy in the ties and struts.

For the beam AC measuring x from A

$$M_x = -Wx + \frac{R}{2}x + [W(x - L)]$$

the term in square brackets only occurring when positive.

$$\frac{\partial M_x}{\partial R} = \frac{x}{2}$$

Hence

$$\left[\frac{\partial U}{\partial R}\right]_{AC} = \frac{1}{EI}\int_0^{2L}\left(\frac{R}{2} - W\right)\frac{x^2}{2}\,.\,dx + \int_L^{2L}\frac{Wx}{2}(x - L)\,dx$$

$$= \frac{2RL^3}{3EI} - \frac{11}{12}\cdot\frac{WL^3}{EI}$$

For AD and DC $\quad P = \frac{R}{\sqrt{2}} \qquad \frac{\partial P}{\partial R} = \frac{1}{\sqrt{2}}$

and length $\quad = \sqrt{2}L$

Hence $\quad \left[\frac{\partial U}{\partial R}\right]_{AD} = \frac{RL}{\sqrt{2}AE} = \left[\frac{\partial U}{\partial R}\right]_{DC}$

For DE $\quad P = R \qquad \frac{\partial P}{\partial R} = 1$

and $\quad$ length $= 2L$

Hence $\quad \left[\frac{\partial U}{\partial R}\right]_{DE} = 2\frac{RL}{AE}$

For the whole structure

$$\frac{\partial U}{\partial R} = 2\left[\frac{\partial U}{\partial R}\right]_{AC} + 4\left[\frac{\partial U}{\partial R}\right]_{AD} + \left[\frac{\partial U}{\partial R}\right]_{DE} = 0$$

Hence $\quad R\left[\frac{4L^3}{3I} + \frac{2\sqrt{2}L}{A} + \frac{2L}{A}\right] = \frac{11}{6}\cdot\frac{WL^3}{I}$

Since I and A are given in mm units L must be taken in mm units also. Substituting the various values gives

$$R\left[\frac{4 \times 125 \times 10^9}{3 \times 10^9} + \frac{2\sqrt{2} \times 5 \times 10^3}{2 \times 10^3} + \frac{2 \times 5 \times 10^3}{2 \times 10^3}\right]$$

$$= \frac{11}{6} \times \frac{100 \times 125 \times 10^9}{10^9}$$

giving $R = 128$ kN.

The loading coming on the beam is then as shown in Fig. 5.9 substituting 128 for R, the bending moment diagram being drawn in Fig. 5.10.

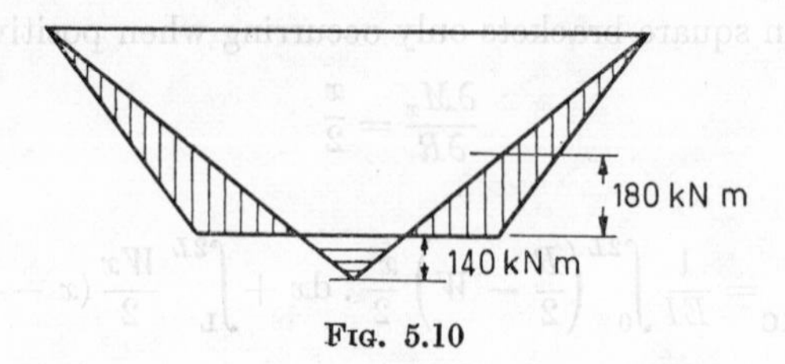

FIG. 5.10

The bending moment under the load, M_{10}, is given by

$$M_{10} = -100 \times 5 + \frac{128}{2} \times 5 = -180 \text{ kN/m}$$

whilst that at the centre of the span, M_{20}, is

$$M_{20} = -100 \times 5 + \frac{128}{2} \times 10 = +140 \text{ kN/m}$$

There is an alternative method of solution for the king-posted beam problem in which the integration is carried out graphically. The procedure is as follows.

Remove the redundant member DE and put loads of 1 kN at D and E in the direction DE. These unit loads produce compressions in AD and EB each equal to $\frac{1}{\sqrt{2}}$, and tensions of $\frac{1}{\sqrt{2}}$ in DC and EC. The load on the beam ACB is then a vertical load of 1 kN upwards at C. The bending moment diagram for the beam due to this load is shown in Fig. 5.11. This diagram will be referred to as the "m" diagram.

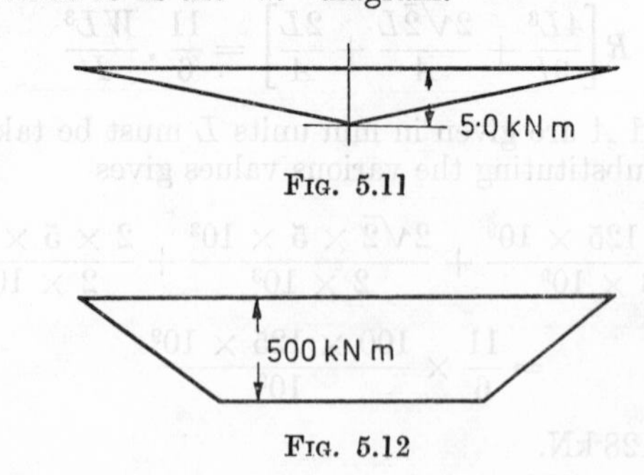

FIG. 5.11

FIG. 5.12

The load on the beam, if the effect of the redundant member is removed, consists of the two 100 kN loads at the quarter points giving a bending moment diagram as shown in Fig. 5.12. This diagram will be referred to as the "M" diagram.

If R is the force in the redundant member, DE, then the application of the principle of least work gives the following equation.

$$\int \frac{Mm\,\mathrm{d}x}{EI} + R\int \frac{m^2\,\mathrm{d}x}{EI} + R\sum \frac{p^2L}{AE} = 0$$

where p = load produced in the bracing members when a unit load replaces the redundancy.

E is common throughout hence it can be cancelled and the equation written

$$\int \frac{Mm\,\mathrm{d}x}{I} + R\int \frac{m^2\,\mathrm{d}x}{I} + R\sum \frac{p^2L}{A} = 0$$

The integrals can be obtained using the M and m diagrams as follows

$$\int Mm\,\mathrm{d}x = -\left[(500 \times 5{\cdot}0 \times \tfrac{1}{2} \times \tfrac{5}{3})2 + 500 \times 10 \times \frac{2{\cdot}5 + 5{\cdot}0}{2}\right]$$

$$= 22\,920$$

$$\int m^2\,\mathrm{d}x = 5 \times 20 \times \tfrac{1}{2} \times \tfrac{2}{3} \times 5 = 500 \times \tfrac{1}{3}$$

$$\sum p^2l = 4 \times 5\sqrt{2} \times \left(\frac{1}{\sqrt{2}}\right)^2 + 10 \times 1^2 = 24{\cdot}14$$

Substituting these values in the equation above and converting to mm gives

$$-\frac{22\,920 \times 10^9}{10^9} + R\frac{500 \times 10^9}{3 \times 10^9} + R\frac{24{\cdot}14 \times 10^3}{2 \times 10^3} = 0$$

leading to $R = 128$ kN as before.

Examples on Redundant Pin-jointed Frameworks

1. Determine the forces in the members of the frame shown in Fig. 5.13 which is pinned to supports A and D and carries loads of 10 kN and 20 kN at B and C respectively. The members AB and CD are 2000 mm² and the remainder 600 mm² in cross-sectional area.

Answer. $AC = BD = 1{\cdot}23$; $BC = 0{\cdot}98$; $DC = 19{\cdot}26$; $AB = 9{\cdot}26$.

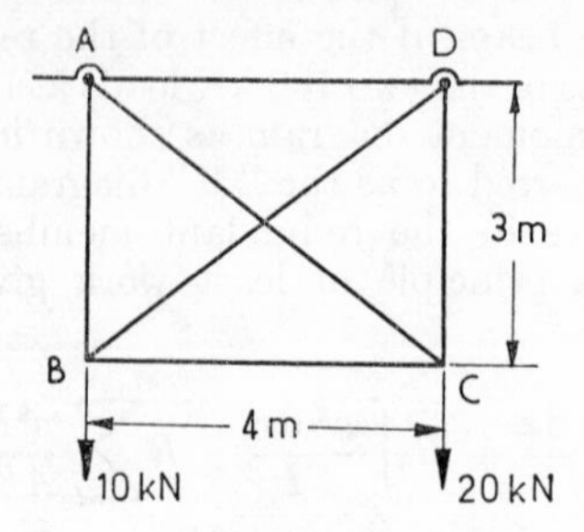

FIG. 5.13

2. The frame ABCD shown in Fig. 5.14 is pinned at A and B and carries a load of 10 kN at D. The members AC and BD are

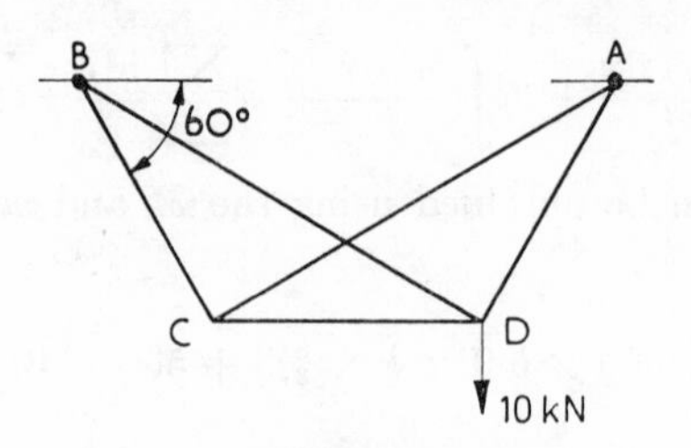

FIG. 5.14

200 mm² and the other members 150 mm² in area: AD = DC = BC = 1 m. Show that the load in CD is approximately 0·37 kN.
Answer. Given in question.

3. The frame shown in Fig. 5.15 is made from bars all having

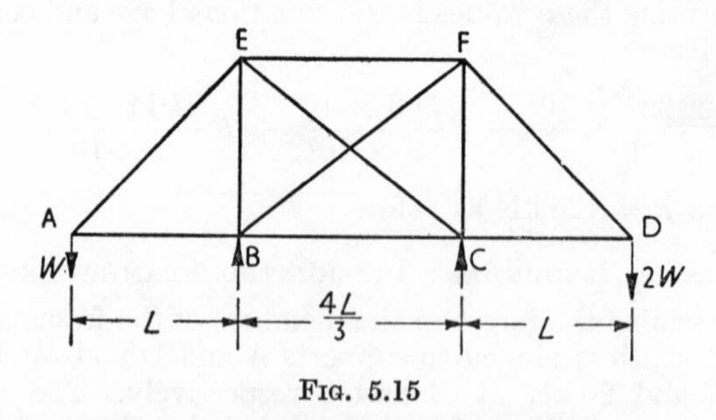

FIG. 5.15

the same extensibility aE. It is supported at B and C and carries loads at A and D. Determine the load in the bar CE.
Answer. $0{\cdot}938W$ compression.

4. The frame shown in Fig. 5.16 is supported at F by a vertical force so that under a load of 10 kN vertically at C there is no

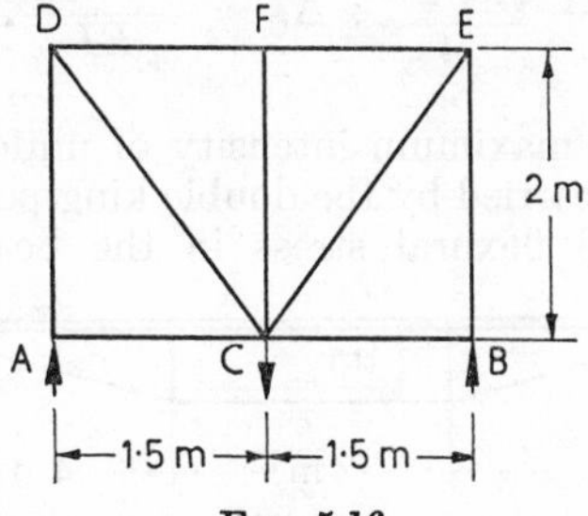

Fig. 5.16

deflexion at F. If the bars all have the same area show that the force at F is 6·28 kN.

Answer. Given in question.

5. A square frame ABCD with two diagonals AC and BD is pinned to supports at A and B and carries loads of 2 kN at C and D in the direction CA and DB. All the members have the same cross-section. Find the forces in CD, CA and CB.

Answer. $CD = 0{\cdot}924$; $CA = 0{\cdot}694$; $CB = 0{\cdot}924$.

6. Tabulate the forces in the members of the frame shown in Fig. 5.17 which is supported on pins at A, B and C and carries a load P at E. All the bars have equal areas.

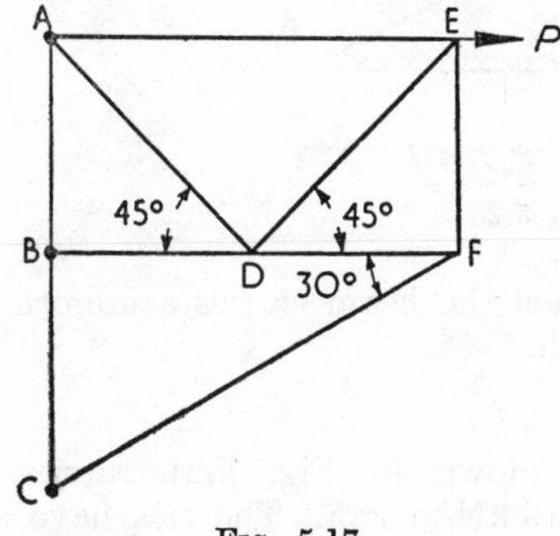

Fig. 5.17

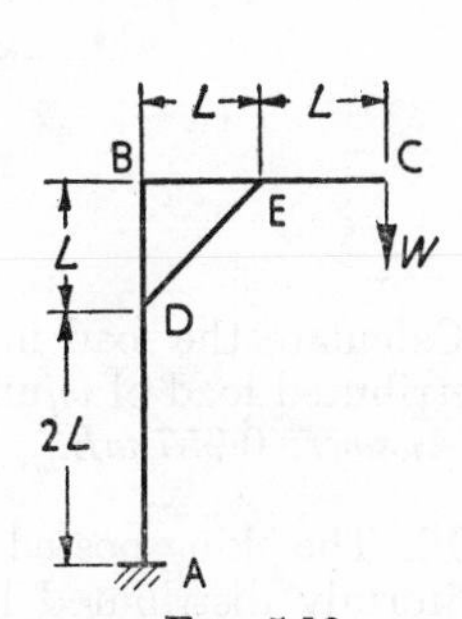

Fig. 5.18

Answer. $AE = +\,0{\cdot}943$; $BD = +\,0{\cdot}214$; $DF = +\,0{\cdot}098$; $EF = -\,0{\cdot}058$; $AD = -\,0{\cdot}081$; $DE = +\,0{\cdot}081$; $FC = -\,0{\cdot}115$.

7. The structure shown in Fig. 5.18 consists of a bent cantilever ABC of constant flexural rigidity and a rigid cross-member DE each end of which is pinned to the cantilever. Calculate the load in DE and the vertical deflexion of C. (*St. Andrews*)

Answer. $DE = +\dfrac{11\sqrt{2}W}{4}$; $\Delta_C = \dfrac{9{\cdot}62WL^3}{EI}$.

8. Calculate the maximum intensity of uniformly distributed load which can be carried by the double king-posted beam shown in Fig. 5.19. The flexural stress in the beam is limited to

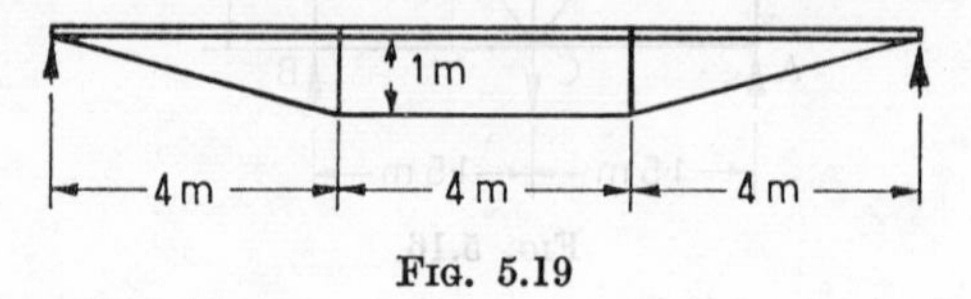

FIG. 5.19

150 N/mm² and the direct stress in the ties and struts to 130 N/mm². I for the beam $= 3{\cdot}5 \times 10^8$ mm⁴ units. A for the struts $= 1500$ mm² and for the ties 1000 mm².

Answer. 9·02 kN/m.

9. The beam shown in Fig. 5.20 is pinned at A and supported by two wires, DC and DB, each of extensibility $50EI/L^2$, where EI is the flexural rigidity of the beam.

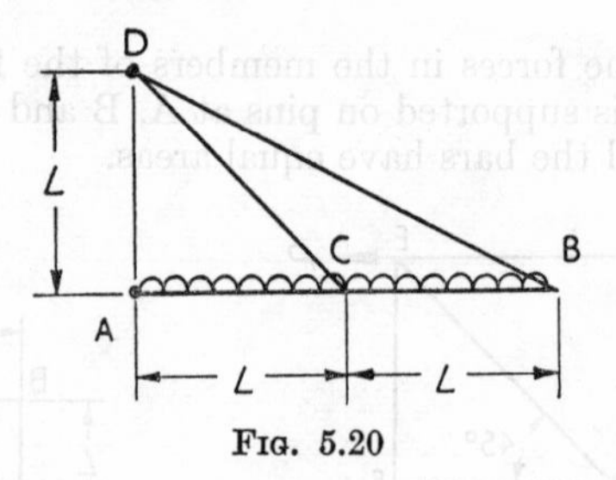

FIG. 5.20

Calculate the load in DB when the beam carries a uniformly distributed load of w/unit length.

Answer. 0·957 wL.

10. The king-posted beam shown in Fig. 5.21 carries a uniformly distributed load of 10 kN/m run. The ties have an

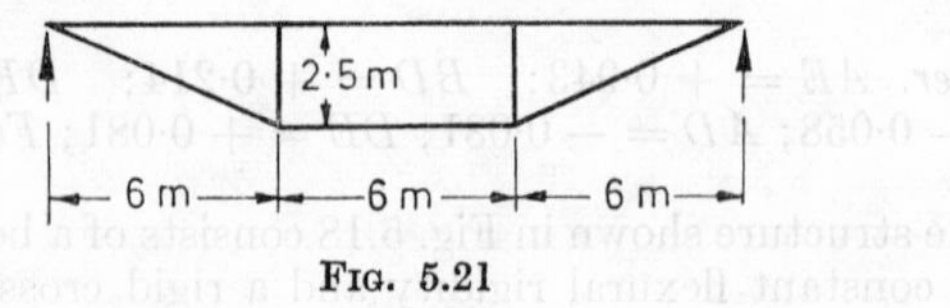

FIG. 5.21

area of 1500 mm² and the struts 3000 mm². The beam is 250 mm deep and has a second moment of area = $5 \cdot 00 \times 10^7$ mm⁴ units. Calculate the stresses in the members.

Answer. $f_b = 75$ N/mm²; $f_s = 21 \cdot 8$ N/mm²; $f_{IT} = 113 \cdot 5$ N/mm²; $f_{LT} = 105$ N/mm².

11. A vertical mast of height $2L$ is subjected to a horizontal pull at the top. It is pinned at the base and held in position by two pairs of guys one attached to the top of the mast and the other at mid-height. Each guy is anchored at a distance L from the base of the mast and placed in plan as shown in Fig. 5.22. Determine the load in the guys if the extensibility of each guy is $50EI/L^2$, where EI is the flexural rigidity of the mast in the plane of P. (*Glasgow*)

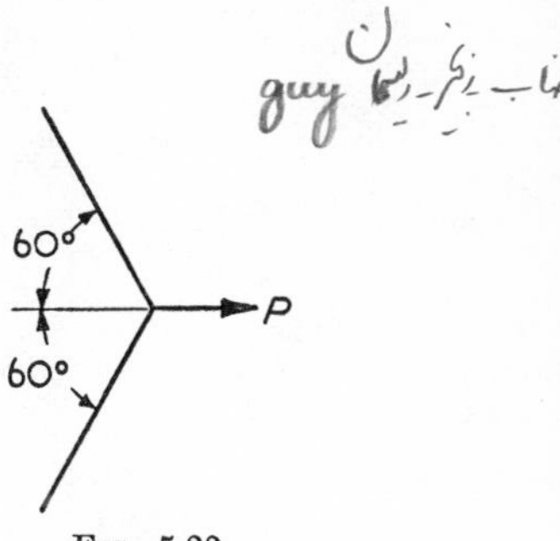

FIG. 5.22

Answer. $1 \cdot 592P$; $0 \cdot 811P$.

12. The structure shown in Fig. 5.23 is hinged to supports at A and F and subjected to a load of 20 kN at C. Assuming the

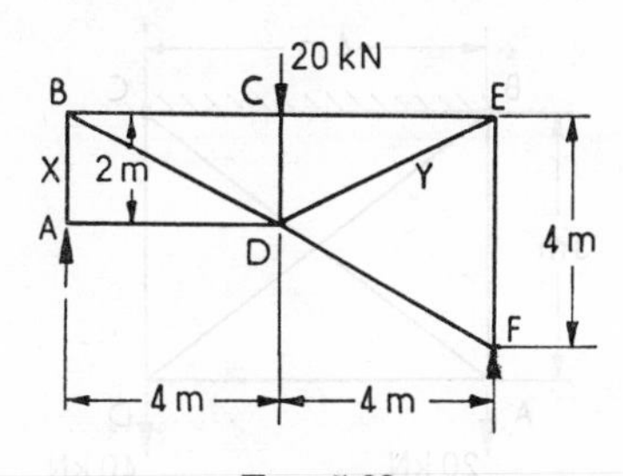

FIG. 5.23

sectional area of all the members to be the same determine the force in each of the members X and Y. (*Aberdeen*)

Answer. $X = 5 \cdot 98$ kN; $Y = 13 \cdot 38$ kN.

13. A vertical pile, AB, of length 10 m is hinged at the base, B, and braced by pin-connected struts, AC and DE, as shown in Fig. 5.24. The pile is loaded with a uniform horizontal load of 10 kN/m run. Determine the force in each of the struts. EI for pile = 50 000 kN·m². EA for struts = 300 000 kN. (*Aberdeen*)

Answer. 92·5 kN; 33·5 kN.

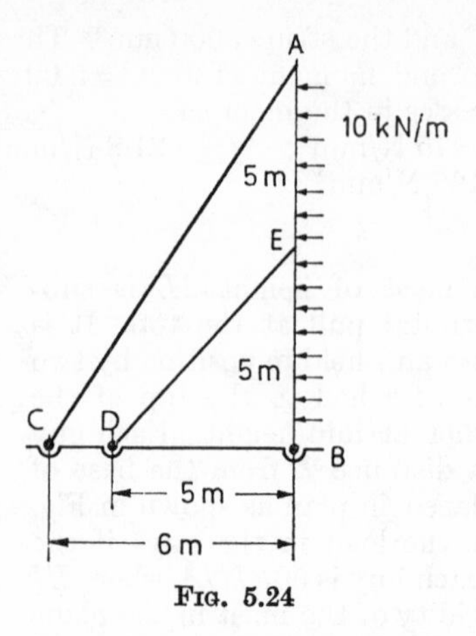

FIG. 5.24

14. The rectangular frame shown in Fig. 5.25 consists of five bars pinned together at A, B, C and D and suspended in a vertical plane from a rigid beam. All the bars are of steel and 1000 mm² in cross-sectional area. Find the force in the member AD due to the two loads at A and D. If the temperature is raised 20°C,

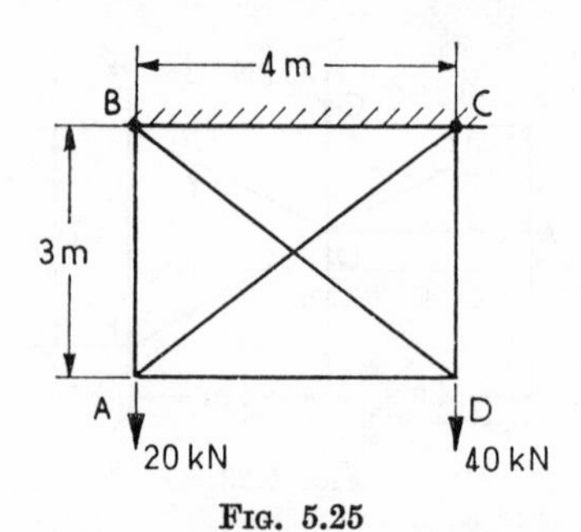

FIG. 5.25

the distance BC remaining unchanged, find the force in each of the bars. $E = 200$ kN/mm². Coefficient of linear expansion $= 11 \times 10^{-6}$ per °C.

Answer. $-7{\cdot}1$ kN; $AB = +22{\cdot}4$; $AD = +0{\cdot}55$; $DC = +41{\cdot}4$; $AC = -0{\cdot}65$; $DB = -0{\cdot}65$.

15. If the right-angled connexion at B in the steel structure shown in Fig. 5.26 were converted into a rigid joint by welding determine the percentage reduction in the load carried by the tie bar. Both angle members are of the same section having

$I = 4{\cdot}0 \times 10^6$ mm^4 units and the frame is unstressed after welding when carrying no load. (*London*)

Answer. 1·5 per cent (approx.).

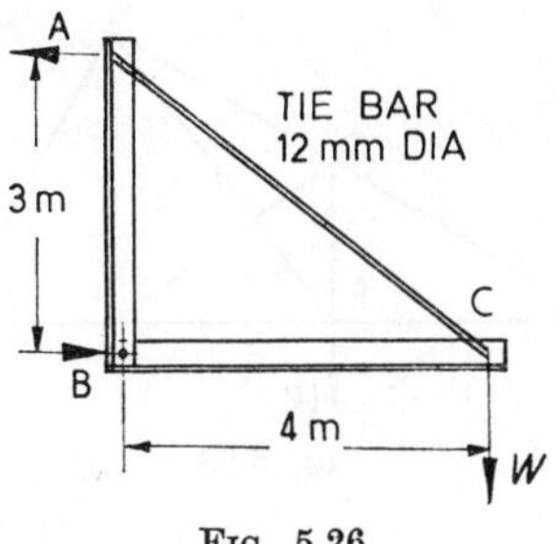

FIG. 5.26

16. The lower section of one side of a braced tower takes the form of a symmetrical trapezium with top and bottom parallel and of lengths $12\frac{1}{2}$ and $19\frac{1}{2}$ m respectively, and of height 12 m. The two top corners carry vertical loads of 20 kN each. The top member has a cross-section of 1200 mm^2, the legs of 2500 mm^2 and the two diagonals of 600 mm^2. The bottoms of the legs are hinged to a rigid concrete base. Find the load in the top member which can be taken as redundant. The trapezium is vertical. (*London*)

Answer. 2·4 kN compression.

17. The frame ABCD (Fig. 5.27) is supported at A and B and carries a load of 100 kN vertically at C. Calculate the loads in the members and the vertical deflexion of C.

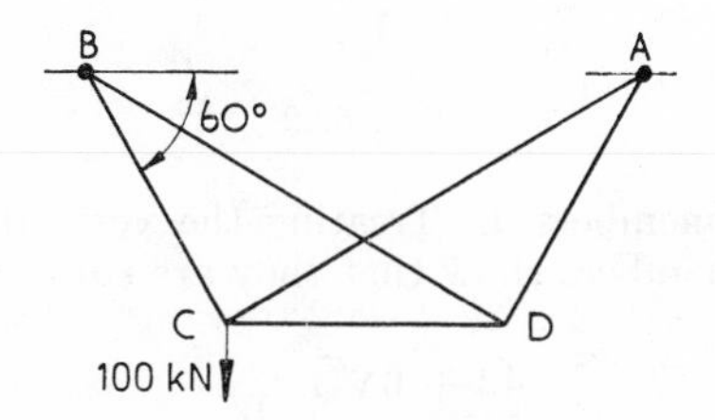

FIG. 5.27

The members AC and BD are 200 mm^2 and the other members are 150 mm^2 in area. The members AD, BC and DC are each 1 m long and $E = 200$ kN/mm^2. (*London*)

Answer. $BC = 88{\cdot}4$; $AC = 46{\cdot}8$; $CD = 3{\cdot}7$; $BD = -3{\cdot}2$; $AD = 1{\cdot}9$; $\Delta = 3{\cdot}3$ mm.

18. The Fig. 5.28 shows a pin-jointed framework carrying a vertical load, W, at E and supported by vertical reactions at A and B. The dimensions of the figure are such that if the line CE

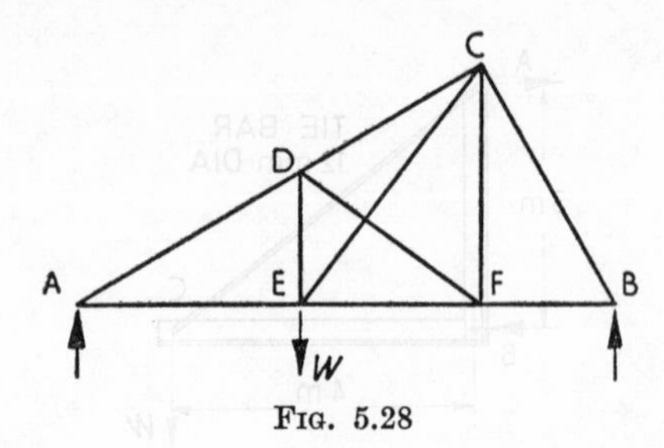

Fig. 5.28

is omitted all the angles are either 30°, 60° or 90°. The members are all of the same material and cross-section. Find the load in the member CE. *(Oxford)*

Answer. $0{\cdot}534W$.

19. In the pin-jointed framework shown in Fig. 5.29 the three triangles ABG, HCF and GDE are equilateral. The members AB, BC, CD, DE, BH, CG and DF have a cross-section $2a$ and

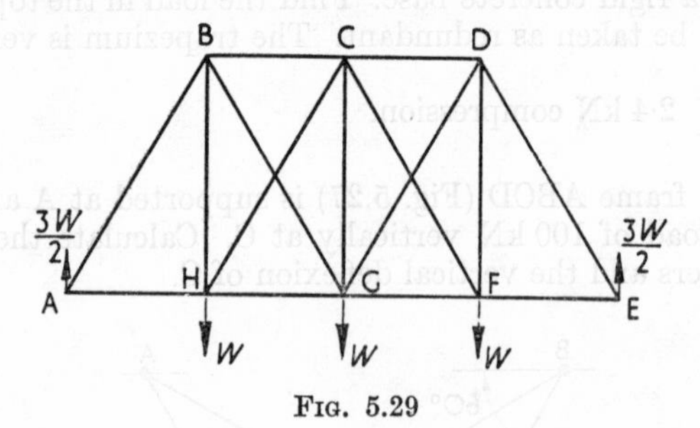

Fig. 5.29

the remaining members a. Treating the verticals BH and DF as redundant members show that they are subjected to tensions of magnitude

$$\frac{42 + 6\sqrt{3}}{35 + 9\sqrt{3}} \cdot W$$

(Cambridge)

Answer. Given in question.

20. The pin-jointed rectangular frame, ABCD, shown in Fig. 5.30 consists of 4 aluminium bars, AB, BC, CD and DA, cross-braced by the steel wires AC and BD. The cross-sectional area of each of the aluminium bars is 2000 mm² and that of the steel wire 150 mm². AB:BC = 2:3. The lengths of the wires are equal when the frame is free from load.

Supported as indicated the frame has to carry a load of 20 kN at B while the temperature can vary between 5° and 25°C. The bracing wires are initially tensioned when the frame is free from external load and the temperature is 15°C. Calculate the least

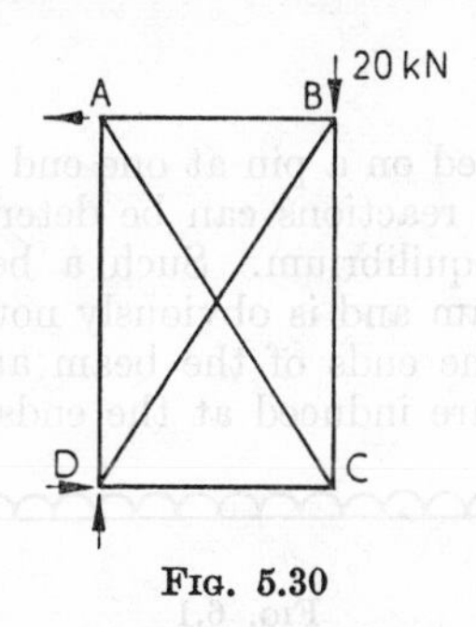

Fig. 5.30

initial tension necessary to prevent the wire BD becoming slack at the minimum temperature. Also determine with this initial tension the greatest tension in AC possible under the given conditions.

Young's Modulus for steel = 200 kN/mm².

Young's Modulus for aluminium = 70 kN/mm².

Temperature coefficient of expansion for steel = 11×10^{-6}.

Temperature coefficient of expansion for aluminium

$= 25{\cdot}5 \times 10^{-6}$ per °C.

(*Cambridge*)

Answer. 20·24 kN; 31·58 kN.

Chapter 6

The Built-in Beam

IF A BEAM is supported on a pin at one end and a roller bearing at the other end the reactions can be determined by using the equations of static equilibrium. Such a beam is known as a simply supported beam and is obviously not a redundant structure. If, however, the ends of the beam are restrained in any way then moments are induced at the ends by these restraints

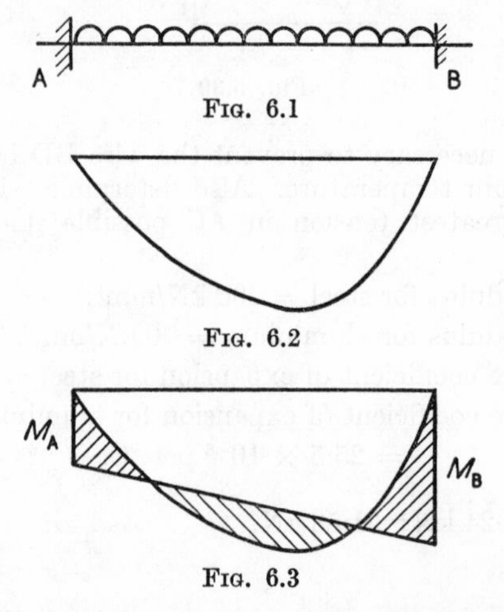

FIG. 6.1

FIG. 6.2

FIG. 6.3

and these end moments make the beam a redundant structure. The commonest type of end restraint is that in which the beam is fully restrained or built-in as shown in Fig. 6.1. The feature of this restraint is that the slope of the beam at the supports is zero.

If this beam had been simply supported then the bending moment diagram would have been as shown in Fig. 6.2. The moments at the ends act in a direction tending to reduce the slope of the beam and are therefore of opposite sign to the moments due to lateral load. The bending moment diagram for the built-in beam is shown in Fig. 6.3, M_A and M_B being the restraining moments at the supports.

The problem with the built-in beam is to determine the values of M_A and M_B.

These values can be obtained by application of the slope deflexion equations as follows.

Let AB be a beam of span l, carrying any lateral load system and having the ends completely built-in. These restraints produce moments M_A and M_B.

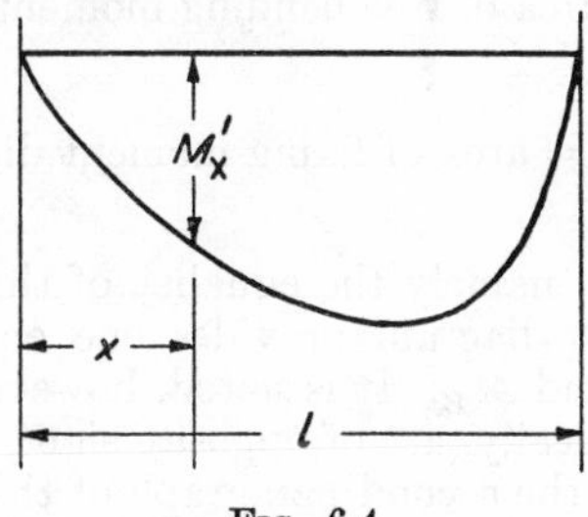

Fig. 6.4

The bending moment diagram due to the lateral loads alone is shown in Fig. 6.4, M'_x being the moment at a point distant x from the left-hand support. The bending moment diagram due to the end fixity alone is shown in Fig. 6.5.

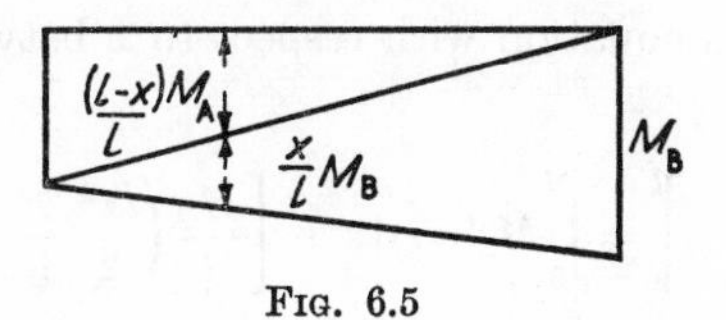

Fig. 6.5

The actual bending moment M_x at a point distant x from the left-hand support is the sum of that due to lateral loads and fixity and is given by

$$M_x = M_x' + \left(\frac{l-x}{l}\right) M_A + \frac{x}{l} M_B$$

or

$$EI \frac{d^2y}{dx^2} = M_x' + \left(\frac{l-x}{l}\right) M_A + \frac{x}{l} M_B$$

Integrating this equation between the limits 0 and l gives

$$EI \left[\frac{dy}{dx}\right]_0^l = \int_0^l M_x' \,.\, dx + \frac{M_A}{l}\left[lx - \frac{x^2}{2}\right]_0^l + \frac{M_B}{l}\left[\frac{x^2}{2}\right]_0^l$$

But if the ends are built-in there is zero slope at the supports, hence both upper and lower limits of the left-hand integral vanish and the equation becomes

$$\int_0^l M_x' \, . \, \mathrm{d}x + \frac{l}{2}(M_A + M_B) = 0$$

But $\int_0^l M_x' \, . \, \mathrm{d}x$ = area of free bending moment diagram (say A),

and $\frac{l}{2}(M_A + M_B)$ = area of fixing moment diagram.

This relationship, namely the equality of the areas of the free and fixing moment diagram provides one equation wherewith to determine M_A and M_B. It is noted, however, that the areas, whilst equal numerically, are of opposite sign. This confirms the statement made in the second paragraph of this chapter.

The second equation required is obtained by multiplying the expression for $EI \frac{\mathrm{d}^2y}{\mathrm{d}x^2}$ by x giving

$$EIx \frac{\mathrm{d}^2y}{\mathrm{d}x^2} = M_x'x + \left(\frac{l-x}{l}\right) x \, . \, M_A + \frac{x^2}{l} M_B$$

Integrating this equation with respect to x between the limits 0 and l gives

$$EI \left[x \, . \, \frac{\mathrm{d}y}{\mathrm{d}x} - y\right]_0^l = \int_0^l M_x'x \, . \, \mathrm{d}x + \left[\frac{M_A}{l}\left(\frac{lx^2}{2} - \frac{x^3}{3}\right) + \frac{x^3}{3l} M_B\right]_0^l$$

But if, in addition to the slopes being zero at each end, there is also no relative deflexion of the two supports then the left-hand side in this equation also vanishes giving

$$0 = \int_0^l M_x'x \, . \, \mathrm{d}x + \left[M_A \frac{l^2}{6} + M_B \frac{l^2}{3}\right]$$

But $\int_0^l M_x'x \, . \, \mathrm{d}x$ = moment of the area of the free bending moment diagram about the left-hand support

$= A\bar{x}$

where $\bar{x}$ = distance of the centre of area of the free bending moment diagram from the left-hand support.

Hence the two equations to determine M_A and M_B are

$$M_A + M_B = -\frac{2A}{l} \quad \text{and} \quad M_A + 2M_B = -\frac{6A\bar{x}}{l^2}$$

The above equations have been derived on the assumption that the slopes at the ends are both zero and that there is no relative sinking of the supports. Equations can be derived in a similar manner if the slopes at the ends are not zero but have values θ_A and θ_B at A and B respectively.

In this case it can be shown that

$$\theta_A = -\frac{l}{6EI}\left\{2M_A + M_B + \frac{6A}{l^2}(l - \bar{x})\right\}$$

$$\theta_B = \frac{l}{6EI}\left(M_A + 2M_B + \frac{6A\bar{x}}{l^2}\right)$$

If in addition the end B sinks relative to A by an amount δ then the equations become

$$\theta_A = -\frac{l}{6EI}\left\{2M_A + M_B + \frac{6A}{l^2}(l - \bar{x})\right\} + \frac{\delta}{l}$$

$$\theta_B = \frac{l}{6EI}\left(M_A + 2M_B + \frac{6A\bar{x}}{l^2}\right) + \frac{\delta}{l}$$

In some cases the best method is a straightforward application of the equation

$$EI\frac{d^2y}{dx^2} = M.$$

Example on Built-in Beams

Example 6.1. An encastré beam of span 6 m weighing 1 kN/m run carries loads of 80 kN at each of the third points. Determine the value of the end fixing moments.

Solution. In this case the loading is symmetrical, consequently the moments at each end are the same and only one equation is required to determine them.

The free bending moment diagram due to the distributed load is parabolic and has an area A_1 given by

$$A_1 = -\frac{2}{3} \times 1 \times \frac{6^3}{8} = 18 \text{ kN m}^2 \text{ units.}$$

The bending moment diagram for the point loads is as shown in Fig. 6.6 the area A_2 being given by

$$A_2 = -(2 \times 160 \times 2) = 640 \text{ kN m}^2 \text{ units}$$

The negative sign appears in each case because the free moments are negative.

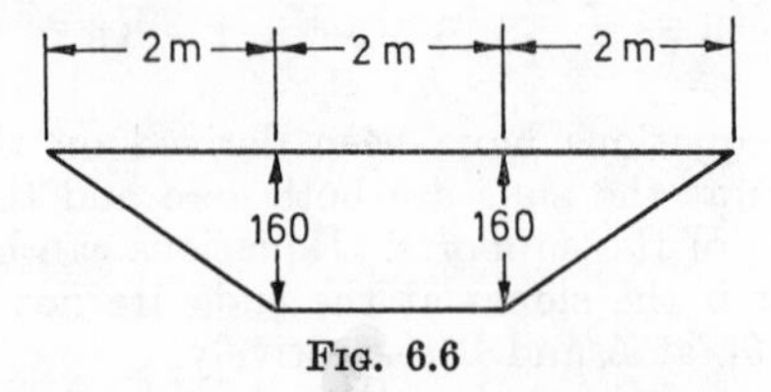

FIG. 6.6

If M_F = fixing moment at the supports then equating the areas of the free and fixing moment diagrams gives

$$M_F \times 6 = 18 + 640 = 658 \text{ kN m}^2 \text{ units}$$

Hence $M_F = 109{\cdot}7$ kN m.

Example 6.2. An encastré beam of span L carries a concentrated load W at a distance $\frac{L}{4}$ from the left-hand support. Calculate the bending moment at mid-span.

Solution. This is an example where the use of the principle of superposition simplifies the work. The stated loading can

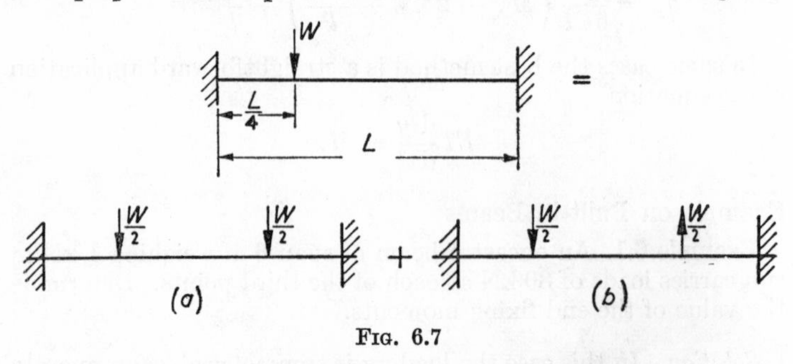

FIG. 6.7

be split as shown in Fig. 6.7 into a symmetrical plus a skew-symmetrical system. The latter system, as has been pointed out, has no central deflexion and no central bending moment. Consequently all that is required to answer the question is to determine the central bending for the loading shown in Fig. 6.7 (a).

This loading is symmetrical, consequently the fixing moment at each end is the same. The area, A_1, of the free bending moment diagram for the loading shown in Fig. 6.7 (a) is given by

$$A_1 = \left\{\frac{WL}{8} \cdot \frac{L}{2} + \frac{WL}{8} \cdot \frac{L}{4}\right\} = -\frac{3WL^2}{32}$$

Hence the fixing moment at each end $= \dfrac{3WL}{32}$

$\therefore$ Central bending moment $= -\dfrac{WL}{8} + \dfrac{3WL}{32} = -\dfrac{WL}{32}$

Example 6.3. A beam AB of uniform section and length, l, is built-in at B and supported by a tie at a point distant r from A. The elasticity of the tie is so designed that when the beam carries a uniformly distributed load, A and B remain at the same level. Show that for the maximum bending moment in the beam to have its smallest value, $r = 0{\cdot}28l$ approx. (*Oxford*)

Solution. This example is shown diagrammatically in Fig. 6.8. The maximum bending moment will occur either at B or C or on

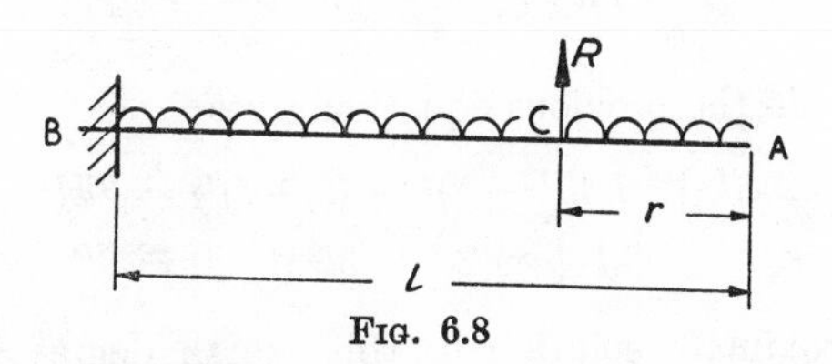

FIG. 6.8

the span between them. For small values of r the moment at C will be small and the moment at B greater. As r is increased so the moment at C becomes larger whilst that at B becomes less. Consequently the value of r which makes the maximum bending moment on the beam have its smallest value is that which gives $M_B = M_C$. Smaller values of r mean that M_B will have a greater value than when the equality relationship is satisfied, and for greater values of r, M_C will be greater.

The problem therefore is to determine the value of r so that $M_B = M_C$.

Let the load in the tie be R. Writing down the relationships between the moments, slopes and deflexions gives

	A to C	C to B
$EI\dfrac{d^2y}{dx^2} =$	$\dfrac{wx^2}{2}$	$\dfrac{wx^2}{2} - R(x - r)$
$EI\dfrac{dy}{dx} =$	$\dfrac{wx^3}{6} + A$	$\dfrac{wx^3}{6} - R\dfrac{(x - r)^2}{2} + A$
$EIy =$	$\dfrac{wx^4}{24} + Ax + B$	$\dfrac{wx^4}{24} - R\dfrac{(x - r)^3}{6} + Ax + B$

When $x = 0,\ y = 0 \qquad \therefore\ B = 0$

$$x = l,\ \frac{dy}{dx} = 0 \qquad \therefore\ A = R\frac{(l-r)^2}{2} - \frac{wl^3}{6}$$

$$x = l,\ y = 0$$

hence $$\frac{wl^4}{24} - R\frac{(l-r)^3}{6} - \frac{wl^4}{6} + \frac{Rl(l-r)^2}{2} = 0$$

or $$4R\{3l(l-r)^2 - (l-r)^3\} = 3wl^4$$

For $M_B = M_C$

$$\frac{wr^2}{2} = \frac{wl^2}{2} - R(l-r)$$

or $$R = \frac{w}{2}(l+r)$$

Substituting in the previous equation gives

$$2w(l+r)\{3l(l-r)^2 - (l-r)^3\} = 3wl^4$$

$$2(l+r)(2l^3 - 3l^2 + r^3) = 3l^4$$

The approximate solution to this fourth degree equation is $r = 0{\cdot}28l$. The moments at B and C are then each equal to $0{\cdot}0392wl^2$. It remains to check that the moment on the span between B and C is not greater than this. The moment on the span $= \left(\frac{wr^2}{8} - 0{\cdot}0392wl^2\right)$. This gives a span moment of $0{\cdot}0258wl^2$ which is less than the moments at B or C.

$\therefore$ For maximum moment to have its smallest value $r = 0{\cdot}28l$.

Example 6.4. A beam, AB, is encastré at A and B and carries a load which varies uniformly from w at A to $2w$ at B. The flexural rigidity of the beam also varies uniformly, its value at B being twice that at A.

Determine the fixing moments at A and B and the maximum deflexion. (*Oxford*)

Solution. The problem is shown diagrammatically in Fig. 6.9.

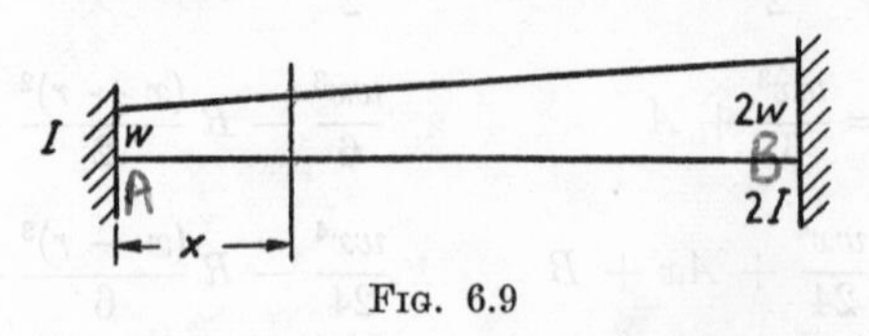

FIG. 6.9

At a distance x from the left-hand support the intensity of loading $w_x = \frac{w}{l}(l + x)$ and the flexural rigidity $I_x = \frac{I}{l}(l + x)$.

Since the flexural rigidity varies throughout, the fundamental relationship from which to start is $\frac{d^2M_x}{dx^2}$ = intensity of loading $= w\left(\frac{l + x}{l}\right)$.

Integrating twice gives $M_x = \frac{w(l + x)^3}{6l} + Ax + B$ where A and B are constants of integration.

If EI_A = flexural rigidity at support A then since

$$M_x = EI_x \frac{d^2y}{dx^2}$$

$$EI_A \frac{d^2y}{dx^2} = \frac{w(l + x)^2}{6} + \frac{Alx}{l + x} + \frac{Bl}{l + x} \quad . \quad . \quad . \quad (1)$$

Integrating gives

$$EI_A \frac{dy}{dx} = \frac{w(l + x)^3}{18} + Alx + l(B - Al)\log_e(l + x) + C$$

$$\frac{dy}{dx} = 0 \text{ at } x = 0$$

$$\therefore \quad C = -\frac{wl^3}{18} - l(B - Al)\log_e l$$

$$\therefore EI_A \frac{dy}{dx} = w\left\{\frac{(l + x)^3 - l^3}{18}\right\} + Alx + l(B - Al)\log_e\left(\frac{l + x}{l}\right) \quad (2)$$

$$EI_A y = \frac{w(l + x)^4}{72} - \frac{wl^3x}{18} + \frac{Alx^2}{2}$$
$$+ l(B - Al)\left\{x \log_e\left(\frac{l + x}{l}\right) - x + l\log_e(l + x)\right\} + D$$

$$y = 0 \text{ at } x = 0$$

$$\therefore \quad D = -\frac{wl^4}{72} - l^2(B - Al)\log_e l$$

$$\therefore EI_A y = \frac{w}{72}\{(l + x)^4 - l^4\} - \frac{wl^3x}{18} + \frac{Alx^2}{2}$$
$$+ l(B - Al)\left\{x \log_e\left(\frac{l + x}{l}\right) - x\right\} + l\log_e\left(\frac{l + x}{l}\right) \quad (3)$$

Using the fact that when $x = l$, $y = 0$ and $\frac{dy}{dx} = 0$ gives

$$\frac{11}{72} wl^4 + \frac{Al^3}{2} + l^2 (B - Al) \{2 \log_e 2 - 1\} = 0$$

$$\frac{7}{18} wl^3 + Al^2 + l (B - Al) \{\log_e 2\} = 0$$

Solving these for A and B gives

$$A = wl \left\{ -\frac{7}{18} + \frac{1}{12} \left(\frac{\log_e 2}{2 - 3 \log_e 2} \right) \right\} \quad . \quad . \quad (4)$$

$$B = wl^2 \left\{ -\frac{7}{18} + \frac{1}{12} \left(\frac{\log_e 2 - 1}{2 - 3 \log_e 2} \right) \right\}$$

The fixing moments M_A and M_B at A and B respectively are given by $M_A = M_x$ when $x = 0$, and $M_B = M_x$ when $x = l$. This gives $M_A = \frac{wl^2}{6} + B$ and $M_B = \frac{4}{3} wl^2 + Al + B$.

Substituting A and B from equation 4 gives

$$M_A = wl^2 \left\{ -\frac{2}{9} + \frac{1}{12} \left(\frac{\log_e 2 - 1}{2 - 3 \log_e 2} \right) \right\} = +\, 0{\cdot}100\, wl^2$$

$$M_B = wl^2 \left\{ +\frac{5}{9} + \frac{1}{12} \left(\frac{2 \log_e 2 - 1}{2 - 3 \log_e 2} \right) \right\} = +\, 0{\cdot}150\, wl^2$$

To find the position of maximum deflexion put $\frac{dy}{dx} = 0$ in equation 2.

If $x = kl$ when $\frac{dy}{dx} = 0$ after substituting the numerical values for A and B the following equation is obtained, the solution of which gives the value of k.

$$0 = (1 + k)^3 - 1 - 20{\cdot}0891\, k + 18{\cdot}8822 \log_e (1 + k)$$

The trial and error solution is $k = 0{\cdot}483$.

Substituting $x = 0{\cdot}483l$ in equation 3 gives y_{max}

$$= 0{\cdot}002706 \frac{wl^4}{EI_A}.$$

Example 6.5. The vertical member AB shown in the Fig. 6.10 is 8 m high and is subject to a load from a crane girder of 40 kN applied at an eccentricity of 0·2 m. Calculate the moments at A and B due to this load, assuming both A and B are encastré.

Solution. The stanchion is equivalent to a beam of span 8 m with a moment of 8 kN m applied 2 m from the left-hand support.

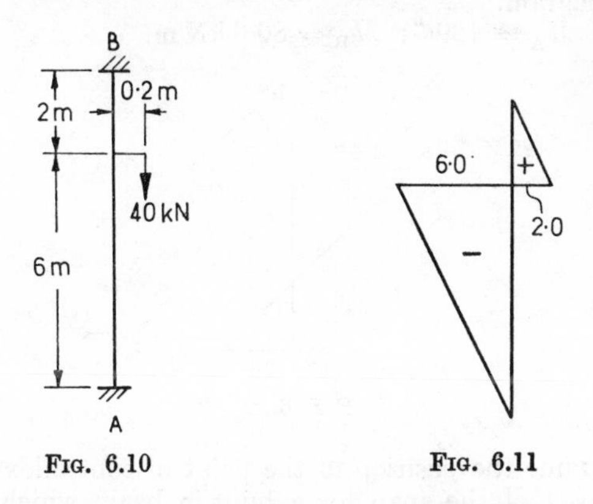

FIG. 6.10 FIG. 6.11

The bending moment diagram due to this moment is shown in Fig. 6.11.

The area, A, of this diagram is given by

$$A = 2 \times 2 \times \tfrac{1}{2} - 6 \times 6 \times \tfrac{1}{2} = -16$$

The moment Ax of this area about the left-hand support is given by

$$Ax = \frac{2{\cdot}0 \times 4}{3} - 18 \times 4 = -69\tfrac{1}{3}$$

If M_A and M_B are the fixing moments at A and B respectively then applying the equations derived earlier in the chapter gives

$$(M_A + M_B) \times 4 = 16$$

$$(2M_A + M_B) \times \frac{8^2}{6} = 69\tfrac{1}{3}$$

The solution of which is $M_A = +2{\cdot}5$ kN m

$$M_B = +1{\cdot}5 \text{ kN m}$$

The bending moment diagram for the member (neglecting all moments in the stanchion due to causes other than the eccentricity of the crane girder load) is as shown in Fig. 6.12.

EXAMPLES FOR PRACTICE ON BUILT-IN BEAMS

1. A built-in beam of 8 m span carries loads of 90 kN and 40 kN at distances of 2 m and 5 m respectively from the left-hand support. Determine the end fixing moments and draw the bending moment diagram.

Answer. $M_A = 129\cdot5$; $M_B = 80\cdot6$ kN m.

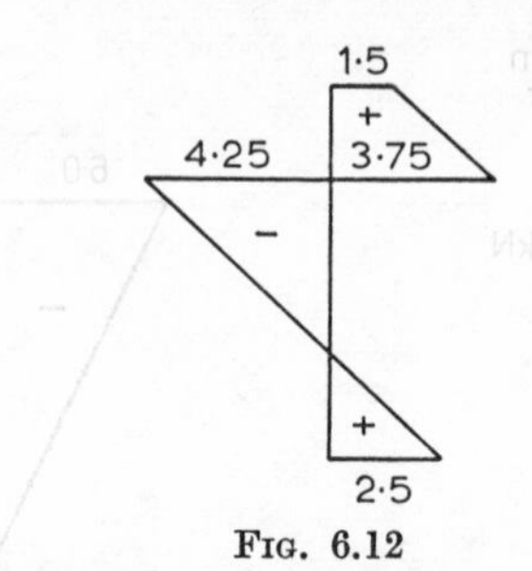

FIG. 6.12

2. Determine the position of the point of contraflexure in the unloaded part of the span for a built-in beam which carries a uniformly distributed load on the left-hand half of the span.

Answer. $0\cdot278L$.

3. A built-in beam carries a load which varies uniformly in intensity from zero at A to $2w$ at B. A prop is placed at mid-span which removes all the deflexion at that point. Calculate the load on the prop. (*St. Andrews*)

Answer. $\dfrac{wL}{2}$.

4. A built-in beam of span $3L$ carries a uniformly distributed load of w per unit length and is propped at a distance L from the left-hand support. If the deflexion of the beam at this point is KR, where R = load on the prop, determine the magnitude of R. (*St. Andrews*)

Answer. $\dfrac{27wL^4}{162EIK + 16L^3}$.

5. A beam ACB of span $2L$ is built-in at A and B and pinned at its mid-point C. It carries a distributed load of w per unit length on AC. The flexural rigidity EI_x of a section of either half distant x from C is kx^2, where k is a constant. Calculate the fixing moments at A and B. (*St. Andrews*)

Answer. $M_A = \frac{3}{8}wL^2$; $M_B = \dfrac{wL^2}{8}$.

6. A beam, AB, of flexural rigidity EI and span L, carries a uniformly distributed load of intensity w per unit length. It is built-in at A and B but support B settles during the application of the load by an amount δ. Show that if

$$\delta = \frac{wL^4}{72EI}$$

there is no fixing moment at B.

Answer. Given in question.

7. A beam, AB, of span l and constant flexural rigidity EI is built-in at B and supported at A so that it rotates through an angle αM_A, where M_A is the moment at A. Calculate the values of the moments at both supports when it carries a load W at a point distant a from A. *(Oxford)*

Answer. $\dfrac{Wa(l-a)^2}{l(l+4EI\alpha)}$; $\dfrac{Wa(l-a)\{al+2EI\alpha(l+a)\}}{l^2(l+4EI\alpha)}$.

8. A beam, AB, is rigidly fixed at both ends of the span L. Show that for a load carried at the centre the deflexion at any point X, when AX $= kL$, is given by

$$\delta = 4(3k^2 - 4k^3)\Delta$$

where $\Delta =$ central deflexion.

A beam of length 20 m is rigidly fixed at the ends and simply supported at mid-length. The left-hand span carries a concentrated load of 100 kN at mid-span, and the right-hand two concentrated loads each 50 kN at the quarter points. Determine the central reaction using the formula established above. *(Aberdeen)*

Answer. 100 kN.

9. A beam AB of constant section and span l is built-in at A and carries a load W at a distance $\dfrac{3l}{4}$ from A. During loading B sinks by an amount δ and rotates through an angle $\dfrac{3\delta}{l}$. Determine the fixing moment at A if the rotation at B is (a) anticlockwise (b) clockwise.

Answer. $\dfrac{12EI\delta}{l} + \dfrac{3Wl}{64}$; $\dfrac{3Wl}{64}$.

10. A uniform beam, fixed horizontally at each end spans an opening of 8 m and carries point loads of 20 kN and 40 kN at distances 2 m and 6 m respectively from one end.

Calculate the values of the fixing couples and reactions at the supports and the deflexion at the centre of the beam. $I = 3 \times 10^7$ mm^4 units. Draw the bending moment diagram. (*Glasgow*)

Answer. 37·5 kN m; 52·5 kN m; 23·12 kN; 36·88 kN; 13·4 mm.

11. A steel beam, AB, 6 m long is fixed horizontally at the ends A and B which are at the same level. The beam carries a load of 50 kN downwards at C, 1 m from A, and a load of 50 kN upwards at D, 2 m from A. Find the end fixing moments and sketch the bending moment and shear force diagrams. Find also the deflexion at D. The moment of inertia of the cross-section is $8{\cdot}0 \times 10^7$ mm^4 units. (*Glasgow*)

Answer. $M_A = 9{\cdot}75$ kN m; $M_B = 15{\cdot}25$ kN m; $\Delta = 1{\cdot}47$ mm upwards.

12. A built-in beam of span l carries a downward load of uniform intensity w over one half of the span measured from one support and an upward concentrated load of $\frac{wl}{4}$ at mid-span. Calculate the deflexion at the centre of the span. (*London*)

Answer. Zero.

13. A horizontal girder, AB, of uniform section has its ends fixed, its free length being l. Find an expression giving the variation in the bending moment at A produced by a load W as it advances along the span. If the load is at a section distant a and b from A and B respectively, a being greater than b, show that for this position of the load the greatest deflexion produced occurs at a section distant $\frac{2al}{2a + l}$ from A and is of magnitude, $\frac{2a^3b^2W}{3(2a + l)^2EI}$. (*Cambridge*)

Answer. Given in question.

14. A steel beam of uniform section, 3 m long, is attached to supports at the ends and carries a load of 10 kN in the centre. The supports are not quite rigid but will yield $1{\cdot}2 \times 10^{-5}$ radians/kN m. The relevant moment of inertia of the joist is 4×10^7 mm^4.

Find the fixing moments at the ends of the beam.

Answer. 35·2 kN m.

Chapter 7

The Continuous Beam

IF LOADS, W_1, W_2, W_3 and W_4, have to be carried in the four bays, AB, BC, CD and DE, of a building there are two general possibilities. Either a number of simply supported beams are used as shown in Fig. 7.1, or one continuous beam as shown in Fig. 7.2.

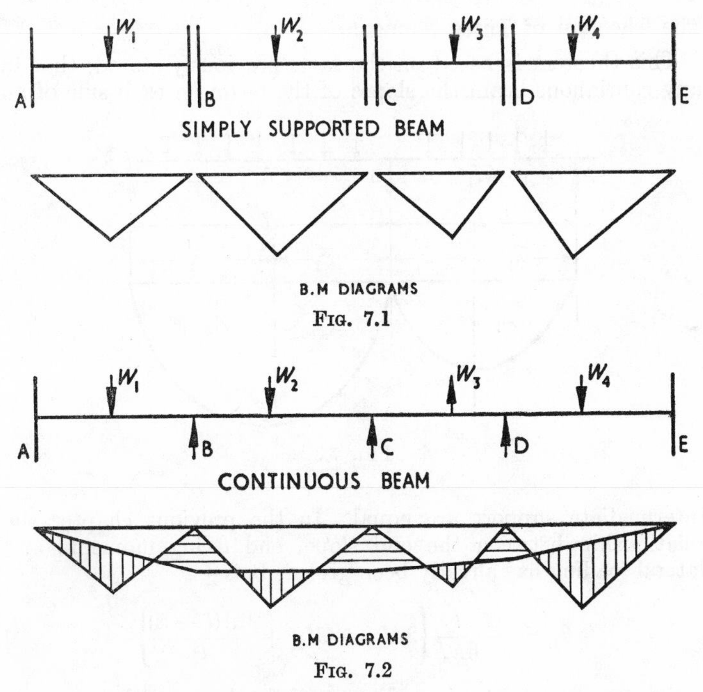

FIG. 7.1

FIG. 7.2

If the first alternative is adopted, then the problem is statically determinate. The reactions can be obtained from the laws of static equilibrium. Each beam is free to take up its own deflected form irrespective of the load in an adjacent span. If, however, one continuous beam is used, then the slope of the beam at a support B must obviously be the same on each side of the support, its magnitude being affected not only by the load in the spans on either side of it but also by all the other loads on the beam. The continuous beam is a redundant structure and the reactive forces at the supports cannot be obtained simply by the application of the laws of static equilibrium. In order to determine the bending moments and shearing forces at all points on the continuous beam either the reactive forces or bending moments at the supports must first be determined.

The usual procedure is to determine the bending moments at the supports. There are many methods for so doing but the two on which the average student concentrates are—

1. The theorem of three moments.
2. The moment-distribution method.

Only these two methods will be used in the chapter.

The Theorem of Three Moments

This theorem is based on the fact, previously stated, that in any continuous beam the slopes of the beam on each side of an

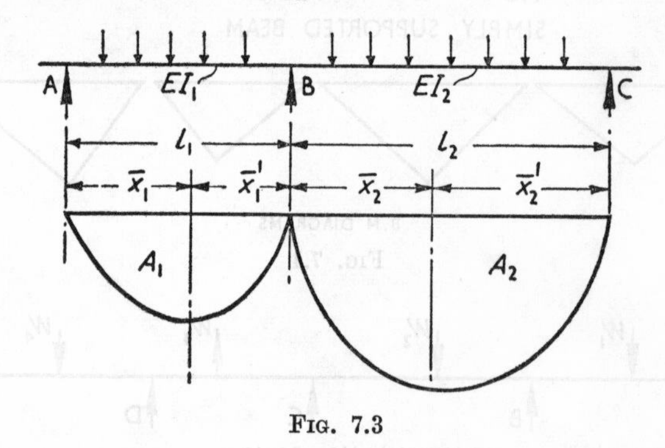

FIG. 7.3

intermediate support are equal. In the previous chapter the relationship between the end slope, end fixing moments and lateral loading has already been given. It is—

$$\theta_L = -\frac{l}{6EI}\left\{2M_L + M_R + \frac{6A(l-\bar{x})}{l^2}\right\}$$

$$\theta_R = \frac{l}{6EI}\left(M_L + 2M_R + \frac{6A\bar{x}}{l^2}\right)$$

where θ_L, M_L are the slope and moment at the left-hand support and θ_R, M_R the corresponding values at the right-hand support.

Let AB and BC as shown in Fig. 7.3 be any two adjacent spans of a continuous beam each subjected to loading so that the areas of the free bending moment diagrams on the two spans are A_1 and A_2. Let the centre of area of A_1 be distant $\bar{x}_1$ from A and $\bar{x}_1'$ from B and the centre of area of A_2 be $\bar{x}_2$ from B and $\bar{x}_2'$ from C. Let AB $= l_1$, BC $= l_2$, and the flexural rigidities EI_1 and EI_2 respectively.

If the beam AB is considered, then the slope at B, the right-hand support, is given by

$$\theta_B = \frac{l_1}{EI_1}\left(M_A + 2M_B + \frac{6A_1\bar{x}_1}{l_1^2}\right)$$

If the beam BC is considered, then B becomes the left-hand support and the slope is given by

$$\theta_B = -\frac{l_2}{EI_2}\left(2M_B + M_C + \frac{6A_2\bar{x}_2'}{l_2^2}\right)$$

Since the beam is continuous these two slopes are equal, hence

$$\frac{l_1}{EI_1}M_A + 2\left(\frac{l_1}{EI_1} + \frac{l_2}{EI_2}\right)M_B + \frac{l_2}{EI_2}M_C + \frac{6A_1\bar{x}_1}{EI_1l_1} + \frac{6A_2\bar{x}_2'}{EI_2l_2} = 0$$

This equation contains three unknowns, the support moments M_A, M_B and M_C, and it is obvious why it is known as the three-moment equation. It was derived by considering a pair of adjacent spans. If the continuous beam has n spans, then considering them in pairs will provide $(n - 1)$ equations. There are, however, $(n + 1)$ supports and therefore $(n + 1)$ unknown support moments. The two further equations necessary for the solution of the problem are obtained from the known end conditions. If the ends are pinned, then the moments at the first and last supports are zero. If they are cantilevers, then the value of the moment is known. If the ends are encastré, this gives a further relationship to determine the moments since an encastré end means a zero slope.

Obviously, if a beam has a large number of spans, solution by the theorem of three moments is lengthy but if there are only two or three spans it is straightforward.

In the case of a number of spans, however, a quicker solution is obtained by the moment-distribution method.

The Moment-distribution Method

The moment-distribution method consists in part of balancing the moments applied at a joint or junction. Thus one is concerned primarily with the effect of the moments on the joints. The general practice in this method, therefore, is to employ a convention for the sign of a moment based on its effect on a joint. The standard convention used is that moments acting in a clockwise direction are positive and those acting in an anticlockwise direction are negative. Thus in the beam shown in Fig. 7.4 the fixing moment

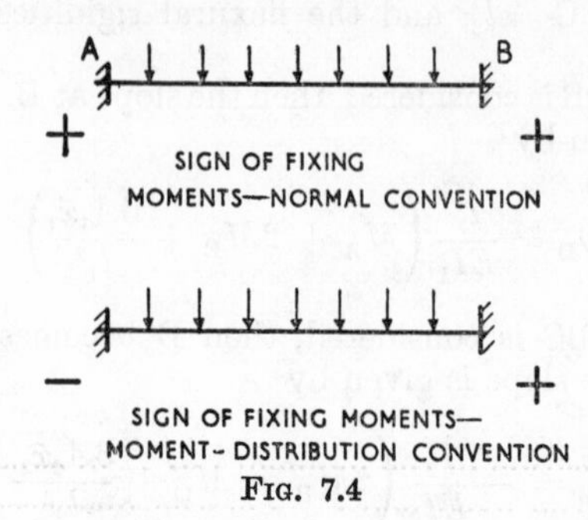

FIG. 7.4

is negative at A but positive at B, although if the standard beam convention were adopted each moment would be positive in sign.

If in a continuous beam the slope at the supports were known, the moments could be calculated. The moment-distribution method starts by taking these slopes as zero, i.e. the joint is

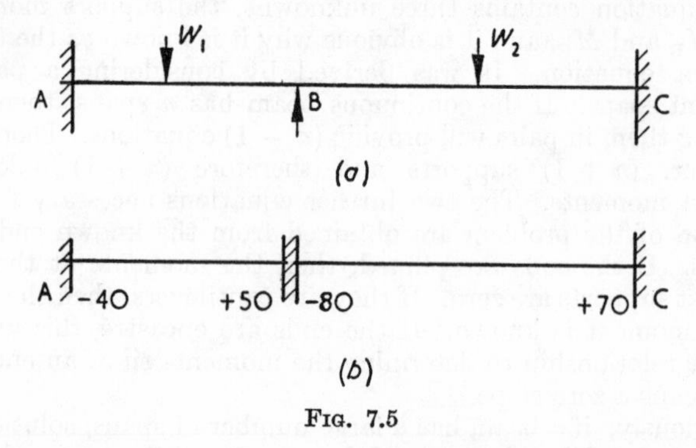

FIG. 7.5

encastré, and then successively releasing and balancing the joint until a state of equilibrium has been obtained. The moments at the ends of a beam when they are encastré can be determined by the methods given in the previous chapter.

Consider the beam, ABC, shown in Fig. 7.5 which is encastré at supports A and C and continuous over B.

Consider this firstly as two separate encastré beams, AB and BC. Let the load on AB be such that the fixing moments (using the moment-distribution convention) are -40 and $+50$ at A and B respectively. Let the load on BC be such that the fixing moments are -80 and $+70$ at B and C respectively. These moments are obtained by taking the beam at B fixed. In the actual problem this beam is free to rotate over support B. If the restraint at B is removed, then there is a moment of 50 tending to rotate the joint in a clockwise direction and a moment of 80 tending to rotate it in an anticlockwise direction. The joint will rotate until the moments on each side of the joint are equal and opposite in sign. When this is so let the moment BA be $(+50 + b)$ and the moment BC will then be $\{-80 + (30 - b)\}$. One problem therefore is the determination of b, the proportion of the out-of-balance moment at B applied to the side BA.

The twisting of the end B of the beam BA due to the application of the moment b also causes an increase in the moment at the fixed end A. Similarly the rotation of the end B of the beam BC due to the application of the moment $(30 - b)$ causes a change in the moment at the fixed end C. Let the increases at these two supports be a and c respectively. Then the final moments are

At A, $-40 + a$

At BA, $+50 + b$

At BC, $-50 - b$

At C, $+70 + c$

If therefore, a, b and c can be determined, the problem is solved. Expressed in words, the factors to be determined are—

1. The distribution of the out-of-balance moment.
2. The amount of moment carried over.

It is not proposed to go more fully into the basic theory but it can be shown—

1. That the out-of-balance moment at any support is distributed amongst the members meeting at it in the ratio of their stiffnesses. The stiffness (denoted by k) of a member is equal to the flexural rigidity divided by its length, i.e. $k = \dfrac{EI}{l}$.

2. The moment carried over is one-half of the moment applied at the far end in the case of beams of uniform section throughout.

If in the simple example shown in Fig. 7.5 the stiffness of AB is twice that of BC, then $\frac{2}{3}$ of the out-of-balance at B go to BA and the remaining $\frac{1}{3}$ to BC. Thus the solution to the problem is as shown in Table 7.1 (*a*).

TABLE 7.1 (*a*)

A	B		C
	$\frac{2}{3}$	$\frac{1}{3}$	
− 40	+ 50	− 80	+ 70
+ 10	+ 20	+ 10	+ 5
− 30	+ 70	− 70	+ 75

The moments at A, B and C are therefore 30, 70 and 75 respectively.

The above problem reached a balance quickly because both ends, A and C, were encastré. If the end A had been a pinned joint then the solution would have been longer since the final moment at A must always equal zero. The work is set out in Table 7.1 (*b*) below.

TABLE 7.1 (*b*)

A	B		C
	$\frac{2}{3}$	$\frac{1}{3}$	
− 40	+ 50	− 80	+ 70
+ 10	+ 20	+ 10	+ 5
+ 30	+ 15		
− 5	− 10	− 5	− 3
+ 5	+ 3		
− 1	− 2	− 1	
+ 1			
0	+ 76	− 76	+ 72

It can be shown, however, that if it is desired to keep one end of a member pinned, that is, not to carry over any moment to that end, the stiffness of that member should be taken as $\frac{3}{4}$ of its actual stiffness. The actual stiffness of AB is twice that of BC but if A is to remain a pin then the equivalent stiffness of AB will be only 1·5 times that of BC and the out-of-balance moment at B will be divided in the ratio $\frac{3}{5}:\frac{2}{5}$.

TABLE 7.2

A	B		C
	$\frac{3}{5}$	$\frac{2}{5}$	
− 40	+ 50	− 80	+ 70
+ 40	+ 20		
	+ 6	+ 4	+ 2
0	+ 76	− 76	+ 72

The work is now much simplified as shown in Table 7.2. It should be noted that since A is a pin the first step is to make the moment at that point zero.

Settlement of the Supports

The theory in both methods given has been based on no relative settlement of the supports. If such settlement does take place both methods can be amended to allow for it.

In the case of the three-moment theorem, if δ_A, δ_B and δ_C are the settlements of the supports A, B and C respectively, then the revised form of the theorem is

$$\frac{l_1}{EI_1}M_A + 2\left(\frac{l_1}{EI_1}+\frac{l_2}{EI_2}\right)M_B + \frac{l_2}{EI_2}M_C + \frac{6A_1\bar{x}_1}{EI_1l_1} + \frac{6A_2\bar{x}_2'}{EI_2l_2} + \frac{6(\delta_B-\delta_A)}{l_1} - \frac{6(\delta_C-\delta_B)}{l_2} = 0$$

It is seen that if

$$\frac{\delta_B-\delta_A}{l_1} = \frac{\delta_C-\delta_B}{l_2}$$

there is no change in the equation for M_A, M_B and M_C The relationship between the δ and l is that which occurs if the supports A, B and C remain collinear after deflexion.

In the moment-distribution method the moments for the initial encastré case are

$$\frac{6EI_1(\delta_B-\delta_A)}{l_1^2}$$

at each support. These moments are added to those due to the loading and the balancing carried out in the standard way. Settlements of supports can also be dealt with by means of Castigliano's theorem as given in Example 7.6.

Examples on Continuous Beams

Example 7.1. A continuous beam, ABCD, is pinned at A and simply supported at B and C, these points being at the same level. AB = 10 m, BC = 12 m and CD = 6 m. It carries a point load of 100 kN at the mid-point of BC and a uniformly distributed load of 10 kN/m run from A to D.

Find the fixing moments at B and C and the reaction on the pin at A.

Solution. The example is shown diagrammatically in Fig. 7.6. In this case the three-moment method will be used for the solution.

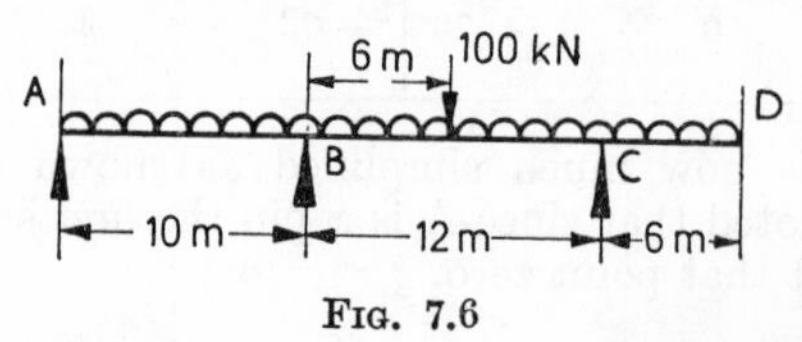

FIG. 7.6

The support A is a pin hence $M_A = 0$. A cantilever of length 6 m overhangs from C. It carries a load of 10 kN/m run hence

$$M_C = 10 \times 6 \times 3 = 180 \text{ kN m}$$

If suffix 1 is used for span AB and 2 for BC then

$$A_1 = \tfrac{2}{3} \times 10 \times 10 \times 10^2 \times \tfrac{1}{8} = \frac{1}{12} \times 10^4 \text{ kN m}^2 \text{ units}$$

$$A_2 = (\tfrac{2}{3} \times 12 \times 10 \times 12^2 \times \tfrac{1}{8}) + (\tfrac{1}{2} \times 12 \times 100 \times 12 \times \tfrac{1}{4})$$
$$= 1440 + 1800$$

$$\bar{x}_1 = 5 \qquad \bar{x}_2 = 6$$

Substituting in the three-moment equation gives

$$2M_B(10 + 12) + 12M_C = \frac{6 \,.\, 1440 \,.\, 6}{12} + \frac{6 \,.\, 1800 \,.\, 6}{12} + \frac{6 \,.\, 10^4 \,.\, 5}{12 \,.\, 10}$$

The EI term cancels throughout

Putting $\qquad M_C = 180$

gives $\qquad 44M_B = 10\,060$

hence $\qquad M_B = 228 \text{ kN m}$

The reaction R_A at A is given by

$$R_A = R_A' + \frac{M_A - M_B}{L}$$

where R_A' is the reaction at A if the beam AB whose span is L were simply supported. Hence

$$R_A = \tfrac{1}{2} \times 10 \times 10 + \frac{0 - 228}{10}$$

$$= 50 - 22{\cdot}8 = 27{\cdot}2 \text{ kN}$$

Example 7.2. A steel beam AD shown in Fig. 7.7 is of constant flexural rigidity EI. It is supported in a horizontal position by steel ties BE, CE and DE whose cross-sectional areas are A_1, A_2 and A_3, respectively. The point A is hinged freely to a rigid abutment and the three tie rods are hinged freely at the point E which is attached to a rigid wall. A uniformly distributed load

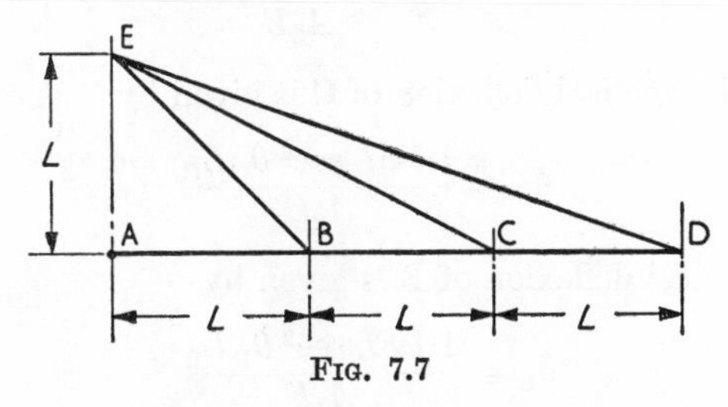

FIG. 7.7

of intensity w is applied to the beam and the deflected form is such that A, B, C and D are collinear. Neglecting the weight of the beam and any direct energy in it show that

$$A_1 : A_2 : A_3 : : 1 : 2 : \tfrac{4}{3} \text{ (approx.)}$$

(Oxford)

Solution. The points of support remain collinear hence the theorem of three moments can be used without the introduction of any terms to allow for the deflexion.

It is also obvious that, due to the symmetry of the loading, $M_B = M_C$. It is therefore only necessary to apply the three-moment equation to the spans AB and BC. Doing so gives

$$4L \, . \, M_B + L \, . \, M_C = \frac{6wL^3}{24} + \frac{6wL^3}{24}$$

giving $$M_B = M_C = \frac{wL^2}{10}$$

$$R_A = R_D = \frac{wL}{2} - \frac{wL}{10} = 0{\cdot}4wL$$

Hence $R_B = R_C = 1{\cdot}1wL$

The force in $DE = R_D \sec \theta_D$

where $\theta_D = \widehat{AED}$.

Hence the extension of DE

$$= \frac{R_D \sec \theta_D L_D}{A_3 E}$$

where L_D = length of DE.

The vertical deflexion of D,

$$\delta_D = \frac{R_D \sec^2 \theta_D L_D}{A_3 E}$$

$$= \frac{0{\cdot}4wL \sec^2 \theta_D L_D}{A_3 E}$$

Similarly the vertical deflexion of C is given by

$$\delta_C = \frac{1{\cdot}1wL \sec^2 \theta_C L_C}{A_2 E}$$

and the vertical deflexion of B is given by

$$\delta_B = \frac{1{\cdot}1wL \sec^2 \theta_B L_B}{A_1 E}$$

where L_C, L_B are the lengths CE and BE respectively and θ_C θ_B are the angles $\widehat{AEC}$ and $\widehat{AEB}$ respectively.

But the beam remains collinear hence

$$\delta_C = 2\delta_B \quad \text{and} \quad \delta_D = 3\delta_B$$

hence
$$\frac{0{\cdot}4wL \sec^2 \theta_D L_D}{A_3 E} = \frac{3{\cdot}3wL \sec^2 \theta_B L_B}{A_1 E}$$

giving
$$\frac{A_3}{A_1} = \frac{0{\cdot}4 \sec^2 \theta_D L_D}{3{\cdot}3 \sec^2 \theta_B L_B}$$

and
$$\frac{1{\cdot}1wL \sec^2 \theta_C L_C}{A_2 E} = \frac{2{\cdot}2wL \sec^2 \theta_B L_B}{A_1 E}$$

But
$$\sec^2 \theta_D = 10 \quad \text{and} \quad L_D = \sqrt{10}L$$

$$\sec^2 \theta_C = 5 \qquad L_C = \sqrt{5}L$$

$$\sec^2 \theta_B = 2 \qquad L_B = \sqrt{2}L$$

Hence $\frac{A_2}{A_1} = \frac{\sec^2 \theta_C L_C}{2 \sec^2 \theta_B L_B} = \frac{5\sqrt{5}}{4\sqrt{2}} = 1{\cdot}98 = 2$ (approx.)

and $\frac{A_3}{A_1} = \frac{0{\cdot}4 \times 10\sqrt{10}}{3{\cdot}3 \times 2\sqrt{2}} = 1{\cdot}36 = \frac{4}{3}$ (approx.)

Example 7.3. Determine the bending moments at the supports for the continuous beam shown in Fig. 7.8, (*a*) when the beam is pinned at A, (*b*) when it merely rests on the support at A.

The flexural rigidity of the beam can be assumed constant.

(*St. Andrews*)

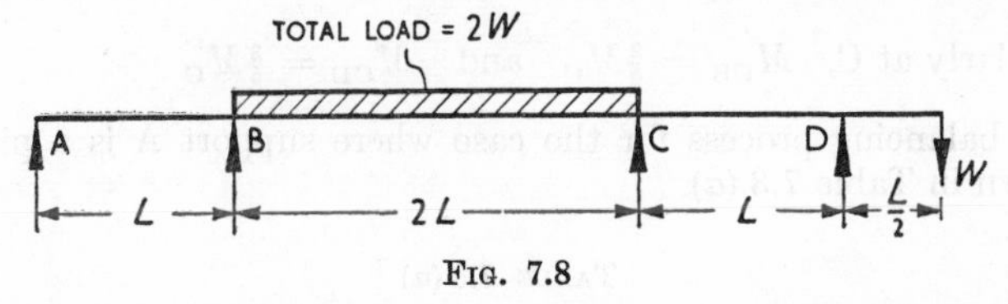

FIG. 7.8

Solution. This problem will be solved by moment-distribution methods.

Before carrying out the balancing process, the end fixing moments, taking the beams as built-in, and the distribution of out-of-balance moment must be determined.

Span BC, $M_{BC} = M_{CB} = \frac{2W \,.\, 2L}{12} = \frac{WL}{3} = 333$ (say)

$M_D = \frac{WL}{2} = 500$ (say)

It has already been mentioned that in the case of a beam with one end pinned it is an advantage to take the equivalent stiffness as three-quarters of the actual stiffness and not to carry over any moment to the pinned end during the balancing process. This method will certainly be applied to span AB but it is also an advantage to apply it to the span CD. The moment at D is fixed at $\frac{WL}{2}$, and any moment carried over from C to D can have no effect on the final moment at D which must be $\frac{WL}{2}$. It therefore saves work in the balancing process if the two sides at D are balanced straight away and thereafter this support not included in any carry over moments. This can be done if span CD is taken as having an equivalent stiffness of three-quarters of its actual stiffness.

The distribution of out-of-balance moment is then given by

$$\text{At support B,}\quad M_{BA} = \left(\frac{\dfrac{3}{4}\cdot\dfrac{I}{L}}{\dfrac{3I}{4L}+\dfrac{I}{2L}}\right) M_B = \tfrac{3}{5}M_B$$

$$M_{BC} = \left(\frac{\dfrac{I}{2L}}{\dfrac{3I}{4L}+\dfrac{I}{2L}}\right) M_B = \tfrac{2}{5}M_B$$

Similarly at C, $M_{CB} = \frac{2}{5}M_C$ and $M_{CD} = \frac{3}{5}M_C$

The balancing process for the case where support A is a pin is shown in Table 7.3 (*a*).

TABLE 7.3 (*a*)

A		B		C		D
	$\frac{3}{5}$	$\frac{2}{5}$	$\frac{2}{5}$	$\frac{3}{5}$		
0	0	− 333	+ 333	0	0	− 500
				+ 250	+ 500	
		− 116	− 233	− 350		
	+ 269	+ 180	+ 90			
		− 18	− 36	− 54		
	+ 11	+ 7	+ 3			
			1	− 2		
0	+ 280	− 280	+ 156	− 156	+ 500	− 500

Hence the bending moments at the supports are

$$M_A = 0 \qquad M_C = 0{\cdot}156WL$$

$$M_B = 0{\cdot}28WL \qquad M_D = 0{\cdot}50WL$$

The reaction at A is equal to $-0{\cdot}28W$, the negative sign implying that the reactive force acts downwards. It is possible for a pin to supply such a reactive force. If, however, the beam merely rests on the supports at A such a reaction is not possible and under the conditions of loading given in the question the beam will raise itself off the support at A. Thus the support, A, virtually becomes non-existent and BC becomes an end span with the support, B, acting as a pin.

The fixed end moments remain as before. There is no distribution of out-of-balance moment at B and that at support C is changed.

At support M_C

$$M_{CB} = \left(\frac{\frac{3}{4} \cdot \frac{I}{2L}}{\frac{3}{4} \cdot \frac{I}{2L} + \frac{3}{4} \cdot \frac{I}{L}} \right) M_C = \frac{M_C}{3}$$

$$M_{CD} = \frac{\frac{3}{4} \cdot \frac{I}{L}}{\frac{3}{4} \cdot \frac{I}{2L} + \frac{3}{4} \cdot \frac{I}{L}} M_C = \frac{2M_C}{3}$$

The balancing is carried out as shown in Table 7.3 (*b*).

TABLE 7.3 (*b*)

B	C		D	
	$\frac{1}{3}$	$\frac{2}{3}$		
− 333	+ 333	0	0	− 500
+ 333	+ 167	+ 250	+ 500	
	− 250	− 500		
0	+ 250	− 250	+ 500	− 500

The moments at the supports are then

$$M_A = M_B = 0$$

$$M_C = 0{\cdot}25WL$$

$$M_D = 0{\cdot}50WL$$

Example 7.4. The continuous beam shown in Fig. 7.9 rests on piers built from a material the safe stress in which is 3·5 N/mm². Calculate a suitable cross-section for each pier. The flexural rigidity of the beam is constant throughout. (*St. Andrews*)

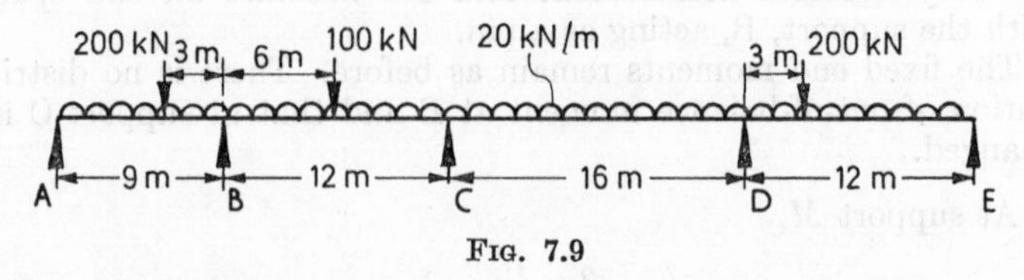

FIG. 7.9

Solution. The answer required is the size of pier. This depends on the load carried. This must therefore be found in order to solve the problem.

Using the moment-distribution method of solution the fixed end moments are—

Span AB,

$$M_{AB} = \frac{200 \times 3^2 \times 6}{9^2} + \frac{20 \times 9^2}{12} = 238 \text{ kN m}$$

$$M_{BA} = \frac{200 \times 3 \times 6^2}{9^2} + \frac{20 \times 9^2}{12} = 402 \text{ kN m}$$

Span BC,

$$M_{BC} = M_{CB} = \frac{100 \times 12}{8} + \frac{20 \times 12^2}{12} = 540 \text{ kN m}$$

Span CD,

$$M_{CD} = M_{DC} = \frac{20 \times 16^2}{12} = 426 \text{ kN m}$$

Span DE,

$$M_{DE} = \frac{200 \times 9^2 \times 3}{12^2} + \frac{20 \times 12^2}{12} = 578 \text{ kN m}$$

$$M_{ED} = \frac{200 \times 3 \times 9^2}{12^2} + \frac{20 \times 12^2}{12} = 353 \text{ kN m}$$

The distribution of the out-of-balance moment at the various supports is

Support B, $$M_{BA} = \frac{\frac{3}{4} \cdot \frac{I}{9}}{\frac{3}{4} \cdot \frac{I}{9} + \frac{I}{12}} M_B = \frac{M_B}{2} = M_{BC}$$

$$\text{Support C,} \qquad M_{CB} = \left(\frac{\dfrac{I}{12}}{\dfrac{I}{12} + \dfrac{I}{16}}\right) M_C = \tfrac{4}{7} M_C$$

$$\text{hence} \qquad M_{CD} = \tfrac{3}{7} M_C$$

$$\text{Support D,} \qquad M_{DC} = \left(\frac{\dfrac{I}{16}}{\dfrac{3}{4} \cdot \dfrac{I}{12} + \dfrac{I}{16}}\right) M_D = \frac{M_D}{2} = M_{DE}$$

The balancing of the moments is shown in Table 7.4 (*a*), the moments being given in kN m units.

TABLE 7.4 (*a*)

A	B		C		D		E
	$\frac{1}{2}$	$\frac{1}{2}$	$\frac{4}{7}$	$\frac{3}{7}$	$\frac{1}{2}$	$\frac{1}{2}$	
− 268	+ 402	− 540	+ 540	− 426	+ 426	− 578	+ 353
+ 268	+ 134					− 176	− 353
	+ 2	+ 2	+ 1	+ 82	+ 164	+ 164	
		− 56	− 112	− 85	− 42		
	+ 28	+ 28	+ 14	+ 10	+ 21	+ 21	
		− 7	− 14	− 10	− 5		
	+ 3	+ 4	+ 2	+ 1	+ 2	+ 3	
			− 2	− 1			
0	+ 569	− 569	+ 429	− 429	+ 566	− 566	0

The reaction at any support is due to the loading and any difference between the two adjacent support moments. These values have been found separately and are shown in Table 7.4 (*b*), the letters (L) and (R) denote respectively the left- and right-hand side of the support.

The total load is the sum of the loads from the left- and right-hand sections This total load is converted into kiloNewtons and divided by $3{\cdot}5 \times 10^3$ to give the size of pier required in square metres.

TABLE 7.4 (b)

Support		Shear Due to Applied Loads kN	Shear Due to Moments kN	Total Load (kN)	Pier Area (m^2)
A		223	− 63	160	$4{\cdot}56 \times 10^{-2}$
B	(L)	157	+ 63	402	$1{\cdot}15 \times 10^{-1}$
	(R)	170	+ 12		
C	(L)	170	− 12	309	$8{\cdot}83 \times 10^{-2}$
	(R)	160	− 9		
D	(L)	160	+ 9	502	$1{\cdot}44 \times 10^{-1}$
	(R)	270	+ 63		
E		170	− 63	107	$3{\cdot}06 \times 10^{-2}$

Example 7.5. A three-span continuous beam, ABCD, of total length $3L$ is encastré at A and freely supported at D. Determine the lengths of AB, BC and CD so that when it carries a uniformly distributed load of w/unit length

$$M_A = M_B = M_C = \frac{wL_1^2}{12}$$

where L_1 = length of AB.

Solution. The beam is shown diagrammatically in Fig. 7.10. The problem will be solved by the theorem of three moments to

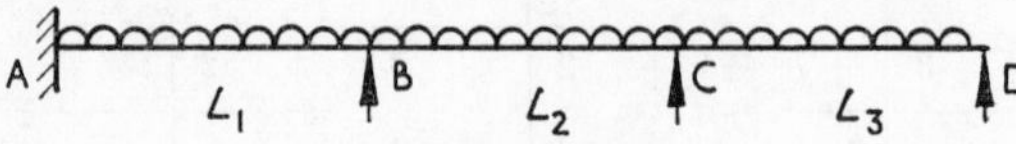

A B C D L_1 L_2 L_3

FIG. 7.10

show the student how to use this method in the case where the end support is encastré.

For each span of length L

$$\frac{6A\bar{x}}{l} = -\frac{wL^2}{4}$$

The slope θ_A at the support A is given by

$$\theta_A = -\frac{L_1}{6EI}\left(2M_A + M_B - \frac{wL_1^2}{4}\right) = 0$$

hence $$M_A = \frac{wL_1^2}{8} - \frac{M_B}{2}$$

THE WORLD'S GREATEST BOOKSHOP

STOCK OF OVER FOUR MILLION VOLUMES

119-125 Charing Cross Road London WC2

01-437 5660 ★ *Open 9-6 (inc. Sats.)*

Sold by SF Dept. 7 Date 29/10.

	£	
	1	30
	3	95
	2	90
£8.15 SF.	8	15

PAID FOYLES

12076 - 19

This bill must be produced in the event of any query regarding purchase.

Considering the spans in pairs gives

$$M_A L_1 + 2M_B(L_1 + L_2) + M_C L_2 = \frac{wL_1^3}{4} + \frac{wL_2^3}{4}$$

$$M_B L_2 + 2M_C(L_2 + L_3) = \frac{wL_2^3}{4} + \frac{wL_3^3}{4}$$

Substituting for M_A gives

$$\frac{wL_1^3}{8} - \frac{M_B L_1}{2} + 2M_B(L_1 + L_2) + M_C L_2 = \frac{wL_1^3}{4} + \frac{wL_2^3}{4}$$

But $$M_B = M_C = \frac{wL_1^2}{12}$$

hence $$\frac{3L_1}{2} + 3L_2 = \frac{w}{M_B}\left(\frac{L_2^3}{4} + \frac{L_1^3}{8}\right) = \frac{3L_2^3}{L_1^2} + \frac{3}{2}L_1$$

$$\therefore \quad L_1 = L_2$$

Also for the second three-moment equation

$$3L_2 + 2L_3 = 3L_2 + \frac{3L_3^3}{L_2^2}$$

giving $$L_3 = \sqrt{\tfrac{2}{3}}L_2 = 0{\cdot}815L_2$$

Also $L_1 + L_2 + L_3 = 3L$

hence $$L_1 = \frac{3}{2{\cdot}815}L = 1{\cdot}068L = L_2$$

and $$L_3 = 0{\cdot}815 \times 1{\cdot}068L = 0{\cdot}864L$$

An alternative and somewhat quicker solution is an argument as follows.

$$M_B = M_A = \frac{wL_1^2}{12}$$

Hence the slope at B which is given by

$$\theta_B = \frac{l}{6EI}\left(M_A + 2M_B + \frac{6A\bar{x}}{l^2}\right) = \frac{L_1}{6EI}\left(\frac{3wL_1^2}{12} - \frac{wL_1^2}{4}\right) = 0$$

The slope at B is, therefore, zero. But if consideration is given to the span BC

$$\theta_B = -\frac{L_2}{6EI}\left(2M_B + M_C - \frac{wL_2^2}{4}\right)$$

and for this to be zero L_2 must equal L_1. This in turn gives zero slope at C.

If consideration is now given to the span CD, the slope at C is given by

$$\theta_C = -\frac{L_3}{6EI}\left(2M_C - \frac{wL_3^2}{4}\right)$$

and for this to be zero

$$\frac{wL_3^2}{4} = \frac{wL_1^2}{6}$$

i.e. $$L_3 = \sqrt{\tfrac{2}{3}}L_1 \text{ as before.}$$

Example 7.6. A bridge of uniform cross-section rests on rigid abutments at the ends and on three equal pontoons as shown in Fig. 7.11, and has a concentrated load, W, at the middle. When

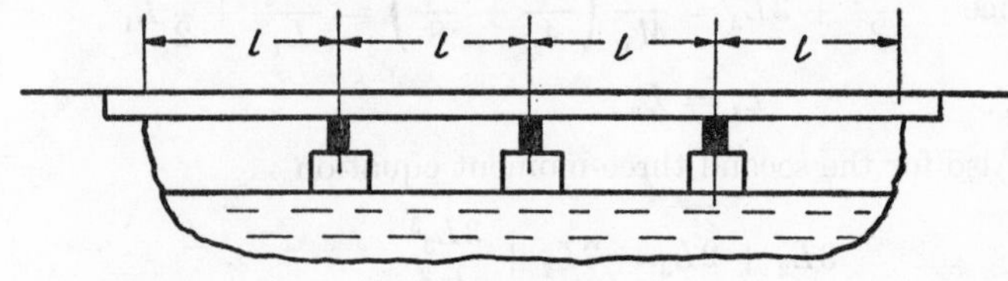

FIG. 7.11

the bridge is unloaded the pontoons just touch it without exerting any force. With the load, W, at the middle and the two end pontoons removed the central deflexion is one half what it would be with no pontoon.

Find the reactions and draw the bending moment diagram for the bridge due to the central load with the three pontoons in position. (*Cambridge*)

Solution. The load carried by a pontoon is proportional to the depth of immersion. Since, when the beam is unloaded, the pontoons touch it without exerting any force, the force which any one of them will exert when it deflects downwards by an amount Δ from this initial position will be $k\Delta$.

The question gives data to determine k from the behaviour of the beam with only a central pontoon. In such a case half the central load is carried by the pontoon and the central deflexion is

$$\frac{1}{2}\cdot\frac{W(4l)^3}{48EI}$$

Hence $$k\cdot\frac{1}{2}\cdot\frac{W(4l)^3}{48EI} = \frac{W}{2}$$

thus $$k = \frac{48EI}{64l^3} = \frac{3EI}{4l^3}$$

The problem is shown diagrammatically in Fig. 7.12. Due to the symmetry of the loading $R_D = R_B = R_1$. There are two

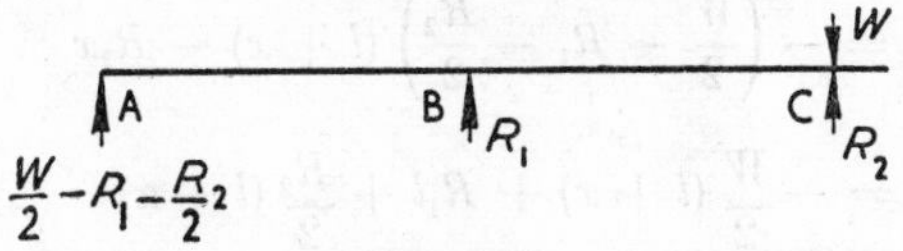

FIG. 7.12

redundancies, the reactions R_1 and R_2, and since the amount by which the pontoon sinks is $\frac{R}{k}$ the equations for the determination of the two unknowns are

$$\frac{\partial U}{\partial R_1} = -\frac{2R_1}{k} \quad \text{and} \quad \frac{\partial U}{\partial R_2} = -\frac{R_2}{k}$$

The total $\frac{\partial U}{\partial R_1}$ gives the deflexion not only of B but also of D, each of these is equal to $\frac{R_1}{k}$, hence

$$\frac{\partial U}{\partial R_1} = -\frac{2R_1}{k}$$

For the portion AB, the moment, M_x, at a section distant x from A is given by

$$M_x = -\left(\frac{W}{2} - R_1 - \frac{R_2}{2}\right) x$$

$$\frac{\partial M_x}{\partial R_1} = x \qquad \frac{\partial M_x}{\partial R_2} = \frac{x}{2}$$

$$\left[\frac{\partial U}{\partial R_1}\right]_{AB} = \frac{1}{EI}\int_0^l \left(R_1 + \frac{R_2}{2} - \frac{W}{2}\right) x^2 \, dx$$

$$= \frac{l^3}{3EI}\left(R_1 + \frac{R_2}{2} - \frac{W}{2}\right)$$

$$\left[\frac{\partial U}{\partial R_2}\right]_{AB} = \frac{1}{2EI}\int_0^l \left(R_1 + \frac{R_2}{2} - \frac{W}{2}\right) x^2 \, dx$$

$$= \frac{l^3}{6EI}\left(R_1 + \frac{R_2}{2} - \frac{W}{2}\right)$$

For the portion of the beam BC the moment, M_x, at a section distant x from B is given by

$$M_x = -\left(\frac{W}{2} - R_1 - \frac{R_2}{2}\right)(l + x) - R_1 x$$

$$= -\frac{W}{2}(l + x) + R_1 l + \frac{R_2}{2}(l + x)$$

$$\frac{\partial M_x}{\partial R_1} = l \qquad \frac{\partial M_x}{\partial R_2} = \frac{(l + x)}{2}$$

$$\left[\frac{\partial U}{\partial R_1}\right]_{BC} = \frac{1}{EI}\int_0^l \left\{R_1 l^2 + \frac{R_2}{2}(l^2 + lx) - \frac{W}{2}(l^2 + lx)\right\} dx$$

$$= \frac{1}{EI}\left(R_1 l^3 + \frac{3R_2 l^3}{4} - \frac{3Wl^3}{4}\right)$$

$$\left[\frac{\partial U}{\partial R_2}\right]_{BC} = \frac{1}{2EI}\int_0^l \left\{R_1(l^2 + lx) + \frac{R_2}{2}(l^2 + 2lx + x^2) - \frac{W}{2}(l^2 + 2lx + x^2)\right\} dx$$

$$= \frac{1}{2EI}\left(\frac{3R_1 l^3}{2}\right) + \left(\frac{7R_2 l^3}{6} - \frac{7Wl^3}{6}\right)$$

There is no need to integrate for the other half of the beam since the expressions will be similar. Hence for the whole structure—

$$\frac{\partial U}{\partial R_1} = \frac{2l^3}{3EI}\left(R_1 + \frac{R_2}{2} - \frac{W}{2}\right) + \frac{2l^3}{EI}\left(R_1 + \frac{3R_2}{4} - \frac{3W}{4}\right)$$

$$= -\frac{8R_1 l^3}{3EI}$$

$$\frac{\partial U}{\partial R_2} = \frac{2l^3}{6EI}\left(R_1 + \frac{R_2}{2} - \frac{W}{2}\right) + \frac{2l^3}{2EI}\left(\frac{3R_1}{2} + \frac{7R_2}{6} - \frac{7W}{6}\right)$$

$$= -\frac{4R_2 l^3}{3EI}$$

giving

$$2R_1 + R_2 - W + 6R_1 + \frac{9}{2}R_2 - \frac{9}{2}W = -8R_1$$

$$R_1 + \frac{R_2}{2} - \frac{W}{2} + \frac{9}{2}R_1 + \frac{7}{2}R_2 - \frac{7}{2}W = -4R_2$$

which simplifies to

$$32R_1 + 11R_2 = 11W$$

$$11R_1 + 16R_2 = 8W$$

the solution of which is

$$R_2 = 0{\cdot}345W \quad \text{and} \quad R_1 = 0{\cdot}225W$$

The reaction at A is

$$(0{\cdot}5 - 0{\cdot}225 - 0{\cdot}173)W = 0{\cdot}102W$$

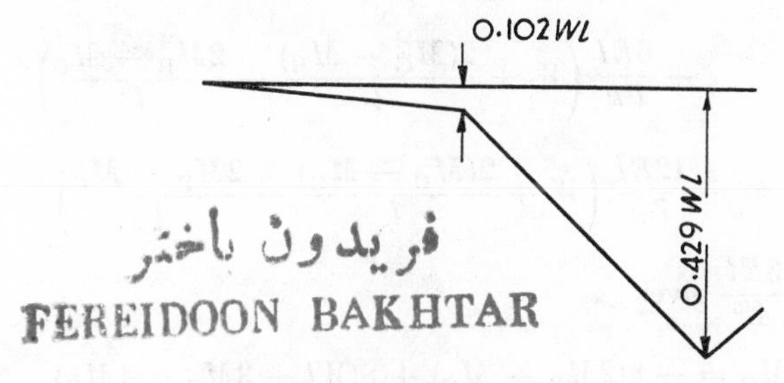

FIG. 7.13

The bending moment at B is $-0{\cdot}102Wl$, whilst at C the bending moment is

$$(-0{\cdot}102 \times 2 - 0{\cdot}225)Wl = -0{\cdot}429Wl$$

giving the bending-moment diagram shown in Fig. 7.13.

A useful check on this strain energy analysis can be obtained by using the theorem of three moments and allowing for the sinking of the supports. The settlement of B and D is $\dfrac{R_1}{k}$ whilst that of C is $\dfrac{R_2}{k}$. The three moment equations for spans AB, BC and BC, CD are respectively

$$0 + 4M_B + M_C = -\frac{6EIR_1}{l^2k} + \frac{6EI(R_2 - R_1)}{l^2k}$$

$$M_B + 4M_C + M_D = -\frac{6EI(R_2 - R_1)}{l^2k} - \frac{6EI(R_2 - R_1)}{l^2k}$$

But $M_D = M_B$, hence

$$2M_B + 4M_C = -\frac{12EI(R_2 - R_1)}{l^2k}$$

But

$$R_1 = \frac{M_B}{l} + \frac{M_B - M_C}{l}$$

and

$$R_2 = W + \frac{2(M_C - M_B)}{l}$$

Substituting these values for R_1 and R_2 gives

$$4M_B + M_C = -\frac{6EI}{l^2k}\left(\frac{2M_B - M_C}{l}\right)$$

$$+ \frac{6EI}{l^2k}\left(W + \frac{2(M_C - M_B)}{l} - \frac{2M_B - M_C}{l}\right)$$

$$2M_B + 4M_C = -\frac{12EI}{l^2k}\left(W + \frac{2(M_C - M_B)}{l} - \frac{2M_B - M_C}{l}\right)$$

Putting $k = \frac{3EI}{4l^3}$ gives

$$4M_B + M_C = -8(2M_B - M_C) + 8(Wl + 3M_C - 4M_B)$$

$$2M_B + 4M_C = -16(Wl + 3M_C - 4M_B)$$

which on simplifying gives

$$52M_B - 31M_C = 8Wl$$

$$62M_B - 52M_C = 16Wl$$

The solution of these equations gives

$$M_C = -0{\cdot}43Wl \text{ and } M_B = -0{\cdot}102Wl$$

A reasonable check with the strain-energy method.

Examples for Practice on Continuous Beams

1. The continuous beam shown in Fig. 7.14 is of constant section throughout. Spans AB and CD carry uniformly distributed loads of w/unit length, span BC of $2w$ and span DE of $2{\cdot}5w$.

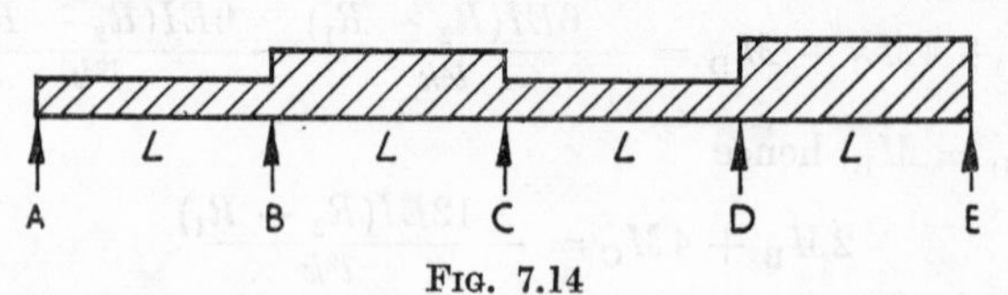

Fig. 7.14

The ends, A and E, are so restrained that the moment and slope at A equal those at E. Determine the moments at the supports.

Answer. $A = 0{\cdot}151wL^2$; $B = 0{\cdot}120wL^2$; $C = 0{\cdot}120wL^2$; $D = 0{\cdot}151wL^2$.

2. The continuous beam shown in Fig. 7.15 carries loads as indicated and weighs 5 kN/m run. If the flexural rigidity of

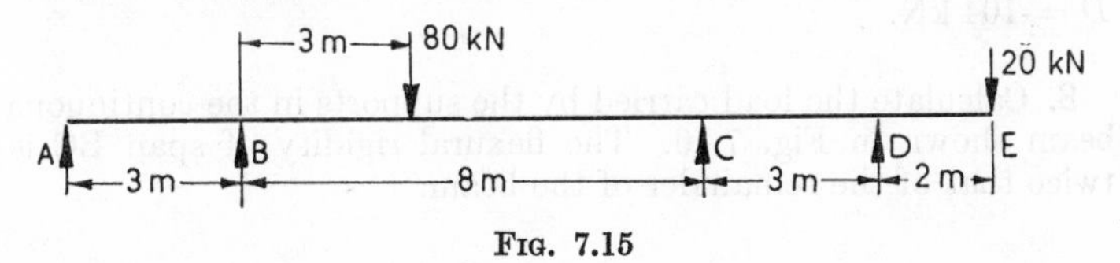

Fig. 7.15

span BC is twice that of the other spans determine the moments at the supports B, C and D.

Answer. $B = 237$ kN m; $C = 209$ kN m; $D = 40$ kN m.

3. A beam, ABC, is pinned to supports at A and C and continuous over B, these points being at the same level. It carries a concentrated load of 50 kN at the mid-point of AB. If the flexural rigidity is constant throughout, and AB = 10 m and BC = 6 m calculate the bending moment and reaction at B.

Answer. 58·5 kN m; 30·85 kN.

4. A beam, ABC, is freely supported at A and C and continuous over B. If the beam carries a load of 5 kN/m run, and AB = 12 m and BC = 6 m determine the value of the fixing moment at B and the reactions at A, B and C.

Answer. $M_B = 67{\cdot}5$ kN m; $A = 24{\cdot}4$ kN; $B = 61{\cdot}9$ kN; $C = 3{\cdot}7$ kN.

5. A continuous beam, ABCDE, is of uniform section throughout and carries a distributed load of 10 kN/m run, and concentrated loads of 200 kN at the mid-point of BC and 100 kN at the overhanging end E. Calculate the reactions at the supports A, B, C and D (which are collinear) if AB = 15 m, BC = 20 m, CD = 9 m and DE = 4 m. (*St. Andrews*)

Answer. $A = 30{\cdot}5$ kN; $B = 323{\cdot}5$ kN; $C = 253{\cdot}0$ kN; $D = 173{\cdot}0$ kN.

6. A beam, ABCD, is encastré at A and supported at B and C, these three points being collinear. It carries a load of 40 kN at 3 m from A, a uniformly distributed load of 15 kN/m run on BC, and a load of 20 kN at D. AB = 12 m, BC = 18 m, CD = 4 m. The flexural rigidity is constant throughout. Determine the moments at the supports.

Answer. $M_A = 114{\cdot}2$ kN m; $M_B = 386$ kN m; $M_C = 80$ kN m.

7. A beam, ABCDE, is supported on four points, A, B, C and D, all at the same level. AB = 12 m, BC = 24 m, CD = 15 m and DE = 3 m. The beam, which is of constant section throughout, carries a uniformly distributed load of 10 kN/m run on its whole length, and point loads of 200 kN at the mid-point of BC and 40 kN at the overhanging end E. Calculate the reactions at the supports.

Answer. $A. = -13$ kN; $B = 357$ kN; $C = 332$ kN; $D = 104$ kN.

8. Calculate the load carried by the supports in the continuous beam shown in Fig. 7.16. The flexural rigidity of span BC is twice that of the remainder of the beam.

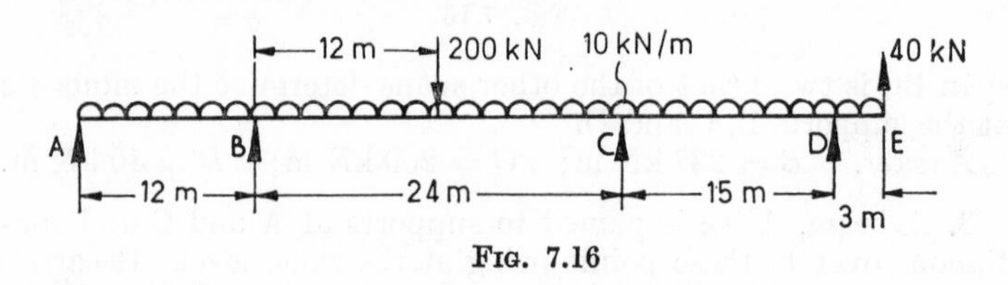

Fig. 7.16

Answer. $A = -0{\cdot}5$ kN; $B = 339{\cdot}5$ kN; $C = 349{\cdot}1$ kN; $D = 10{\cdot}9$ kN.

9. A continuous beam, ABCD, is of constant flexural rigidity throughout. AB = BC = L and CD = $L/4$. It is encastré at A and supported at B and C, the three supports being collinear. It carries a uniformly distributed load of w/unit length on BC and a point load at D of $\frac{wL}{2}$. Determine the slope of the beam at C in terms of w, L and EI. (*St. Andrews*)

Answer. $0{\cdot}00597 \dfrac{wL^3}{EI}$.

10. The continuous beam shown in Fig. 7.17 is pinned at A and supported at B, C and D, the supports being at the same level. It carries a point load of 4 kN at E, distributed loads of 6 kN/m run on AB and CD, and a point load of 30 kN at F.

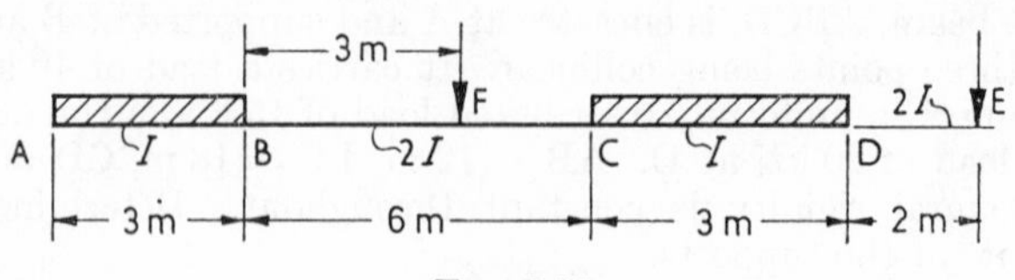

Fig. 7.17

If the second moments of area are as shown, determine the values of the bending moment and reaction at each support.

Answer. $A = 0$, 5·7 kN; $B = 9{\cdot}9$ kN m, 27·8 kN; $C = 7{\cdot}1$ kN m, 23·2 kN; $D = 8{\cdot}0$ kN m, 13·3 kN.

11. A uniform beam, ABCD, is pinned at A and is continuous over supports B and C. AB = BC = CD = L. If a load, W, acts vertically at D, draw the bending moment and shearing force diagrams for the beam

(*a*) if the beam is attached to the supports,

(*b*) if the beam merely rests on the supports before the load is applied.

Answer. (*a*) $M_B = -\frac{Wl}{4}$, $R_{BL} = -\frac{W}{4}$; (*b*) Unstable.

12. A beam of constant flexural rigidity B and length $2L$ carries a uniformly distributed load w/unit length and is supported on three springs, one at its centre and one at each end. When the load is applied to the beam the end springs are found to deflect $L/100$ and the centre spring to deflect $3L/100$. Find the deflexion constants of the springs in terms of B, L and w.

(*Oxford*)

Answer. Centre $41{\cdot}63W - \frac{4B}{L^3}$; Outer $37{\cdot}5W + \frac{6B}{L^3}$.

13. A continuous beam, ABCD, is encastré at A and D. It carries a load of 10 kN/m run on AB, 40 kN at the mid-point of BC and 15 kN/m run on CD. AB = 9 m, BC = 12 m and CD = 8 m. The beam is of steel (E = 200 kN/mm²) and I is constant throughout at $2{\cdot}0 \times 10^8$ mm⁴ units.

Determine the moments at the joints if, due to the loading, B sinks by 25 mm and C by 10 mm, A and D remaining fixed.

Answer. $M_A = 128$ kN m; $M_B = 24$ kN m; $M_C = 74$ kN m; $M_D = 108$ kN m.

14. Draw the bending moment and shearing force diagram for the beam shown in Fig. 7.18. A is fully fixed, B and C are free supports.

(*Durham*)

Answer. A: 296 kN m, 134 kN; B: 127 kN m, 118 kN; C: 150 kN m, 88 kN.

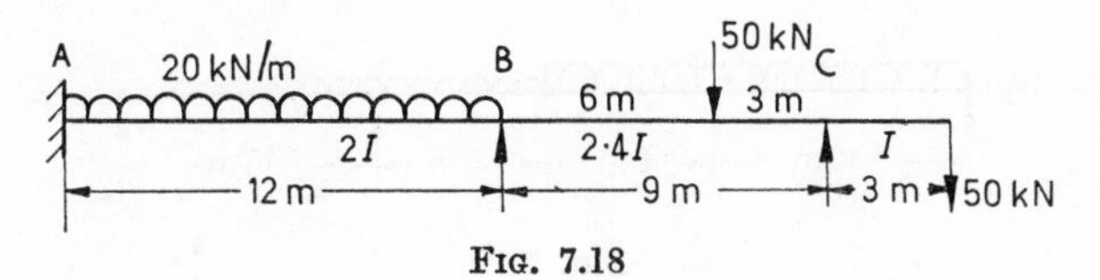

FIG. 7.18

15. ABCD is a straight uniform beam of length $4l$. It is freely supported at its ends A and D, and at two intermediate supports B and C distant l from either end. The supports at A and D are rigid but those at B and C are such that they deflect by an amount λ for each unit of load which is placed upon them. The beam carries a uniformly distributed load w/unit length along its entire length.

Show that the reactions at the supports are

$$\frac{wl}{8}\left[\frac{7l^3 + 48EI\lambda}{4l^3 + 3EI\lambda}\right] \quad \text{and} \quad \frac{3wl}{8}\left[\frac{19l^3}{4l^3 + 3EI\lambda}\right]$$

(Cambridge)

16. Define "stiffness" and "carry-over factor" as applied to moment-distribution methods for solving continuous beam problems. A beam of span L is of uniform E but has a value of I which varies uniformly from I_0 at the centre to $I_0(1 + c)$ at the ends. Prove that the stiffness and carry-over factor are given by

$$\frac{cEI_0}{L}\left\{\frac{1}{\log_e(1+c)} + \frac{2c^2}{c^2 - 2c + 2\log_e(1+c)}\right\}$$

and

$$\frac{c^2 - 2c - 2(c^2 - 1)\log_e(1+c)}{c^2 - 2c + 2(c^2+1)\log_e(1+c)}$$

respectively and obtain numerical values when $c = 0$ and $c = 1$. *(London)*

Answer. $6{\cdot}62\,\dfrac{EI_0}{L}$, $-0{\cdot}565$.

17. A beam 8 m long is simply supported on rigid supports at its ends and at the centre rests across the free end of a cantilever of length 2 m. The flexural rigidity of the beam is twice that of the cantilever. Obtain the influence line for the central reaction when a concentrated load of 1 kN rolls over the beam.

Answer. Max. reaction $= \frac{2}{3}$. *(London)*

18. Fig. 7.19 shows two similar simply supported beams which are connected together with a hinge at H. A uniformly distributed load of 100 kN/m covers the left-hand beam and a uniformly

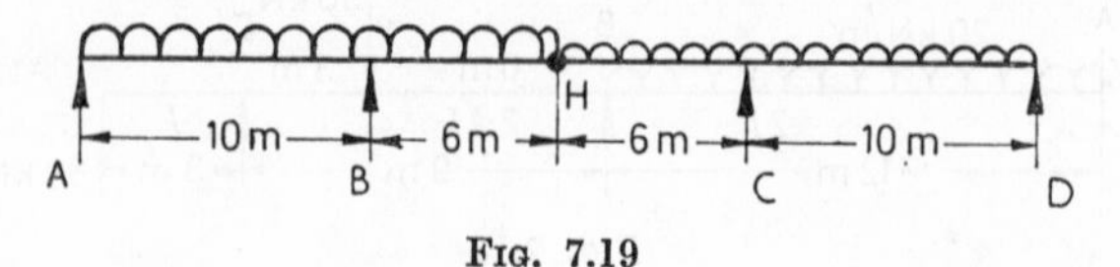

Fig. 7.19

distributed load of 40 kN/m covers the right-hand beam. Draw the shearing force diagram for each beam and calculate the vertical displacement at H. Take $E = 200$ kN/mm² and $I = 600 \times 10^7$ mm⁴ units. (*London*)

Answer. $\Delta_H = 16$ mm; $A = 345$ kN, $B = 1212$ kN, $C = 580$ kN, $D = 102$ kN.

19. The uniform beam shown in Fig. 7.20 is supported at three points A, B and C which are at the same level when the beam carries no load. When the beam is loaded the reaction at each of the points of support is $\frac{1}{k}$ times the deflexion at that point.

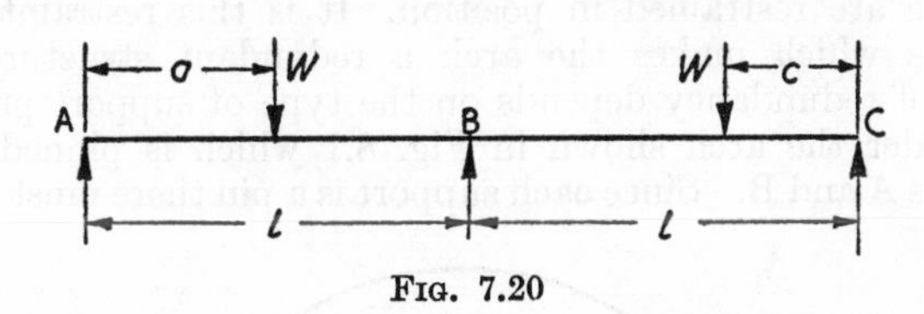

FIG. 7.20

Prove that the reaction at B due to the two equal loads, W, applied at the points indicated is

$$\frac{12kEI + 3l^2(a + c) - a^3 - c^3}{18kEI + 2l^3} W$$

where I is the relevant moment of inertia of a cross-section of the beam. (*Cambridge*)

20. A beam of length $2a$ and flexural rigidity EI carries a uniformly distributed load w/unit length and rests on three supports, one at each end and one in the middle. Assuming that the beam was straight before loading, show that, for the greatest bending moment to be as small as possible, the central support must be $(8\sqrt{2} - 11)wa^4/24EI$ lower than the end supports which are at the same level. (*Cambridge*)

Chapter 8

The Arch

AN ARCH can be defined as a curved structural member whose supports are restrained in position. It is this restraint at the supports which makes the arch a redundant structure. The degree of redundancy depends on the type of support provided.

Consider the arch shown in Fig. 8.1 which is pinned to the supports A and B. Since each support is a pin there must be both

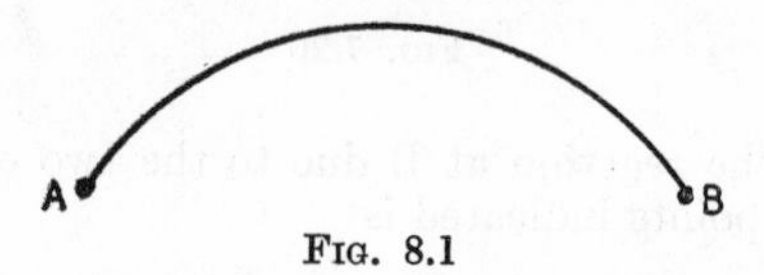

FIG. 8.1

a horizontal and vertical reaction at each. Thus there are in all four reactive forces. But the structure is uniplanar, consequently only three equations of static equilibrium can be used and there is therefore one extra force constituting a redundancy. In general one of the horizontal reactions is taken as redundant and its magnitude determined by the application of Castigliano's theorems.

If, however, a pin is inserted at any point in the arch shown in Fig. 8.2 (such a pin is generally inserted at the crown), then this

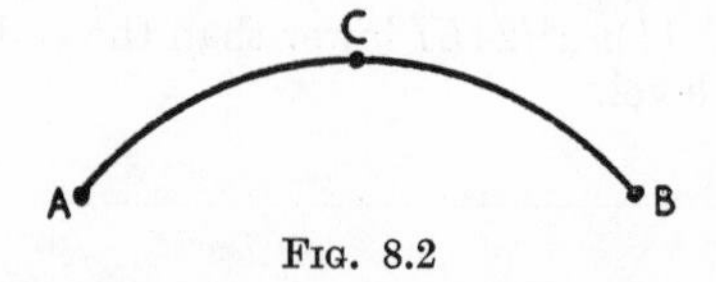

FIG. 8.2

pin means that there is no bending moment at that point. This provides an additional equation to the three static equilibrium equations thus giving sufficient to determine the four reactive forces from the principles of statics. The three-pinned arch is therefore not a redundant structure.

The arch shown in Fig. 8.3 has its supports built-in, consequently there are fixing moments at each support and in all six reactive forces, two vertical, two horizontal and two moments. There are therefore three redundancies which must be solved from three simultaneous equations, produced by the application of Castigliano's theorems to the arch.

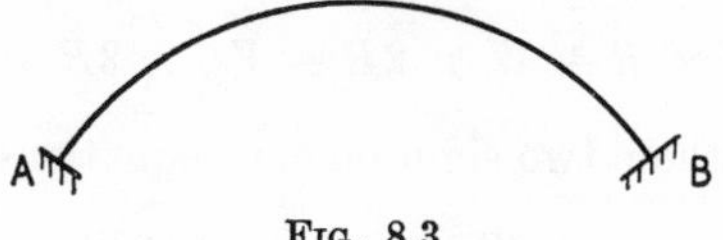

Fig. 8.3

It is in problems of the arch that the principle of superposition is useful. The skew-symmetrical loading mentioned in Chapter III produces no horizontal thrust, thus eliminating one redundancy straight away.

Examples on Three-pinned Arches

Example 8.1. A three-pinned arch consists of two quadrants of circles as shown in Fig. 8.4. Determine the maximum bending

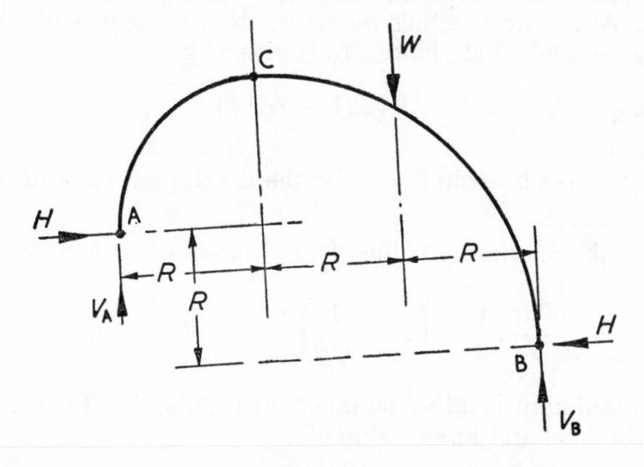

Fig. 8.4

moment on the arch and the vertical and horizontal forces on the pin at the crown when it carries a concentrated load as shown.

Solution. Since there is no applied horizontal load on the arch the horizontal reactions at A and B must be equal and opposite. Let each of these reactions be H, and let V_A, V_B be the vertical reactions at A and B respectively.

Taking moments about C for the left-hand portion gives

$$H \times R = V_A \times R$$

$$\therefore \quad V_A = H$$

Moments about C for the right-hand portion gives

$$H \times 2R - V_B \times 2R + WR = 0$$

Taking moments about A for the whole arch gives

$$H \times R + W \times 2R - V_B \times 3R = 0$$

The solution of these two simultaneous equations gives

$$H = \frac{W}{4} \quad \text{and} \quad V_B = \frac{3W}{4}$$

hence $$V_A = \frac{W}{4}$$

The horizontal force on the pin at the crown must be equal to H, i.e. $\frac{W}{4}$, whilst the vertical force must be equal to V_A, i.e. $\frac{W}{4}$.

To obtain the maximum bending moment consider firstly the portion AC. The bending moment, M_X, at any point X subtending an angle θ at the centre is given by

$$M_X = HR \sin\theta - V_A R(1 - \cos\theta)$$

$$\frac{dM_X}{d\theta} = \cos\theta - \sin\theta = 0 \text{ for the maximum value of } M_X$$

Hence the maximum value of M_X occurs when $\theta = 45°$ and is

$$\frac{WR}{4} \cdot \frac{1}{2} - \left(1 - \frac{1}{\sqrt{2}}\right) = + 0{\cdot}1035WR$$

The maximum bending moment on the portion BC will occur under the load and has a value of

$$H \times \sqrt{3}R - V_B \times R = \frac{WR}{4}(\sqrt{3} - 3) = - 0{\cdot}319WR$$

Hence the maximum bending moment on the span is $0{\cdot}319WR$. It occurs under the load and is negative in sign.

Example 8.2. Draw the influence line for bending moment at the quarter point of a three-pinned parabolic arch of span $2L$ and rise d, the third pin being at the crown. Use this line to determine the maximum bending moment which can occur at this point in an arch of 40 m span and 4 m rise when the live load consists of two loads of 100 kN and 150 kN at 4 m centres.

Solution. The diagram for the example is shown in Fig. 8.5, X denoting the quarter point and a the variable distance of the load from the left-hand support.

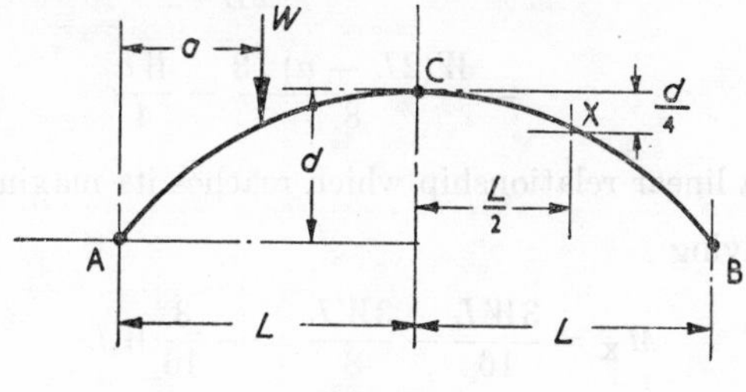

FIG. 8.5

Considering the arch as a whole and taking moments about A and B gives

$$V_B = \frac{a}{2L} . W \quad \text{and} \quad V_A = \frac{2L - a}{2L} . W$$

Consider now the right-hand portion of the arch and take moments about C. This gives

$$Hd = \frac{Wa}{2L} . L = \frac{Wa}{2}$$

hence
$$H = \frac{Wa}{2d}$$

The moment M_X at X when the load is on the portion AC is given by

$$M_X = H \times \tfrac{3}{4}d - \frac{Wa}{2L} . \frac{L}{2}$$

Substituting for H gives M_X

$$M_X = \frac{Wa}{2}\left(\frac{1}{d} . \frac{3d}{4} - \frac{1}{2}\right)$$

$$= \frac{Wa}{8}$$

This is a linear relationship reaching its maximum value when $a = L$, giving $M = \dfrac{WL}{8}$.

When the load is between C and X, H is obtained by considering the left-hand portion and taking moments about C. This gives

$$H = \frac{W(2L - a)}{2d}$$

The bending moment M_X at X is given by

$$M_X = H \times \tfrac{3}{4}d - \frac{Wa}{2L} \cdot \frac{L}{2}$$

$$= \frac{W(2L - a) \,.\, 3}{8} - \frac{Wa}{4}$$

This is a linear relationship which reaches its maximum when $a = \dfrac{3L}{2}$, giving

$$M_X = \frac{3WL}{16} - \frac{3WL}{8} = -\frac{3}{16}WL$$

When the load is between X and B the value of M_X is given by

$$M_X = H \times \tfrac{3}{4}d - \frac{Wa}{2L} \cdot \frac{L}{2} + W\left(a - \frac{3L}{2}\right)$$

This is a linear relationship which vanishes when $a = 2L$.

The complete influence line with the algebraic values is shown in Fig. 8.6. Substituting for L gives the maximum positive and negative values as 2·5 and 3·75 kN/m respectively.

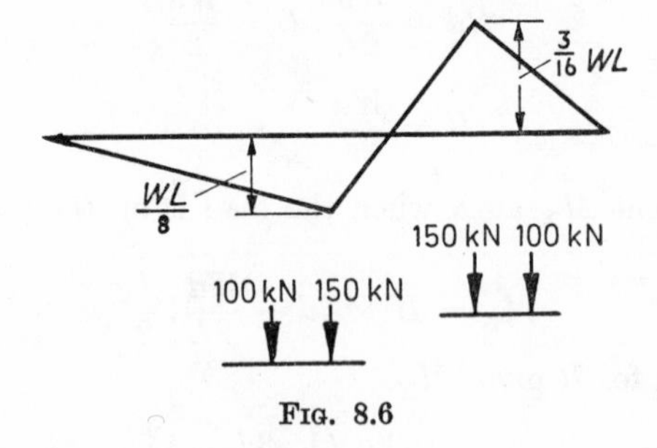

Fig. 8.6

The positions of the load train for maximum positive and negative bending moments are also indicated on this figure.

The maximum negative moment

$$= 150 \times 3{\cdot}75 + 100 \times 2{\cdot}25 = 787{\cdot}5 \text{ kN m}$$

The maximum positive moment

$$= 150 \times 2{\cdot}5 + 100 \times 2{\cdot}0 = 575 \text{ kN m}$$

Examples on Two-pinned Arches

This is a redundant structure and must be solved by the application of Castigliano's theorems. This will involve either a summation or integration. The only arches which can be dealt with therefore are those in which such summation and integration can be carried out in a straightforward manner. There are three main types of arch for which this is possible—

1. *The Braced Arch.* Here the loads are direct and the summation is as for a redundant frame.
2. *The Parabolic Arch.* If the centre line of the arch is parabolic, then integration with respect to x can only be carried out if the flexural rigidity EI_X at any point X is equal to $EI_0 \sec \alpha$, where α is the angle of inclination to the horizontal of the tangent at X, and EI_0 is the flexural rigidity at the crown.
3. *The Segmental Arch.* If the flexural rigidity is constant throughout, then the segmental arch can be integrated trigonometrically.

Examples will be given of each of these three main types.

Example 8.3. The two-pinned braced arch shown in Fig. 8.7 carries a central point load of 10 kN. If all the members have the same cross-sectional area determine the load carried by each. (*London*)

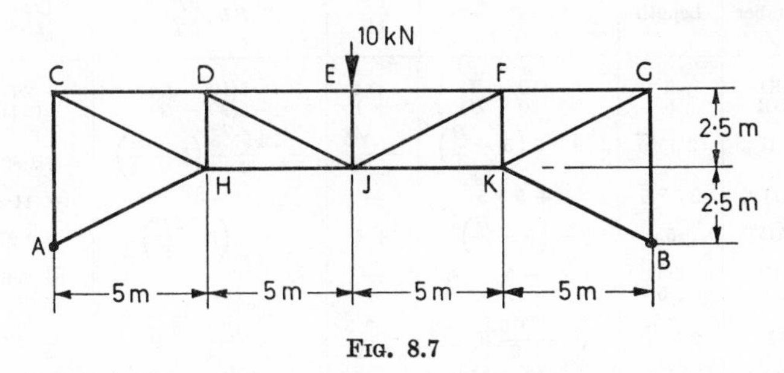

FIG. 8.7

Solution. The pins at A and B mean that in addition to the vertical load of 5 kN there is a horizontal thrust H acting at each pin. This horizontal thrust constitutes the redundancy and its value can be obtained by the solution of the equation

$$\frac{\partial U}{\partial H} = 0$$

The loads in the members are either direct tensions or compressions hence

$$\frac{\partial U}{\partial H} = \sum \frac{PL}{AE} \cdot \frac{\partial P}{\partial H}$$

The load, P, in the members has been obtained by inspection, the results being given in Fig. 8.8. Only half the arch has been stressed since the load is symmetrical.

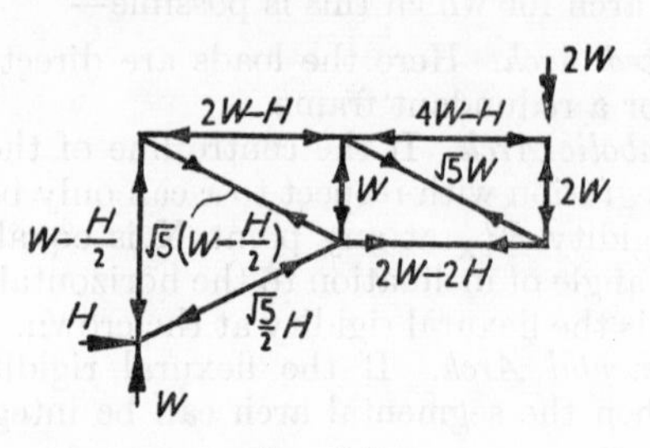

FIG. 8.8

The work for summation is set out in the table below.

TABLE 8.3

Member	Length	P	$\frac{\partial P}{\partial H}$	$PL.\frac{\partial P}{\partial H}$	Load (kN)
CD	5	$-(10-H)$	$+1$	$-5(10-H)$	$-2{\cdot}50$
DE	5	$-(20-H)$	$+1$	$-5(20-H)$	$-12{\cdot}50$
CH	$2{\cdot}5\sqrt{5}$	$+\sqrt{5}\left(5-\frac{H}{2}\right)$	$-\frac{\sqrt{5}}{2}$	$-\frac{12{\cdot}5\sqrt{5}}{2}\left(5-\frac{H}{2}\right)$	$+2{\cdot}80$
DJ	$2{\cdot}5\sqrt{5}$	$+5\sqrt{5}$	—	—	$+11{\cdot}2$
AC	5	$-\left(5-\frac{H}{2}\right)$	$+\frac{1}{2}$	$-\frac{5}{2}\left(5-\frac{H}{2}\right)$	$-1{\cdot}25$
DH	2·5	-5	—	—	$-5{\cdot}0$
EJ	2·5	-10	—	—	$-10{\cdot}0$
AH	$2{\cdot}5\sqrt{5}$	$-\frac{\sqrt{5}}{2}H$	$-\frac{\sqrt{5}}{2}$	$+\frac{12{\cdot}5\sqrt{5}}{4}H$	$-8{\cdot}4$
HJ	5	$+(10-2H)$	-2	$-10(10-2H)$	$-5{\cdot}0$

The summation and equating to zero of $PL\frac{\partial P}{\partial H}$ gives $H = 7{\cdot}50$ kN and the substitution of this value in the previously calculated value for P gives the loads in the members which are tabulated in the last column.

Example 8.4. A two-pinned parabolic arch of rise d and span L carries a vertical load of $2W$ at a point distant $L/3$ from the left-hand support. A support is placed at the mid-span of the arch which removes the central deflexion. If $EI_x = EI_0 \sec \alpha$ calculate the load on the prop. *(St. Andrews)*

Solution. The arch with the load of $2W$ can be split into a symmetrical and a skew-symmetrical load as shown in Fig. 8.9. The skew-symmetrical load has no central deflexion, consequently no prop is required and the problem therefore consists of the determination of the force V in Fig. 8.9.

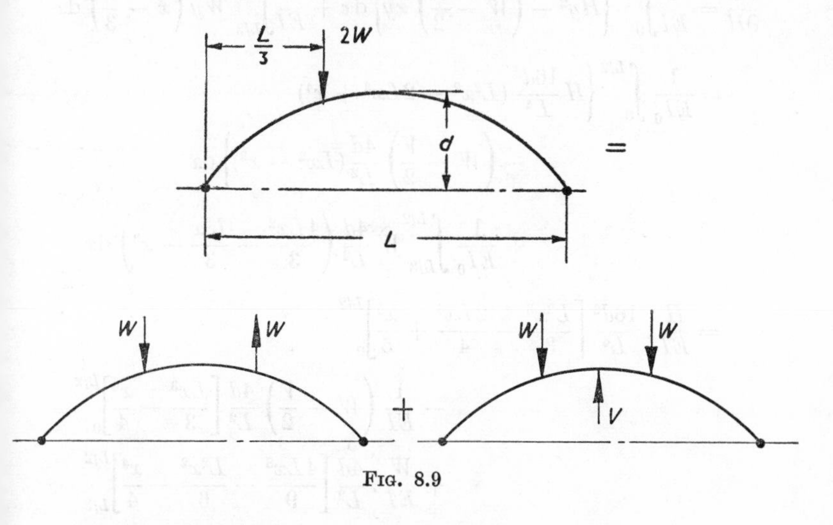

FIG. 8.9

The first problem is to find H in terms of W and V. Then putting $\frac{\partial U}{\partial V} = 0$ gives the value of V.

Since $EI_x = EI_0 \sec \alpha$ the integration has to be carried out with respect to x and not along the arch line.

H is obtained from the fact that

$$\frac{\partial U}{\partial H} = 0$$

The strain energy is that due to bending, consequently

$$\frac{\partial U}{\partial H} = \int \frac{M_x}{EI_x} \cdot \frac{\partial M_x}{\partial H} \,.\, ds = \frac{1}{EI_0} \int M_x \frac{\partial M_x}{\partial H} \,.\, ds$$

Since the load is symmetrical only half the arch need be considered. The moment M_x at any point distance x from the left-hand support is given by

$$M_x = Hy - \left(W - \frac{V}{2}\right) x + \left[W \left(x - \frac{L}{3}\right)\right]$$

the term in square brackets occurring only when positive

$$\frac{\partial M_x}{\partial H} = y \quad \text{and} \quad y = \frac{4\,\mathrm{d}x\,(L - x)}{L^2}$$

$$\frac{\partial U}{\partial H} = \frac{1}{EI}\int_0^{L/2}\left\{Hy^2 - \left(W - \frac{V}{2}\right)xy\right\}\mathrm{d}x + \frac{1}{EI}\int_{L/3}^{L/2} Wy\left(x - \frac{L}{3}\right)\mathrm{d}x$$

$$= \frac{1}{EI_0}\int_0^{L/2}\left\{H\frac{16d^2}{L^4}(L^2x^2 - 2Lx^3 + x^4)\right.$$

$$\left. - \left(W - \frac{V}{2}\right)\frac{4d}{L^2}(Lx^2 - x^3)\right\}\mathrm{d}x$$

$$+ \frac{1}{EI_0}\int_{L/3}^{L/2} W\frac{4d}{L^2}\left(\frac{4Lx^2}{3} - \frac{L^2x}{3} - x^3\right)\mathrm{d}x$$

$$= \frac{H}{EI}\cdot\frac{16d^2}{L^4}\left[\frac{L^2x^3}{3} - \frac{2Lx^4}{4} + \frac{x^5}{5}\right]_0^{L/2}$$

$$- \frac{1}{EI}\left(W - \frac{V}{2}\right)\frac{4d}{L^2}\left[\frac{Lx^3}{3} - \frac{x^4}{4}\right]_0^{L/2}$$

$$+ \frac{W}{EI}\cdot\frac{4d}{L^2}\left[\frac{4Lx^3}{9} - \frac{L^2x^2}{6} - \frac{x^4}{4}\right]_{L/3}^{L/2}$$

Putting in the limits and equating the sum to zero gives

$$H = 0{\cdot}195\frac{L}{d}V + 0{\cdot}34\frac{L}{d}W$$

Since H is now expressed as a function of V the moment M_x can be expressed in terms of V and W giving

$$M_x = (0{\cdot}195V + 0{\cdot}34W)\frac{L}{d}y - \left(W - \frac{V}{2}\right)x + \left[W\left(x - \frac{L}{3}\right)\right]$$

$$= (0{\cdot}195V + 0{\cdot}34W)\frac{4x}{L}(L - x) - \left(W - \frac{V}{2}\right)x$$

$$+ \left[W\left(x - \frac{L}{3}\right)\right]$$

$$= V\left(1{\cdot}28x - \frac{0{\cdot}78x^2}{L}\right) - W\left(\frac{1{\cdot}32x^2}{L} - 0{\cdot}36x\right)$$

$$+ \left[W\left(x - \frac{L}{3}\right)\right]$$

$$\frac{\partial M_x}{\partial V} = 1{\cdot}28x - \frac{0{\cdot}78x^2}{L}$$

$$EI\,\frac{\partial U}{\partial V} = V\int_0^{L/2}\left(1{\cdot}28x - \frac{0{\cdot}78x^2}{L}\right)^2 dx$$

$$- W\int_0^{L/2}\left(1{\cdot}28x - \frac{0{\cdot}78x^2}{L}\right)\left(\frac{1{\cdot}32x^2}{L} - 0{\cdot}32x\right) dx$$

$$+ W\int_{L/3}^{L/2}\left(x - \frac{L}{3}\right)\left(1{\cdot}28x - \frac{0{\cdot}78x^2}{L}\right) dx$$

Equating the sum to zero gives

$$V\int_0^{L/2}\left(1{\cdot}64x^2 - 2{\cdot}01\,\frac{x^3}{L} + 0{\cdot}61\,\frac{x^4}{L^2}\right) dx$$

$$= W\int_0^{L/2}\left(1{\cdot}97\,\frac{x^3}{L} - 0{\cdot}46x^2 - 1{\cdot}06\,\frac{x^4}{L^2}\right) dx$$

$$- W\int_{L/3}^{L/2}\left(1{\cdot}54x^2 - 0{\cdot}43Lx - 0{\cdot}78\,\frac{x^3}{L}\right) dx$$

Integrating and substituting the limits gives

$$0{\cdot}04082V = -\,0{\cdot}0003W$$

or

$$V = -\,0{\cdot}0074W$$

Hence the force on the prop is $0{\cdot}0074W$ acting in a downwards direction. The negative value indicates that the force which the arch exerts on the support is a tension. This means that the support would have to take the form of a tie bar in order to be effective. A simple prop would not be able to remove any deflexion at the centre since this deflexion is upwards.

This example has been worked out by slide rule. The solution involves the difference of two quantities which are nearly equal and would have been carried out on a calculating machine if accuracy to three significant figures was required.

Example 8.4. A semicircular arch of constant section and span $2R$ is pinned at both supports. Find what part of the span must be covered by a uniformly distributed load w/unit length so as to produce the maximum sagging bending moment at mid-span. (*Oxford*)

Solution. The approach to this problem is to draw the influence line for the bending moment at mid-span. The arch is shown diagrammatically in Fig. 8.10 with symmetrical loads of $\frac{W}{2}$ placed at an angle ψ from the centre. The horizontal thrust from

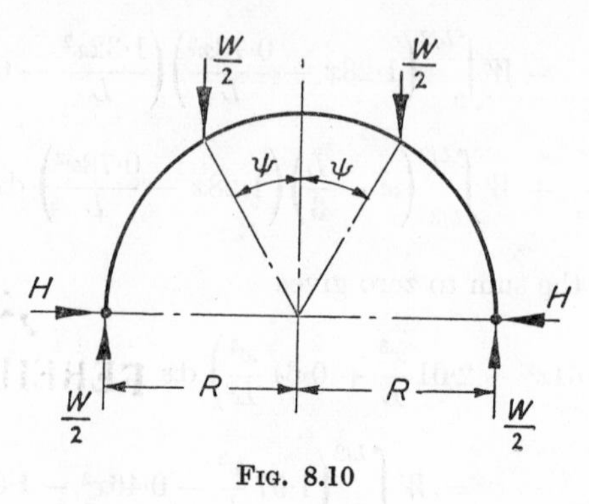

FIG. 8.10

this loading system is, by the principle of superposition, the same as that produced by a single load W placed at a point subtending an angle ψ at the centre.

The moment M_X at any point X subtending an angle α at the centre is given by

$$M_X = -\frac{W}{2}R(1 - \sin\alpha) + HR\cos\alpha + \left[\frac{W}{2}R(\sin\psi - \sin\alpha)\right]$$

the term in square brackets occurring only when positive.

$$\frac{\partial M_X}{\partial H} = R\cos\alpha, \quad ds = R\,d\alpha$$

hence

$$\frac{\partial U}{\partial H} = \frac{R^3}{EI}\int_0^{\pi/2}\left\{H\cos^2\alpha - \frac{W}{2}\left(\cos\alpha - \frac{\sin 2\alpha}{2}\right)\right\}d\alpha$$

$$+\frac{R^3}{EI}\int_{\psi}^{\pi/2}\frac{W}{2}\cos\alpha\,(\sin\psi - \sin\alpha)\,d\alpha$$

Integrating and substituting the limits and equating to zero gives

$$H = \frac{W(\cos 2\psi + 1)}{2\pi}$$

This expression gives the variation in horizontal thrust with the load position. The central bending moment M_C is given by

$$M_C = -\frac{WR}{2} + HR + \frac{WR}{2} \sin \psi$$

Substituting for H gives

$$M_C = \frac{WR}{2\pi} (\cos 2\psi + 1 + \pi \sin \psi - \pi)$$

This expression gives the influence line for M_C. It has the form shown in Fig. 8.11. The maximum sagging bending moment due to a distributed load will occur when the load covers the portion between the two points at which the central bending moment is zero and if ψ_0 = angle subtended at centre by $W/2$ when the central bending moment is zero then

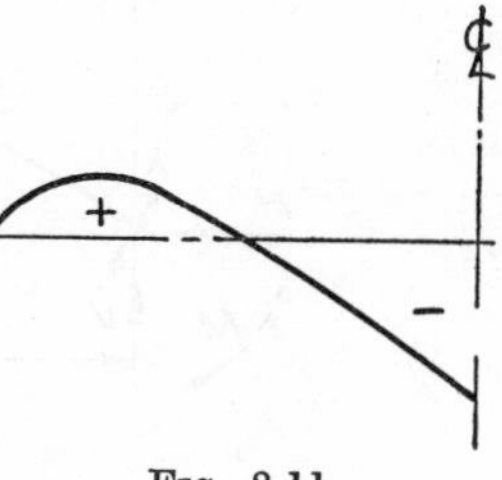

FIG. 8.11

$$\cos 2\psi_0 + \pi \sin \psi_0 - 2{\cdot}142 = 0$$

the solution of which is $\sin \psi_0 = 0{\cdot}571$

i.e. $\psi_0 = 34° \; 50'$

Hence the load between $\psi = 34° \; 50'$ and $-34° \; 50'$ gives the position for maximum sagging bending moment at mid-span.

Examples on Built-in Arches and Rings

Encastré arches can be analysed only by straightforward integration if they are of the three types already specified for the two-pinned arch. Examples will be given of parabolic and segmental arches in this section and also one on a ring. This latter type of problem is similar to the encastré arch.

Example 8.5. An encastré parabolic arch rib with $EI_x = EI_0 \sec \alpha$ has a span of 20 m and rise of 4 m. It carries loads of 25 and 30 kN at 5 m and 10 m respectively from the left-hand springing. Find the horizontal thrust, fixing moments and vertical reaction at each support.

Solution. The arch is shown diagrammatically in Fig. 8.12. Since it is encastré there are three redundant reactions and those at A will be taken as redundant. They can be obtained from the solution of the simultaneous equations

$$\frac{\partial U}{\partial H} = \frac{\partial U}{\partial V} = \frac{\partial U}{\partial M} = 0$$

the preliminary work use algebraic symbols and take

$$y = \frac{4ax(L - x)}{L^2}.$$

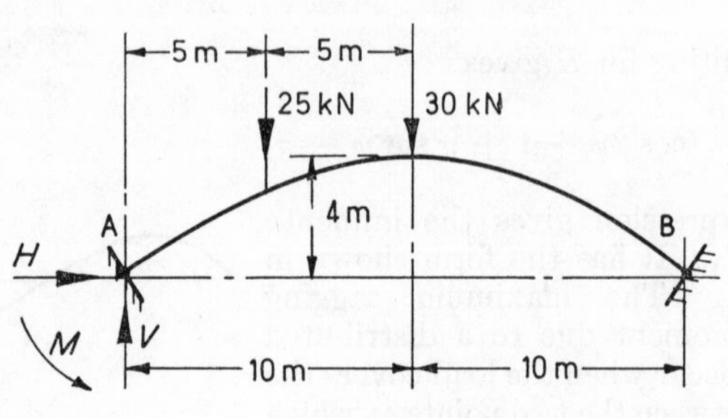

FIG. 8.12

The bending moment, M_x, at any point distant x from the left-hand support is given by

$$M_x = -Vx + Hy + M + \left[25\left(x - \frac{L}{4}\right)\right] + \left[30\left(x - \frac{L}{2}\right)\right]$$

$$= -Vx + \frac{H \,.\, 4ax(L - x)}{L^2} + M + \left[25\left(x - \frac{L}{4}\right)\right]$$

$$+ \left[30\left(x - \frac{L}{2}\right)\right]$$

$$\frac{\partial M_x}{\partial M} = 1$$

$$\therefore EI\frac{\partial U}{\partial M} = \int_0^L \left\{M + H\frac{4ax(L - x)}{L^2} - Vx\right\} \mathrm{d}x$$

$$+ 25\int_{L/4}^L \left(x - \frac{L}{4}\right) \mathrm{d}x + 30\int_{L/2}^L \left(x - \frac{L}{2}\right) \mathrm{d}x$$

$$= \left[Mx + \frac{4aH}{L^2}\left(\frac{Lx^2}{2} - \frac{x^3}{3}\right) - \frac{Vx^2}{2}\right]_0^L$$

$$+ 25\left[\frac{x^2}{2} - \frac{Lx}{4}\right]_{L/4}^L + 30\left[\frac{x^2}{2} - \frac{Lx}{2}\right]_{L/4}^L$$

$$= ML + \frac{4aL}{6} - \frac{VL^2}{2} + 25\,\frac{9L}{32} + 30\,\frac{L}{8}$$

$$\frac{\partial M_x}{\partial H} = \frac{4ax(L-x)}{L^2}$$

$$EI_0 \frac{\partial U}{\partial H} = \frac{4a}{L^2} \int_0^L \left\{ Mx(L-x) + H \frac{4ax^2}{L^2} (L^2 - 2Lx + x^2) \right.$$

$$\left. - Vx^2(L-x) \right\} \mathrm{d}x + \frac{4a}{L^2} . 25 \int_{L/4}^L x(L-x)\left(x - \frac{L}{4}\right) \mathrm{d}x$$

$$+ \frac{4a}{L^2} 30 \int_{L/2}^L x(L-x)\left(x - \frac{L}{2}\right) \mathrm{d}x$$

$$= \frac{4a}{L^2} \left[M \left(\frac{Lx^2}{2} - \frac{x^3}{3} \right) + H \frac{4a}{L^2} \left(\frac{L^2x^3}{3} - \frac{2Lx^4}{4} + \frac{x^5}{5} \right) \right.$$

$$\left. - V \left(\frac{Lx^3}{3} - \frac{x^4}{4} \right) \right]_0^L + \frac{4a}{L^2} . 25 \left[\frac{5Lx^3}{12} - \frac{x^4}{4} - \frac{L^2x^2}{8} \right]_{L/4}^L$$

$$+ 30 \frac{4a}{L^2} \left[\frac{Lx^3}{2} - \frac{x^4}{4} - \frac{L^2x^2}{4} \right]_{L/2}^L$$

$$= \frac{4a}{L^2} \left(\frac{ML^3}{6} + \frac{H . 4aL^3}{30} - \frac{VL^4}{12} + 25 \frac{135L^4}{3072} + 30 \frac{L^4}{64} \right)$$

$$= 4a \left(\frac{ML}{6} + \frac{H . 2aL}{15} - \frac{VL^2}{12} + \frac{25 . 135L^2}{3072} + \frac{30L^2}{64} \right)$$

$$\frac{\partial M_x}{\partial V} = -x$$

$$\therefore EI_0 \frac{\partial U}{\partial V} = \int_0^L \left\{ Vx^2 - Mx - \frac{H . 4a}{L^2} (Lx^2 - x^3) \right\} \mathrm{d}x$$

$$- 25 \int_{L/4}^L \left(x^2 - \frac{Lx}{4} \right) \mathrm{d}x - 30 \int_{L/2}^L \left(x^2 - \frac{Lx}{2} \right) \mathrm{d}x$$

$$= \left[\frac{Vx^3}{3} - \frac{Mx^2}{2} - \frac{4aH}{L^2} \left(\frac{Lx^3}{3} - \frac{x^4}{4} \right) \right]_0^L$$

$$- 25 \left[\frac{x^3}{3} - \frac{Lx^2}{8} \right]_{L/4}^L - 30 \left[\frac{x^3}{3} - \frac{Lx^2}{4} \right]_{L/2}^L$$

$$= \frac{VL^3}{3} - \frac{ML^2}{2} - H \frac{4aL^2}{12} - 25 \frac{81L^3}{384} - 30 \frac{5L^3}{48}$$

Simplifying and equating each expression to zero gives the three following simultaneous equations

$$M + \frac{2a}{3} H - \frac{L}{2} V + \frac{225L}{32} + \frac{15L}{4} = 0$$

$$M + \frac{4a}{5} H - \frac{L}{2} V + \frac{25 \,.\, 135}{512} L + \frac{45}{16} L = 0$$

$$M + \frac{2a}{3} H - \frac{2L}{3} V + \frac{25 \,.\, 81}{192} L + \frac{150}{24} L = 0$$

The solution of these equations gives

$$H = \frac{15L}{2a} \cdot \frac{705}{512}$$

and putting $L = 20$, $a = 4$, gives

$$H = 51{\cdot}6 \text{ kN}$$

$$V = 36{\cdot}1 \text{ kN}$$

and $$M = +\,8{\cdot}0 \text{ kN m}$$

These are the reactive forces at A. Those at B are as follows—

$$H_B = H_A = 51{\cdot}6 \text{ kN}$$

$$V_B = 55 - V_A = 18{\cdot}9 \text{ kN}$$

$$M_B = M_A - 20V_A + 15 \times 25 + 10 \times 30 = -\,39 \text{ kN m}$$

Example 8.6. A circular arched rib of uniform cross-section is encastred at A and B and is subjected to vertical loads as shown

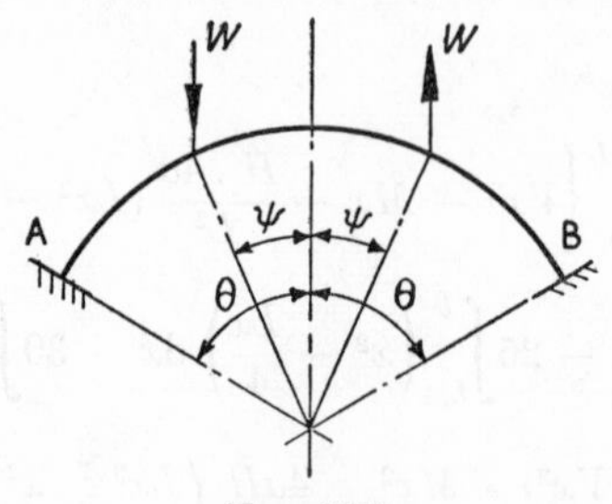

FIG. 8.13

in Fig. 8.13. Find the magnitudes of the vertical reactions at A and B. (*Cambridge*)

Solution. This problem has a skew-symmetrical loading system, consequently the deflexion and bending moment at mid-span are zero. There is also no resultant load on the span, hence there is no horizontal reaction at the supports. The arch can therefore be split into two halves as shown in Fig. 8.14.

If the work is carried out from the centre of the span, then the only unknown is F, the shear force at mid-span, and the vertical reactions at the supports will be $W - F$.

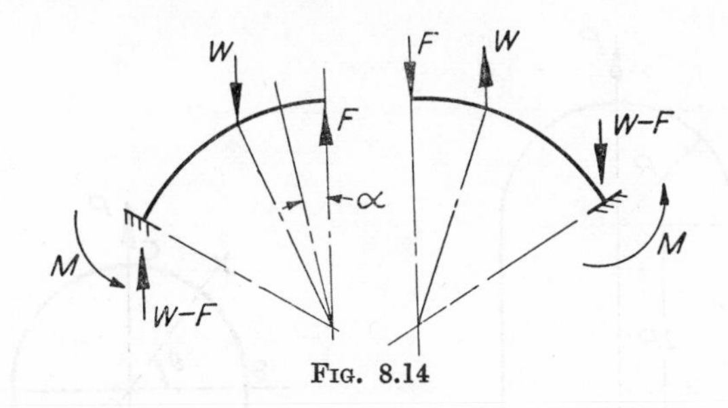

FIG. 8.14

The bending moment M_X, at a point X where the arc between X and mid-span subtends an angle α at the centre, is given by

$$M_X = -FR \sin \alpha + [WR(\sin \alpha - \sin \psi)]$$

$$\frac{\partial M_X}{\partial F} = -R \sin \alpha$$

hence

$$EI \frac{\partial U}{\partial F} = FR^3 \int_0^\theta \sin^2 \alpha \, d\alpha - WR^3 \int_\psi^\theta (\sin^2 \alpha - \sin \alpha \sin \psi) \, d\alpha$$

$$= FR^3 \left[\alpha - \frac{\sin 2\alpha}{4}\right]_0^\theta - WR^3 \left[\alpha - \frac{\sin 2\alpha}{4} + \cos \alpha \sin \psi\right]_\psi^\theta$$

Equating to zero gives

$$F = W \frac{\theta - \psi + 2 \cos \theta \sin \psi - \dfrac{\sin 2\theta}{2} - \dfrac{\sin 2\psi}{2}}{\theta - \dfrac{\sin 2\theta}{2}}$$

The vertical reaction at each support is $W - F$

i.e. $$\frac{2\psi + \sin 2\psi - 4 \cos \theta \sin \psi}{2\theta - \sin 2\theta}$$

Example 8.7. Determine the maximum bending moment which occurs on the ring shown in Fig. 8.15 when it is subjected to an axial pull P. The ring is of constant section throughout. (*St. Andrews*)

Solution. If the ring is split into two halves as shown, then since the loading is axial the vertical reactions on each half must be $P/2$. The moments M_0 will be equal and opposite. If, however, an inward force H is placed on the upper half, then to maintain equilibrium an outward force H must act on the lower half.

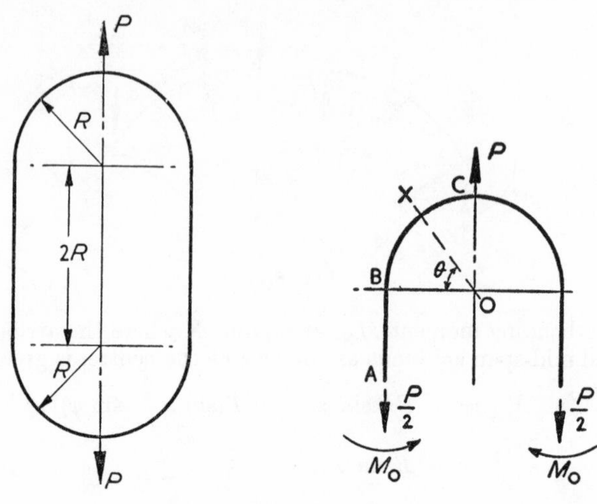

FIG. 8.15

But each half is subjected to the same loading and this symmetry can only be maintained if $H = 0$. The problem therefore reduces to one redundancy, the moment M_0. The value of this moment can be obtained in the usual way by putting $\dfrac{\partial U}{\partial M_0} = 0$.

Consider first the straight portion AB. The bending moment, M_x, at a point distant x from A is given by

$$M_x = M_0 \quad \therefore \quad \frac{\partial M_x}{\partial M_0} = 1$$

and

$$EI\left[\frac{\partial U}{\partial M_0}\right]_{AB} = \int_0^R M_0 \, dx = M_0 R$$

The bending moment, M_X, at a point X on the circular portion BC where the angle BOX $= \theta$ is given by

$$M_X = M_0 + \frac{PR}{2}(1 - \cos\theta)$$

$$\frac{\partial M_X}{\partial M_0} = 1$$

hence $$EI\left[\frac{\partial U}{\partial M_0}\right]_{BC} = R\int_0^{\pi/2}\left\{M_0 + \frac{PR}{2}(1-\cos\theta)\right\}d\theta$$

$$= R\left\{M_0\frac{\pi}{2} + \frac{PR}{2}\left(\frac{\pi}{2}-1\right)\right\}d\theta$$

For the whole structure

$$\frac{\partial U}{\partial M_0} = 4\left(\left[\frac{\partial U}{\partial M_0}\right]_{AB} + \left[\frac{\partial U}{\partial M_0}\right]_{BC}\right)$$

$$= 4R\left\{M_0\left(1+\frac{\pi}{2}\right) + \frac{PR}{2}\left(\frac{\pi}{2}-1\right)\right\}$$

Equating this expression to zero gives

$$M_0 = -\frac{PR}{2}\left(\frac{\pi-2}{\pi+2}\right) = -0{\cdot}111PR$$

The moment under the load

$$= M_0 + \frac{PR}{2}$$

$$= -0{\cdot}111PR + 0{\cdot}500PR = +0{\cdot}389PR$$

$\therefore$ Maximum moment on ring $= 0{\cdot}389PR$ occurring under the load.

EXAMPLES FOR PRACTICE ON THE ARCH

1. An elliptical arch rib of span $2L$ is pinned at the abutments and at the crown where the rise is d.

It carries a load of intensity w/unit length on the left-hand half of the span. Determine the bending moment in the rib at a distance $L/4$ from the right-hand abutment.

Answer. $0{\cdot}1025wL^2$.

2. A three-pinned arch is formed as shown in Fig. 8.16 from two segments. Determine analytically or graphically the maximum bending moment on the rib when it carries the load shown. (*St. Andrews*)

Answer. $0{\cdot}1546WR$.

3. An arch in the form of a parabola with axis vertical has hinges at the abutments and at the vertex. The abutments are at different levels, the horizontal span being l and the heights of the vertex above the abutments being h_1 and h_2.

Show that the horizontal thrust due to a load w/unit length uniformly distributed across the span is

$$\frac{wl^2}{2(\sqrt{h_1} + \sqrt{h_2})^2}$$

(*Cambridge*)

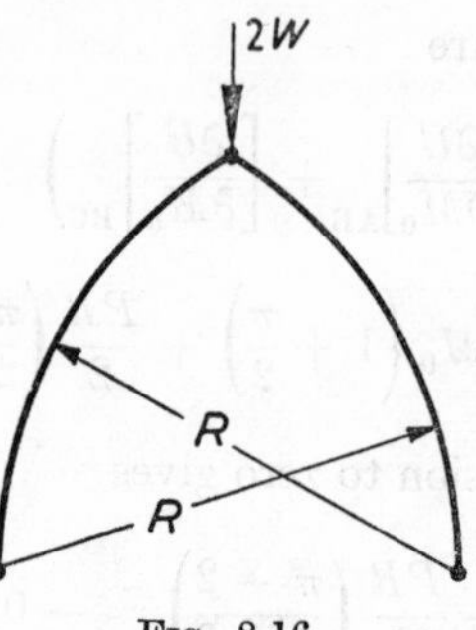

FIG. 8.16

4. Fig. 8.17 shows a three-pinned arch truss. Determine the horizontal and vertical components of the reactions at A and B

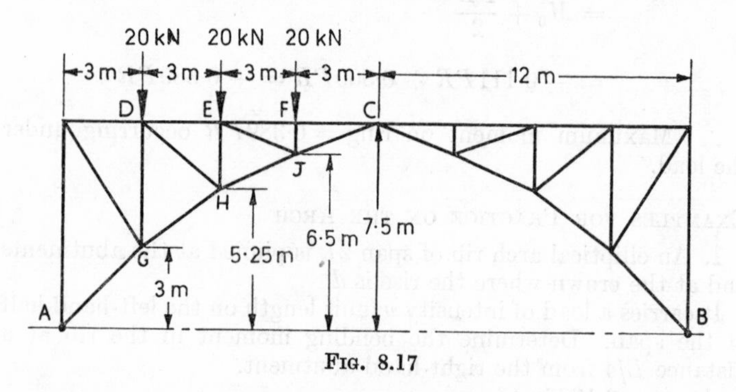

FIG. 8.17

and the forces in the members DE, DH and GH, due to loads of 20 kN placed at each of the panel points D, E and F.

(*Cambridge*)

Answer. $V_A = 45$ kN, $V_B = 15$ kN, $H = 24$ kN, $DE = -37{\cdot}3$ kN, $DH = -29{\cdot}2$ kN, $GH = -12$ kN.

5. Fig. 8.18 shows one half of a symmetrical three-pinned arch the lower chord joints of which lie on a parabola. The dead load is 20 kN/m of span and the live load is 25 kN/m and may be of any length from 0 to 12·5 m. All loads may be assumed to act at the upper chord joints. Determine by drawing influence lines the maximum tensile and/or compressive forces set up in the members AB and AC. (*London*)

Answer. $AB = \pm 99{\cdot}3$; $AC = +42{\cdot}0, -23{\cdot}3$.

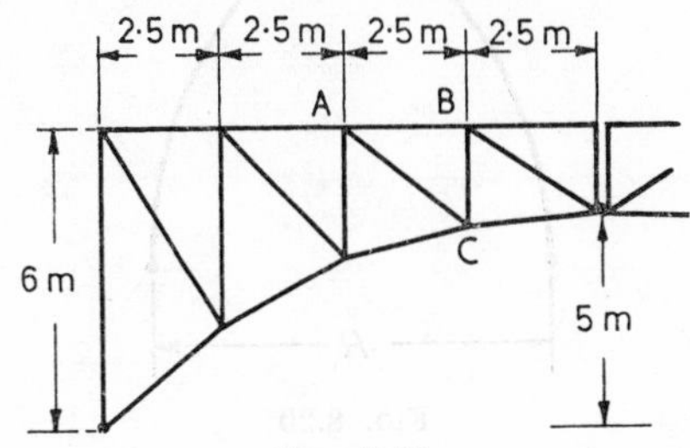

FIG. 8.18

6. The deck of a three-hinged segmental arch rib is carried on spandrel columns. The span of the arch is 30 m, the rise at the crown 6 m and the spandrel columns are spaced at 5 m centres. Determine the bending moment, radial shear and normal thrust at the left-hand quarter span section when a concentrated load of 160 kN is placed over the section. (Assume the deck simply supported between spandrel columns.) (*London*)

Answer. 193·6 kN m; 107 kN; 9·3 kN.

7. Draw the influence line for B.M. at the point, D, of the three-pinned arch ABC, Fig. 8.19. Find the maximum moment at this section when a uniform load of 50 kN/m longer than the

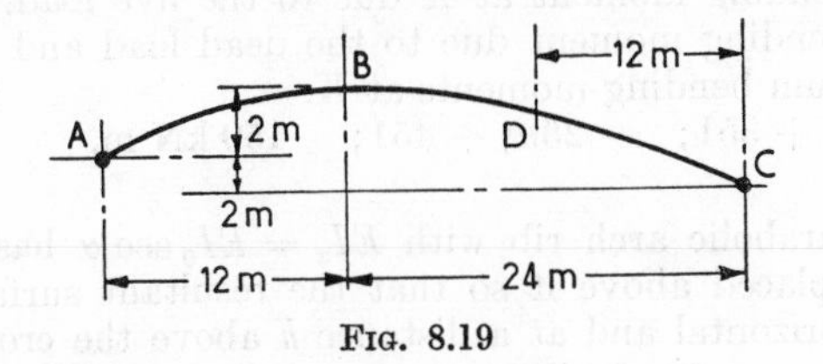

FIG. 8.19

span and which can be broken up into lengths of any size, crosses the structure. The arch segments are parabolic.

For the load in the position found above state the reactions at the hinge B.

Answer. 216 kN m; $H = 1151$ kN; $V = 192$ kN.

8. The two-pinned arch shown in Fig. 8.20 consists of two segments each of radius R. It is of constant EI throughout. Determine the bending moment under the load.

Answer. $0{\cdot}0737\,WR$.

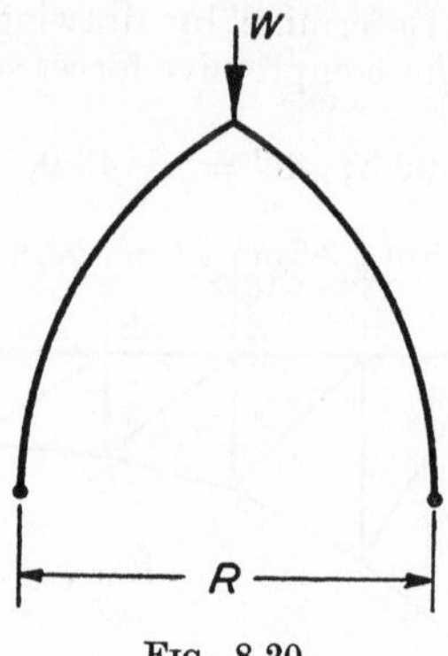

FIG. 8.20

9. A two-hinged arch, parabolic in outline, is of 40 m span and 8 m central rise. The abutment hinges, A and B, are at the same level. The load from the roadway is transmitted to the arch by vertical columns at A and B and at 10 m intervals. The longitudinal roadway girders are supported by these columns and may be taken as simply supported at each column top.

The dead load is uniform at 40 kN/m run, the live load consists of two loads of 100 kN and 200 kN at a fixed distance of 6 m apart. Draw for the section X, 15 m from the left-hand hinge, the influence line for bending moment; also for each column, influence lines of the loads transmitted by them to the arch.

Use these lines to determine the maximum positive and negative bending moment at X due to the live load. Combine with the bending moment due to the dead load and determine the maximum bending moments at X. *(Aberdeen)*

Answer. + 351; − 259; + 451; − 159 kN m.

10. A parabolic arch rib with $EI_x = EI_0 \sec \alpha$ has filling of density σ placed above it so that the resultant surface of the filling is horizontal and at a distance h above the crown of the arch.

If the span of the arch is L and the central rise a, determine the horizontal thrust and central bending moment per foot width of the arch.

Answer. $H = \sigma L^2 \left(\frac{1}{42} + \frac{h}{8a}\right)$; $M = \frac{\sigma a L^2}{336}$.

11. A semicircular arch rib of span $2R$ is pinned to two stanchions as shown in Fig. 8.21. The flexural rigidity, EI, is the same for both stanchions and rib. If the coefficient of linear expansion for the material is β, show that the maximum bending moment in the stanchions due to a rise in temperature, t, is

$$\frac{12EIRh\beta t}{3\pi R^3 + 4h^3}$$ (*Oxford*)

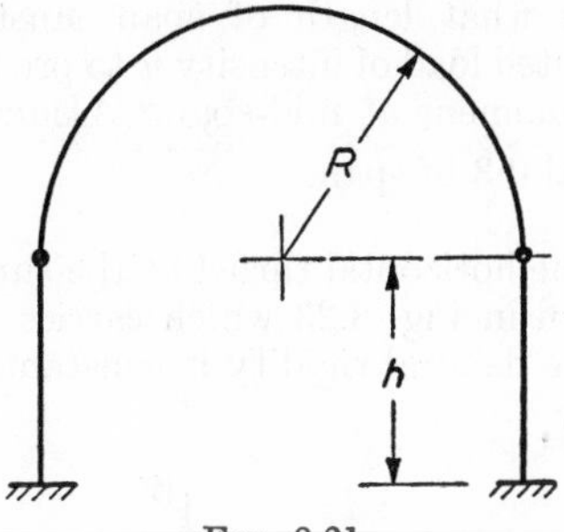

FIG. 8.21

12. The segmental rib ACB shown in Fig. 8.22 is of constant section throughout and carries a load W at its mid-point C.

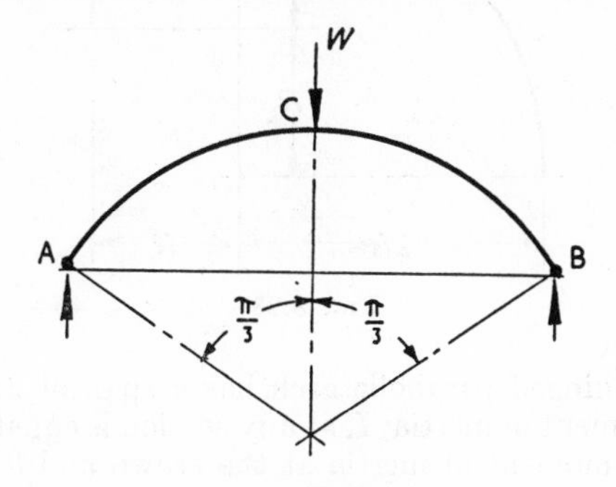

FIG. 8.22

The ends A and B are tied together with a tie whose extensibility

$$aE = \frac{50EI}{R^2}$$

where EI is the flexural rigidity of the rib AB. Calculate the load in the tie. (*St. Andrews*)

Answer. $0{\cdot}561W$.

13. A steel semicircular two-hinged arch rib of 16 m span has a uniform section whose $I = 2 \times 10^8$ mm^4 units. If

$E = 200 \text{ kN/mm}^2$, find the maximum bending moment induced in the rib by a temperature variation of 40°C.

Take the coefficient of expansion per °C as 0·000012.

Answer. 3·06 kN m.

14. A two-pinned parabolic arch has a span of $2L$ and a rise of d. Determine what length of span must be loaded by a uniformly distributed load of intensity w to produce the maximum sagging bending moment at mid-span. Assume $I_x = I_0 \sec \alpha$. (*St. Andrews*)

Answer. Central 0·3 of span.

15. Calculate the horizontal thrust at the supports for the two-pinned arch shown in Fig. 8.23 which carries a point load of W at the crown. The flexural rigidity is constant throughout. (*St. Andrews*)

Answer. 0·28 W.

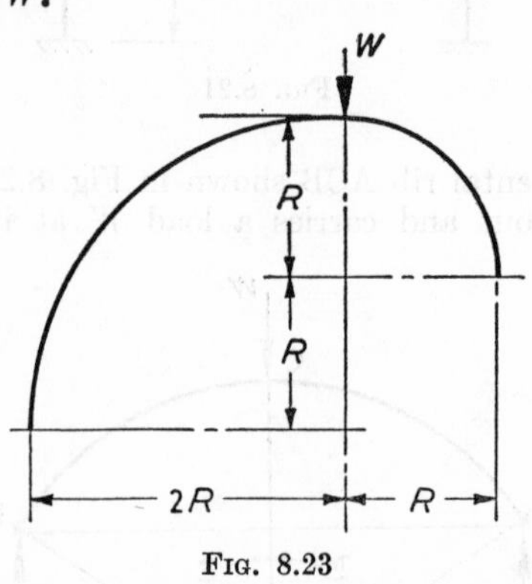

FIG. 8.23

16. A two-hinged parabolic arch has a span of $2L$ and a rise of h. The moment of inertia, I, at any section is equal to $I_C \sec \theta$, I_C being the moment of inertia at the crown and θ the slope of the section. Determine the horizontal movement between the constraints so that under a central point load, W, the bending moments at the quarter points and crown are numerically equal. Compare the final vertical deflexion of the load with the imposed horizontal movements. (*Glasgow*)

Answer. $\Delta_H = 0{\cdot}0405 \dfrac{WhL^2}{EI}$; $\Delta_V = 0{\cdot}135 \dfrac{L}{h} \Delta_H$.

17. Calculate the maximum bending moment on a semi-circular arch rib, of radius R and of uniform section throughout, when it carries a load of $2wR$ uniformly distributed across the span.

Answer. $0{\cdot}0895 wR^2$.

18. A two-pinned parabolic arch has a span of $2L$ and a central rise of $L/2$. Calculate the horizontal thrust at the supports when a load P acts in a direction inclined at 45° to the vertical at a point on the arch rib distant $L/2$ from the centre. Assume $EI_x = EI_0 \sec \alpha$.
Answer. $0{\cdot}690P$ and $0{\cdot}017P$.

19. A parabolic arch rib of span and rise each $= L$ is pinned at the supports and is used as a roof for an area. Determine the horizontal thrust at the supports when a wind pressure of intensity w/unit height acts on one half of the arch.
It may be assumed that $I_x = I_0 \sec \alpha$.
Answer. $0{\cdot}71wL$ and $0{\cdot}29wL$.

20. A parabolic arch rib of span 40 m and rise 4 m carries a uniformly distributed horizontal load of 25 kN/m run on the left-hand half of the span. Determine the bending moment at the crown and calculate also the change in this bending moment if one of the supports yields horizontally by 0·06 mm/kN of horizontal thrust. Take $I_C = 8 \times 10^8$ mm^4 units, and $I_X = I_C \sec \alpha$.
Answer. Zero; 64 kN m.

21. The ordinates z to the reaction locus for a two-hinged arch measured vertically upwards from the springings A and B at distances kL from the support A are as follows—

k	0	0·1	0·2	0·3	0·4	0·5	0·6	0·7	0·8	0·9	1·0
z (m)	45	43·8	40·8	37·1	34·1	33	34·1	37·1	40·8	43·8	45

For the spandrel braced arch shown in Fig. 8.24 determine the forces in the members X, Y and Z, for the given panel point loads.

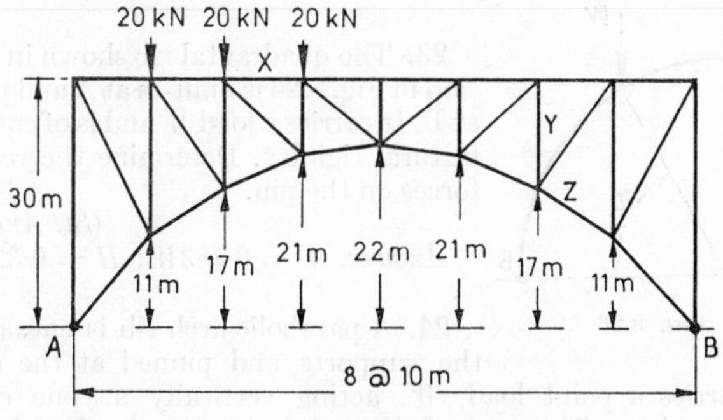

FIG. 8.24

Note. The reaction locus for an arch is a line giving the direction of the resultant force at the abutments when the load is in a given position.

Thus if the curve A′C′B′ is the reaction locus for a given arch, the reactions at A and B when the load is at C act in the directions AC′ and BC′ respectively. Since the directions of the reactions are known their magnitude can be determined.

Answer. $X = -30{\cdot}8$, $Y = -3{\cdot}5$, $Z = -32{\cdot}2$ kN.

22. The members of the arched frame shown in Fig. 8.25 all have the same ratio of length to cross-sectional area (1·5 mm^{-1}).

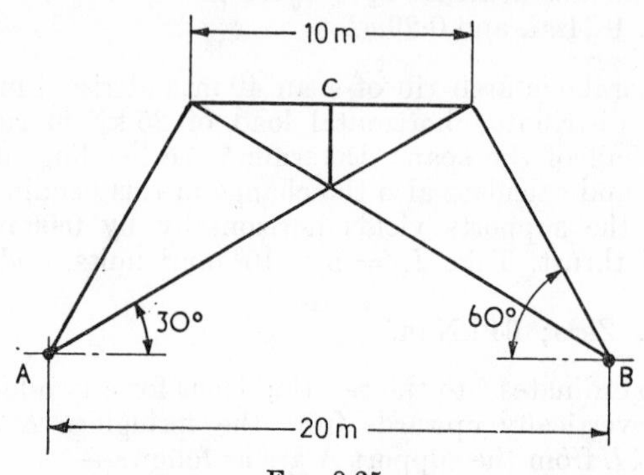

FIG. 8.25

If the supports at A and B are pinned so that no lateral movement is allowed, find the horizontal thrust at the abutments occasioned by a rise in temperature of 30°C. Calculate also the total upward movement of C. Use a coefficient of expansion of 12×10^{-6} per °C and $E = 200$ kN/mm². (*London*)

Answer. H = 43·6 kN; $\Delta_C = 8{\cdot}2$ mm.

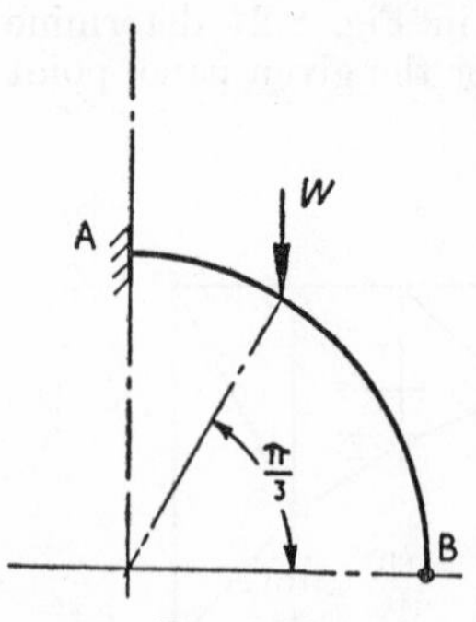

FIG. 8.26

23. The quadrantal rib shown in elevation in Fig. 8.26 is built-in at A and pinned at B. It carries a load W and is of constant flexural rigidity. Determine the reactive forces on the pin.

(*St. Andrews*)

Answer. $V = 0{\cdot}783W$; $H = 0{\cdot}339W$.

24. A parabolic arch rib is encastré at the supports and pinned at the crown and carries a point load, W, acting vertically at one of the quarter points. The span of the arch is $2L$, the rise d, and it can be assumed that the second moment of area at any section is proportional to the secant of the angle of slope at that section. Calculate the end fixing moments. *(Oxford)*

Answer. $0{\cdot}123WL$ and $0{\cdot}065WL$.

25. The horizontal thrust, H, and the vertical reaction, V, at the left-hand springing for a fixed parabolic arch (span L, rise h and $I = I_C \sec\theta$) for a load W at $x = kL$ from the left-hand is given by

$$H = \frac{15}{4}\cdot\frac{WL}{h}k^2(1-k)^2 \quad \text{and} \quad V = W(1-k)^2(1+2k)$$

Using these values, determine the normal shear, normal thrust and bending moment at the left-hand quarter point for a fixed parabolic arch loaded with a vertical load, $W = 20$ kN, at 6 m from the left-hand end if the span is 20 m and the central rise 2 m. *(Aberdeen)*

Answer. 8·94 kN; 35·48 kN; − 14·0 kN m.

26. Find an expression for the horizontal thrust in an arch of span $2l$, whose centre line is given by the curve

$$y = h\left(2\frac{x}{l} - \frac{x^2}{l^2}\right)$$

where the arch is subjected to a central load, W. The arch has a uniform flexural rigidity and h is much less than l. The arch is built in to rigid abutments.

Discuss the trend of the horizontal force as h tends to zero, so that the arch becomes a built-in beam. *(Oxford)*

Answer. $H = \dfrac{15Wl}{32h}$.

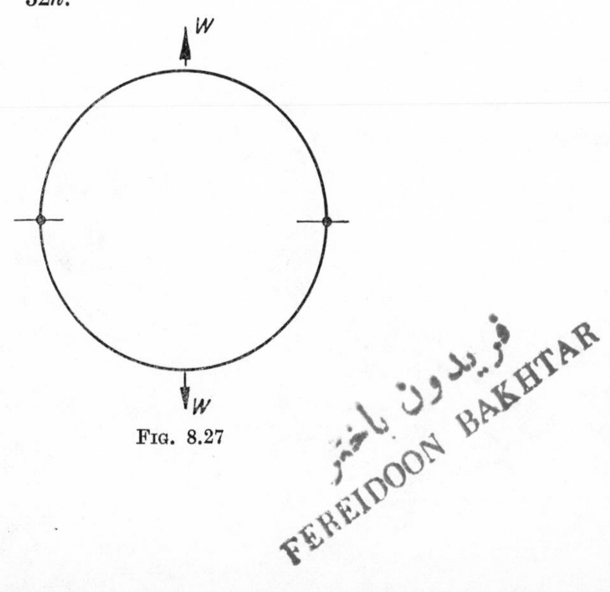

Fig. 8.27

27. A circular ring of mean diameter D has a horizontal diametrical tie which may be considered inextensible. The ring is subjected to equal comprehensive forces, P, along the vertical diameter. Show that the tension in the tie is

$$\frac{2(4 - \pi)}{\pi^2 - 8} P$$

(London)

28. A circular ring of constant section throughout is pinned at two diametrically opposite points as shown in Fig. 8.27. Calculate the maximum bending moment in the ring when it carries a diametrical pull of W. *(St. Andrews)*

Answer. $\frac{WR}{2}$.

Chapter 9

Stiff-jointed Frames

THE STIFF-JOINTED FRAME is in some respects similar to the arch. The latter is a single structural member with a curved centre line. The stiff-jointed frame or portal consists of a number of straight members so jointed as to be able to transmit a bending moment from one member to the next. The similarity between the arch and the portal is that in each case the redundancy is caused by restraint at the supports.

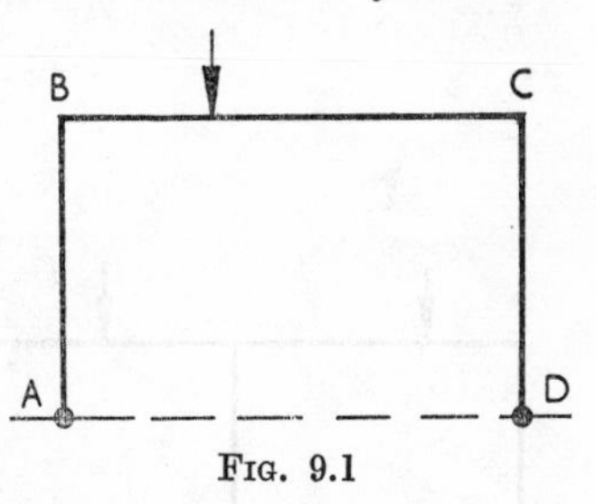

FIG. 9.1

The portal shown in Fig. 9.1 has the supports pinned and at the same level, and this has only one redundant force. The types illustrated in Figs. 9.2 and 9.3 also have only one redundancy. The solution is, however, not quite so simple as the problem of 9.1 owing to the sloping member in 9.2 and in 9.3 because the feet are not at the same level.

If the supports of any of these portals are encastré, then the structure has three redundancies.

Figs. 9.4 and 9.5 illustrate the double portal. This has obviously more redundancies than the single portal, having three if the

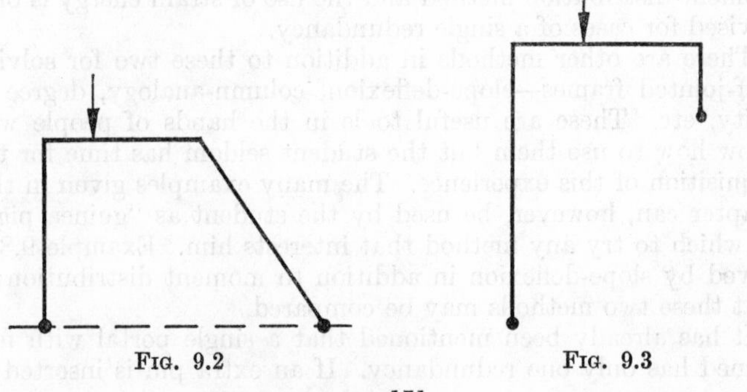
FIG. 9.2 FIG. 9.3

supports are pinned (Fig. 9.4) and six if they are encastré (Fig. 9.6).

The student may also meet the two storeyed portal typified in Fig. 9.6. This consists of a single portal on the top of which another portal has been placed, consequently the number of redundancies is greater than the single storeyed portal.

There are two main methods for analysing stiff-jointed frames, by considering strain energy and by moment distribution. The

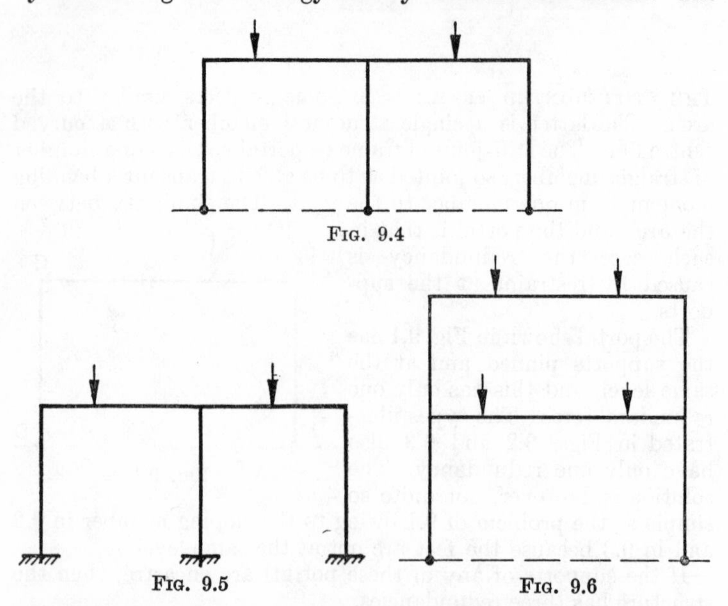

Fig. 9.4

Fig. 9.5

Fig. 9.6

former is straightforward when there is only one redundancy and in such problems is probably the better method, but as the number of redundancies increases so does the usefulness of the moment-distribution method and the use of strain energy is only advised for cases of a single redundancy.

There are other methods in addition to these two for solving stiff-jointed frames—slope-deflexion, column-analogy, degree of fixity, etc. These are useful tools in the hands of people who know how to use them but the student seldom has time for the acquisition of this experience. The many examples given in this chapter can, however, be used by the student as "guinea pigs" on which to try any method that interests him. Example 9.8 is solved by slope-deflexion in addition to moment distribution so that these two methods may be compared.

It has already been mentioned that a single portal with feet pinned has only one redundancy. If an extra pin is inserted in the structure, a three-pinned portal is obtained, which like the three-pinned arch is statically determinate.

The various examples now given will be typical of each of the cases shown in Figs. 9.1–9.6. An example will also be given of a three-pinned portal.

The Three-pinned Portal

Example 9.1. Draw the influence line for horizontal thrust for the three-pinned portal shown in Fig. 9.7. Determine the

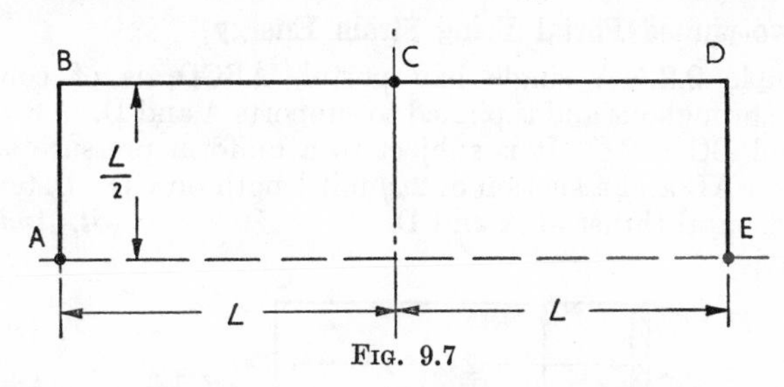

FIG. 9.7

maximum bending moment on this portal when it carries uniformly distributed loads of $2w$/unit length on BC and w on CD. *(Oxford)*

Solution. Let a load W be at a distance x from B on the member BC. Then the vertical reactions at A and E are

$$\left(\frac{2L - x}{2L}\right) W \quad \text{and} \quad \frac{x}{2L} W$$

respectively.

Taking moments about C the right-hand portion of the portal CDE gives

$$H \times \frac{L}{2} = \frac{x}{2L} W \times L$$

hence

$$H = \frac{x}{L} W$$

For an influence line, W = unity, hence the required diagram is as shown in Fig. 9.8.

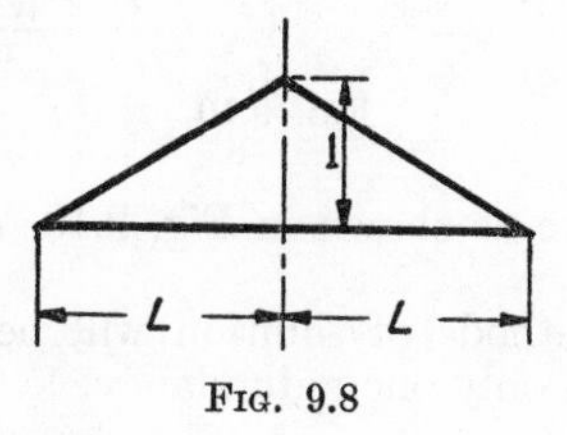

FIG. 9.8

The area under the influence line for the portion $BC = 1 \times L \times \frac{1}{2} = L/2$ = area under influence line for portion CD.

$$\text{Thrust due to } 2w \text{ on BC} = wL$$

$$\text{Thrust due to } w \text{ on CD} = \frac{wL}{2}$$

$$\therefore \quad \text{Total thrust due to given loading} = \frac{3wL}{2}$$

The Two-pinned Portal Using Strain Energy

Example 9.2. A single bay portal, ABCD, is of constant section throughout and is pinned to supports A and D. AB = CD = L and BC = $2L$. It is subject to a uniform pressure w/unit length on AB and a suction of $2w$/unit length on CD. Determine the horizontal thrust at A and D. (*St. Andrews*)

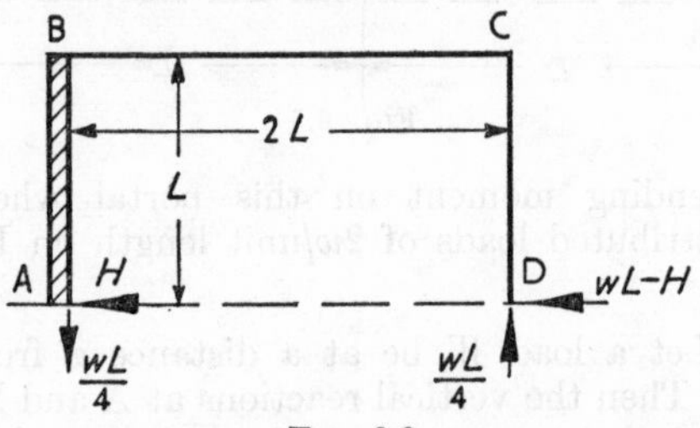

FIG. 9.9

Solution. The portal is shown diagrammatically in Fig. 9.9. Only the effects of the pressure load of w will be analysed as, because of the symmetry of the portal, the effects of the suction

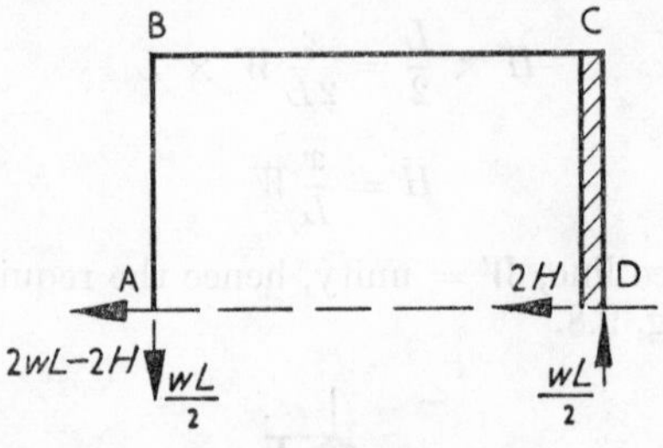

FIG. 9.10

of $2w$ on CD will be as shown in Fig. 9.10, H having the same value as in Fig. 9.9.

Strain-energy methods of solution will be used in this case since the portal has only one redundancy.

If H is taken as the redundancy then

$$\frac{\partial U}{\partial H} = \int \frac{M_x}{EI} \cdot \frac{\partial M_x}{\partial H} \, dx = 0$$

The vertical reactions, V_A and V_D, are obtained by taking moments. They are

$$V_A = -\frac{WL}{4} = -V_D$$

For AB the bending moment M_x at a section distant x from A is given by

$$M_x = \frac{wx^2}{2} - Hx \qquad \frac{\partial M_x}{\partial H} = -x$$

$$\therefore \left[\frac{\partial U}{\partial H}\right]_{AB} = \frac{1}{EI}\int_0^L \left(Hx^2 - \frac{wx^3}{2}\right) dx = \frac{1}{EI}\left(\frac{HL^3}{3} - \frac{wL^4}{8}\right)$$

For BC the bending moment, M_x, at a section distant x from B is given by

$$M_x = -HL + \frac{wLx}{4} + \frac{wL^2}{2} \qquad \frac{\partial M_x}{\partial H} = -L$$

$$\left[\frac{\partial U}{\partial H}\right]_{BC} = \frac{1}{EI}\int_0^{2L} \left(HL^2 - \frac{wL^2x}{4} - \frac{wL^3}{2}\right) dx$$

$$= \frac{1}{EI}\left(2HL^3 - \frac{3wL^4}{2}\right)$$

For CD the bending moment M_x at a section distant x from D is given by

$$M_x = +(wL - H)x \qquad \frac{\partial M_x}{\partial H} = -x$$

$$\therefore \left[\frac{\partial U}{\partial H}\right]_{CD} = \frac{1}{EI}\int_0^L (Hx^2 - wLx^2)\, dx = \frac{1}{EI}\left(\frac{HL^3}{3} - \frac{wL^4}{3}\right)$$

For the whole structure

$$\frac{\partial U}{\partial H} = 0$$

and summing for the various members and equating to zero gives

$$H = \frac{47}{64} wL = 0{\cdot}735wL$$

Therefore the horizontal reactions are—

At A, due to pressure $= 0{\cdot}735wL$

due to suction $= 0{\cdot}530wL$

giving a resultant of $1{\cdot}265wL$.

At B, due to pressure $= 0{\cdot}265wL$

due to suction $= 1{\cdot}470wL$

giving a resultant of $1{\cdot}735wL$.

Example 9.3. A 300 × 150 mm joist ($I = 8{\cdot}6 \times 10^7$ mm^4 units, $A = 5{\cdot}7 \times 10^3$ mm^2) is used to form a single bay portal 5 m high and of 10 m span, the supports being pinned. Calculate the maximum increase in stress due to a temperature rise of 50°C if the coefficient of expansion of steel = 0·000012 per °C.

(*St. Andrews*)

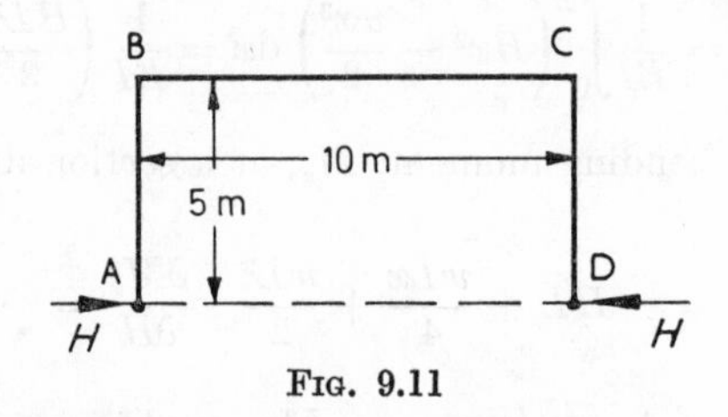

FIG. 9.11

Solution. The portal is shown diagrammatically in Fig. 9.11.

Let H be the horizontal thrust induced by the rise in temperature. Then for the whole structure.

$$\frac{\partial U}{\partial H} = 10 \times 1{\cdot}2 \times 10^{-5} \times 50 \text{ m}$$

For each of the members AB and CD

$$M_x = Hx$$

$$\therefore \quad \frac{\partial U}{\partial H} = \frac{1}{EI}\int_0^5 Hx^2\, dx = \frac{125H}{3EI}$$

For the member BC

$$M_x = 5H$$

$$\frac{\partial U}{\partial H} = \frac{1}{EI}\int_0^{10} 25H\, dx = \frac{250H}{EI}$$

$\therefore$ For the whole structure

$$\frac{\partial U}{\partial H} = \frac{875H}{3EI} = 0{\cdot}6 \times 10^{-2} \text{ m}$$

Substituting for EI (taking care to convert the units) gives

$$H = 0{\cdot}358 \text{ kN}$$

The maximum bending moment due to this thrust is 1·79 kN m and the stresses induced are

$$\text{Stress due to bending, } \frac{1{\cdot}79 \times 10^6 \times 150}{8{\cdot}6 \times 10^7} = 3{\cdot}12 \text{ N/mm}^2$$

$$\text{Stress due to thrust, } \frac{0{\cdot}358 \times 10^3}{5{\cdot}7 \times 10^3} = 0{\cdot}063 \text{ N/mm}^2$$

$\therefore$ Total stress resulting from temperature change

$$= 3{\cdot}18 \text{ N/mm}^2$$

Example 9.4. The frame shown in Fig. 9.12 is pinned to the supports A and E and has stiff joints at B, C and D. Determine (*a*) the bending moments at the joints, (*b*) the maximum bending

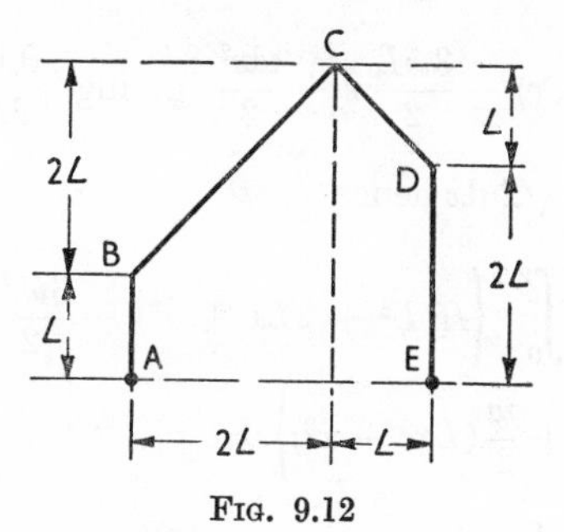

FIG. 9.12

moment of the frame when it carries a uniformly distributed load of w/unit of horizontal length on BC and CD. The flexural rigidity is constant throughout. (*St. Andrews*)

Solution. Frames of this type are probably better solved by strain energy than other methods. If moment-distribution methods are used, the sloping members BC and CD necessitate a sway correction, because the deflected form due to the load is such that B and D move horizontally relative to A and E. The average student does not find the application of this sway correction easy. If strain energy methods are used, four-member frames with sloping ridges are not appreciably more difficult than the three-member frames dealt with in the previous example.

In the example in question the horizontal reaction at each pin will be taken as the redundancy. There is no horizontal load, consequently the horizontal thrusts are equal and opposite.

The vertical reactions are obtained by taking moments giving $V_A = V_E = 3wL/2$.

Bending energy only will be considered, hence

$$\frac{\partial U}{\partial H} = \int \frac{M_x}{EI} \cdot \frac{\partial M_x}{\partial H}\, dx$$

For the member AB, the bending moment, M_x, at a section distant x from A, is given by

$$M_x = Hx \quad \text{hence} \quad \frac{\partial M_x}{\partial H} = x$$

$$\left[\frac{\partial U}{\partial H}\right]_{AB} = \frac{1}{EI}\int_0^L Hx^2\, dx = \frac{HL^3}{3EI}$$

For the member BC, the bending moment M_x, at a section distant x measured vertically from B, is given by the equation

$$M_x = H(L + x) - \frac{3wL}{2}x + \frac{wx^2}{2} \quad \text{giving} \quad \frac{\partial M_x}{\partial H} = (L + x)$$

but for BC $ds = \sqrt{2}\, dx$ hence

$$\left[\frac{\partial U}{\partial H}\right]_{BC} = \frac{\sqrt{2}}{EI}\int_0^{2L}\left\{H(L^2 + 2Lx + x^2) - \frac{3wL}{2}(Lx + x^2) + \frac{w}{2}(Lx^2 + x^3)\right\} dx$$

$$= \frac{\sqrt{2}}{EI}\left[H\left(L^2x + Lx^2 + \frac{x^3}{3}\right) - \frac{w}{2}\left(\frac{3L^2x^2}{2} + Lx^3 - \frac{Lx^3}{3} - \frac{x^4}{4}\right)\right]_0^{2L}$$

$$= \frac{\sqrt{2}}{EI}\left(\frac{26HL^3}{3} - \frac{11wL^4}{3}\right)$$

For the member DE the moment, M_x, at a section distant x from E, is given by

$$M_x = Hx \quad \frac{\partial M_x}{\partial H} = x$$

$$\left[\frac{\partial U}{\partial H}\right]_{DE} = \frac{1}{EI}\int_0^{2L} Hx^2\, dx = \frac{8HL^3}{3EI}$$

For the member DC the moment, M_x, at a section distant x measured vertically from D, is given by

$$M_x = H(2L + x) - \frac{3wL}{2}x + \frac{wx^2}{2} \quad \frac{\partial M_x}{\partial H} = (2L + x)$$

$$\left[\frac{\partial U}{\partial H}\right]_{DC} = \frac{\sqrt{2}}{EI}\int_0^L \left\{H(4L^2 + 4Lx + x^2) - \frac{3wL}{2}(2Lx + x^2) + \frac{w}{2}(2Lx^2 + x^3)\right\} dx$$

$$= \frac{\sqrt{2}}{EI}\left[H\left(4L^2x + 2Lx^2 + \frac{x^3}{3}\right) - \frac{3wL}{2}\left(Lx^2 + \frac{x^3}{3}\right) + \frac{w}{2}\left(\frac{2Lx^3}{3} + \frac{x^4}{4}\right)\right]_0^L$$

$$= \frac{\sqrt{2}}{EI}\left(\frac{19HL^3}{3} - \frac{37wL^4}{24}\right)$$

For the whole structure

$$\frac{\partial U}{\partial H} = 0 = \frac{L^3}{EI}\left\{H(3 + 15\sqrt{2}) - wL\left(\frac{11}{3} + \frac{37}{24}\right)\sqrt{2}\right\}$$

giving $H = 0{\cdot}303wL$

$$M_B = HL = 0{\cdot}303wL^2$$

$$M_C = 3HL - \frac{3wL^2}{2} + \frac{wL^2}{2} = -0{\cdot}091wL^2$$

$$M_D = 2HL = 0{\cdot}606wL^2$$

These are the joint moments. The question also asks for the maximum bending moment on the frame and an analysis must now be made of the moments at sections away from the joints to determine if at any point the moment exceeds M_D.

The moment M_x on the section BC is given by

$$M_x = 0{\cdot}303wL(L + x) - 1{\cdot}5wLx + 0{\cdot}5wx^2$$

Differentiating with respect to x and equating to zero gives as the condition for a maximum value of M_x

$$0{\cdot}303wL - 1{\cdot}5wL + wx = 0$$

i.e.

$$x = 1{\cdot}197L$$

giving $M_{max} = -0{\cdot}417wL^2$ which is numerically less than M_D.

The moment, M_x, on section CD is given by

$$M_x = 0{\cdot}303wL(2L + x) - 1{\cdot}5wLx + 0{\cdot}5wx^2$$

stanchion = an upright bar, post or support

Differentiating with respect to x and equating to zero gives as a condition for a maximum $x = 1{\cdot}197L$. But the expression for M_x only holds for values of x not greater than L, hence the solution is invalid.

Therefore the greatest bending moment on the frame occurs at joint D and has the value $0{\cdot}606wL^2$.

Analysis of Portals by Moment-distribution

It has already been seen from the work on built-in beams that moments at the ends of such beams may occur from two causes—(i) load carried by the beam, and (ii) a movement of one end of the beam relative to the other.

Both of these have been dealt with in problems on continuous and built-in beams and both play a part in the analysis of portals by moment-distribution.

If the single portal, ABCD, shown in Fig. 9.1. is symmetrical and is symmetrically loaded, then the deflected form of the portal is such that there is no relative horizontal movement between A and B or C and D and no relative vertical movement between B and C. Consequently the moments on the members of the portal are due only to joint rotation.

If, however, there is a lack of symmetry either in the loading or in the make-up of the portal, then the deflection under load is such that there is relative horizontal movement between A and B, and C and D. As a result moments are introduced in the structure. The moments so introduced are known as sway moments and the main difficulty with the application of moment distribution to portal frames is in dealing with these moments.

There are two main methods of applying the sway correction. The first makes use of the equilibrium of the horizontal shear in the legs with the horizontally applied loads whilst the second uses an equation of structural equilibrium formed from a consideration of the moments at the top and bottom of the stanchions. The latter method will be used in the examples given. The fundamental theory behind it is as follows.

Consider any stanchion, AB, as shown in Fig. 9.13 which forms part of a portal. Let the horizontal reaction at A due to the external loads on the portal be H_A and let M_{AB} be the fixing moment. Let a load, W_1, be applied horizontally to the stanchion at a distance a_1 above A. Then if M_{BA} is the moment at the top of the stanchion and the moment distribution convention (clockwise moments positive, anticlockwise negative) is used—

$$M_{BA} = -M_{AB} - H_A h_1 + W_1(h_1 - a_1)$$

Thus for the stanchion AB the relationship between the moments and the loadings can be expressed as

$$M_{AB} + M_{BA} = -H_A h_1 + W_1(h_1 - a_1)$$

Similarly for stanchions CD, EF, GH the following relationships hold

$$M_{CD} + M_{DC} = -H_C h_2 + W_2(h_2 - a_2)$$
$$M_{EF} + M_{FE} = -H_E h_3 + W_3(h_3 - a_3)$$
$$M_{GH} + M_{HG} = -H_G h_4 + W_4(h_4 - a_4)$$

Also in order to satisfy the conditions of horizontal equilibrium it can be said that

$$H_A + H_C + H_E + H_G = W_1 + W_2 + W_3 + W_4$$

From these five equations the four unknown H forces can be eliminated thus giving a relationship between the moments at the top and bottom of the stanchions and the applied horizontal loads, W_1, W_2, W_3 and W_4.

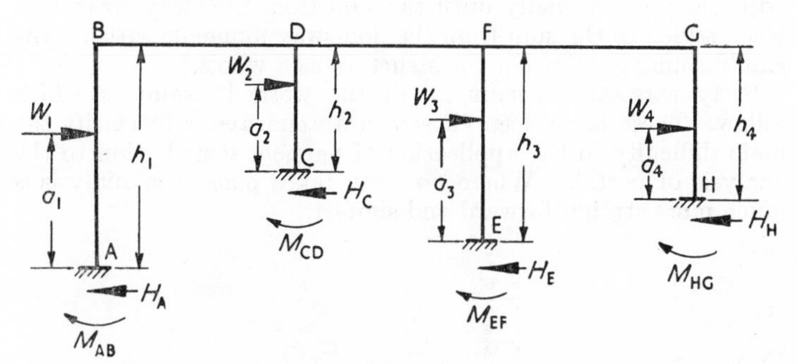

FIG. 9.13

The above analysis has been carried out for the general condition. In many cases simpler expressions can be obtained.

For example if the stanchions are all of the same height h_1, then the sum of the moments at the top and bottom of the stanchions, which will here and elsewhere in this chapter be denoted by $\Sigma M_B{}^T$, has the following relationship

$$\Sigma M_B{}^T = -\Sigma Wa$$

If, in addition to the stanchions being of equal height, there is also no horizontal load on the portal, then $\Sigma M_B{}^T = 0$.

This relationship between the sum of the moments at the top and bottom of the stanchions and the external loading on the structure can be referred to as the equilibrium equation of the structure. The final analysis must produce moments which satisfy it.

The two causes of joint moments, namely external loading and horizontal deflexion, are dealt with separately. The moments due to the external loads are calculated, and in balancing up the joints the structure is held by an applied horizontal force so that no sway takes place. After this distribution $\Sigma M_B{}^T$ is obtained and the lack of balance on the structure as a whole found by comparing this sum with the equilibrium equation produced as previously described. This difference gives the sway correction.

When the portal shown in Fig. 9.13 moves horizontally by a given amount moments are produced in the columns, the moment in any particular column being dependent on the I/h^2 value of that column relative to the others. Therefore to determine the sway moments any arbitrary set of moments in the columns is taken which satisfies the relationship that the moments in any particular column relative to those in another column are in the ratio of the I/h^2 value of those two columns. After balancing the joints the column moments are summed and then adjusted proportionally until the sum from the sway correction when added to the sum from the non-sway moments satisfies the equilibrium equation for the structure as a whole.

Sway correction occurs in all the worked examples which follow. These have been chosen in an endeavour to clarify the main difficulty in the application of moment distribution to the analysis of portals. Where no sway takes place, the analysis is much more straightforward and simpler.

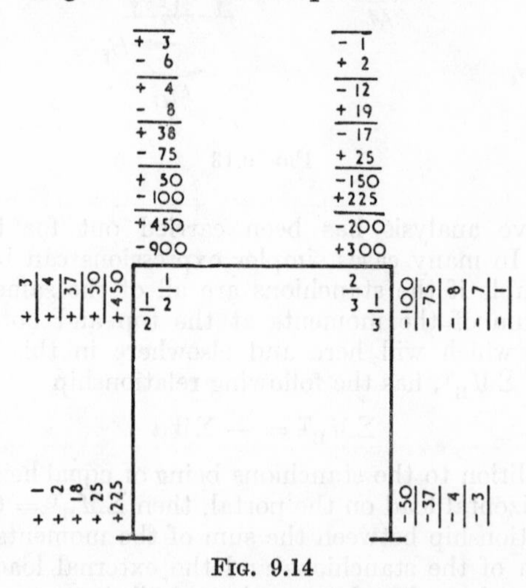

Fig. 9.14

Moment distribution is essentially an arithmetical method and the question of the accuracy required must arise. Engineers are accustomed to slide-rule accuracy, i.e. three significant figures. A comparative accuracy can be obtained in moment distribution if the moments before balancing are in the neighbourhood of 1,000 and the balance is carried down to the nearest whole number. This gives an accuracy comparable with a slide rule.

There are two methods of writing down the balancing; one is to write it against the joints as shown in Fig. 9.14 and the other is to tabulate it.

The latter is neater and will be used in the examples given.

Example 9.5. Use moment distribution methods to draw the bending moment diagram for the frame, shown in Fig. 9.15, which carries a load, W, at a distance $\frac{L}{4}$ from B.

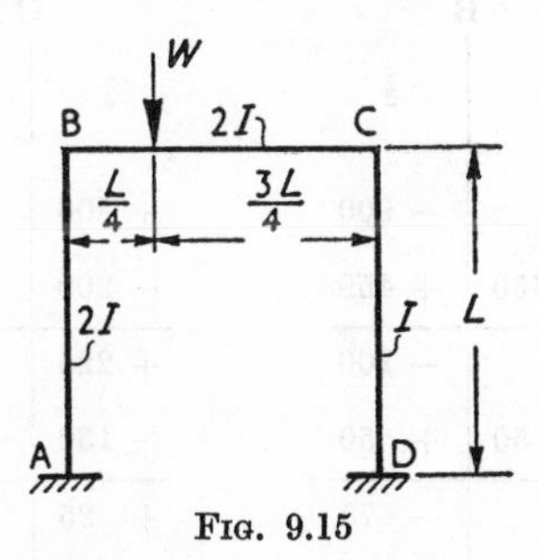

FIG. 9.15

Determine also the horizontal force which would have to be applied at B to prevent side sway. (*London*)

Solution. This example will be solved as follows—

The fixed end moments are

$$M_{BC} = -\frac{9WL}{64} = -900 \text{ (say)}$$

$$M_{CB} = \frac{3WL}{64} = 300 \text{ (say)}$$

The distribution of out-of-balance at the joints is as follows—

$$M_{BA} = \left(\frac{\frac{2I}{L}}{\frac{2I}{L} + \frac{2I}{L}}\right) M_B = \frac{M_B}{2} = M_{BC}$$

$$M_{CB} = \left(\frac{\frac{2I}{L}}{\frac{2I}{L} + \frac{I}{L}}\right) M_C = \tfrac{2}{3} M_C$$

$$M_{CD} = \left(\frac{\frac{I}{L}}{\frac{2I}{L} + \frac{I}{L}}\right) M_C = \tfrac{1}{3} M_C$$

The initial distribution with the structure held against side sway is set out in Table 9.5 (*a*).

TABLE 9.5 (*a*)

A	B		C		D
	½	½	⅔	⅓	
		− 900	+ 300		
+ 225	+ 450	+ 450	− 200	− 100	− 50
		− 100	+ 225		
+ 25	+ 50	+ 50	− 150	− 75	− 37
		− 75	+ 25		
+ 18	+ 37	+ 38	− 17	− 8	− 4
		− 8	+ 19		
+ 2	+ 4	+ 4	− 12	− 7	− 3
		− 6	+ 2		
+ 1	+ 3	+ 3	− 1	− 1	
+ 271	+ 544	− 544	+ 191	− 191	− 94

From this distribution of the moments the sum at the top and bottom of the columns $\Sigma M_B{}^T$ is given by

$$\Sigma M_B{}^T = +271 + 544 - 191 - 94 = +530$$

There is, however, no horizontal load on the structure; consequently in the final analysis $\Sigma M_B{}^T = 0$.

The column stiffnesses are in the ratio 2 : 1, so a set of arbitrary moments in this ratio is taken and distributed as shown in Table 9.5 (*b*).

TABLE 9.5 (b)

A	B		C		D
	$\frac{1}{2}$	$\frac{1}{2}$	$\frac{2}{3}$	$\frac{1}{3}$	
− 200	− 200			− 100	− 100
+ 50	+ 100	+ 100	+ 67	+ 33	+ 17
		+ 33	+ 50		
− 8	− 17	− 16	− 33	− 17	− 8
		− 17	− 8		
+ 4	+ 8	+ 9	+ 5	+ 3	+ 2
		+ 2	+ 4		
	− 1	− 1	− 3	− 1	
− 154	− 110	+ 110	+ 82	− 82	− 93

From this distribution

$$\Sigma M_B{}^T = -154 - 110 - 82 - 93 = -439$$

Therefore sway moments must be multiplied by $\frac{530}{439}$ to balance the structure giving the final moments as below.

TABLE 9.5 (c)

	A	B		C		D
Non-sway . .	+ 271	+ 544	− 544	+ 191	− 191	− 94
Sway Correction .	− 187	− 134	+ 134	+ 99	− 99	− 113
FINAL . .	+ 84	+ 410	− 410	+ 290	− 290	− 207

As a check take

$$\Sigma M_B{}^T = +84 + 410 - 290 - 207 = -3$$

These moments are obtained by taking $\frac{WL}{64} = 100$.

Converting back in terms of WL gives

$$M_A = 0{\cdot}0137WL$$

$$M_B = 0{\cdot}0647WL$$

$$M_C = 0{\cdot}0461WL$$

$$M_D = 0{\cdot}0325WL$$

Let P be the force at the top to prevent side sway. Then

$$PL = \Sigma M_B{}^T \text{ of first distribution}$$

Hence $$P = \frac{530W}{6{,}400} = 0{\cdot}083W$$

The bending moment diagram is shown in Fig. 9.16.

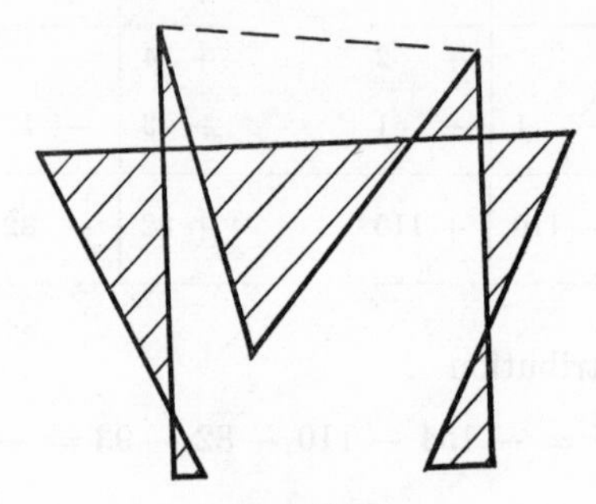

Fig. 9.16

Example 9.6. The stiff-jointed frame shown in Fig. 9.17 is supported on roller bearings at A and D. Draw the bending

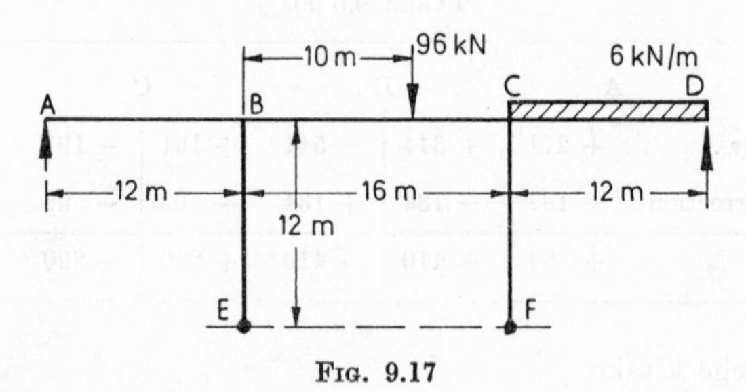

Fig. 9.17

moment diagram for the frame when it carries the loading shown. The flexural rigidity is constant throughout. E and F are pinned joints. *(St. Andrews)*

Solution. Since A and D are roller bearings no horizontal force acts at these points. There is therefore only one redundancy (the horizontal force at E and F) and strain energy analysis would be straightforward.

Moment-distribution analysis is used for the following reasons.

1. It is as straightforward as strain energy.
2. In order to show the reader how to apply sway correction when the columns are pinned at the base.
3. The question asks for a bending moment diagram and moment distribution gives straight away the necessary information for drawing it. Strain-energy methods would have given the value of H. The various bending moments at the joints would have to be calculated from this before the bending moment diagram could be drawn.

The fixed end moments are as follows

$$M_{CB} = \frac{96 \times 10^2 \times 6}{16^2} = 225 \text{ kN m}$$

$$M_{BC} = \frac{96 \times 6^2 \times 10}{16^2} = 135 \text{ kN m}$$

$$M_{CD} = M_{DC} = \frac{6 \times 12^2}{12} = 72 \text{ kN m}$$

These values will be multiplied by 10 for distribution purposes.

The distribution of the out-of-balance moments at each joint is obtained as follows.

At joint B

$$M_{BA} = M_{BE} = \left(\frac{\frac{3I}{48}}{\frac{3I}{48} + \frac{3I}{48} + \frac{I}{16}} \right) M_B = \frac{M_B}{3} = M_{BC}$$

In a similar way it can be shown that at joint C

$$M_{CD} = M_{CF} = M_{CB} = \frac{M_C}{3}$$

The first distribution with the structure held against side sway is carried out in Table 9.6 (*a*).

There is no horizontal load on the structure consequently $\Sigma M_B{}^T = 0$ in the final analysis.

In the first non-sway analysis, however

$$\Sigma M_B{}^T = +530 - 479 = +51$$

This represents the moment introduced into the structure by holding it against side sway. The correction must therefore apply

TABLE 9.6 (*a*)

A		B			C		D
	$\frac{1}{3}$	$\frac{1}{3}$	$\frac{1}{3}$	$\frac{1}{3}$	$\frac{1}{3}$	$\frac{1}{3}$	
0	0		− 1350	+ 2250		− 720	+ 720
						− 360	− 720
			− 195	− 390	− 390	− 390	
	+ 515	+ 515	+ 515	+ 258			
			− 43	− 86	− 86	− 86	
	+ 14	+ 15	+ 14	+ 7			
				− 2	− 3	− 2	
0	+ 529	+ 530	− 1059	+ 2037	− 479	− 1558	0

a moment of − 51 to the tops of the stanchions. The moment must be applied in the ratio of their stiffnesses, i.e. equally to both. But as the bases are pinned the moment can only be applied to the tops of the stanchions. There is, therefore, no need to assume an arbitrary set of moments and then balance up the joints, because the only moments which can balance the structure are $-\frac{51}{2}$ at both BE and CF. These moments (to the nearest whole number) with the corresponding balancing moments are shown in Table 9.6 (*b*).

TABLE 9.6 (*b*)

		B			C	
Sway Correction .	+ 12	− 25	+ 13	+ 13	− 26	+ 13
Non-sway Distribution .	+ 529	+ 530	− 1059	+ 2037	− 479	− 1558
FINAL MOMENTS .	+ 541	+ 505	− 1046	+ 2050	− 505	− 1545

The bending moment diagram is drawn and shown in Fig. 9.18.

The attention of the reader should be drawn to the fact that sway could take place in the above example only because the supports at A and D were rollers. If they had been pinned or

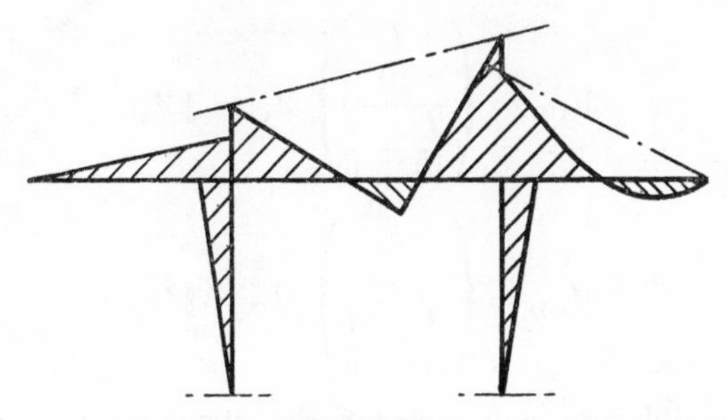

FIG. 9.18

encastré, no sway would have been possible and the first distribution would have given the final moments in the case of pins at A and D.

Example 9.7. A stiff-jointed frame ABCD is of constant section throughout and has the following dimensions, AB = 15 m, BC = 10 m, CD = 6 m. It carries loads of 90 kN at each of the third points of the horizontal beam, BC. The legs, AB and CD, are vertical.

Calculate the moments at the joints if the feet are fully fixed.

(*St. Andrews*)

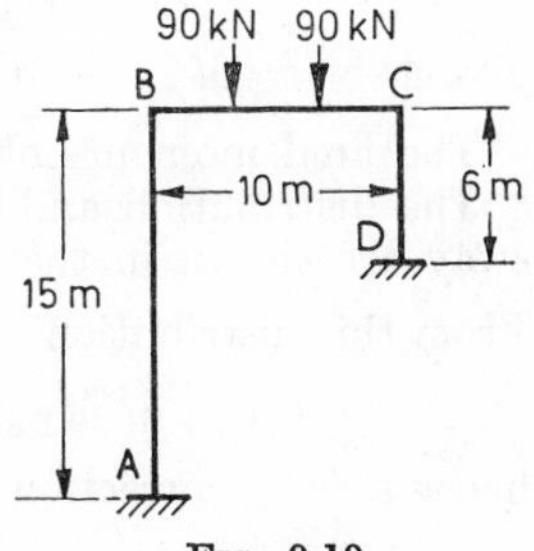

FIG. 9.19

Solution. The frame is shown diagrammatically in Fig. 9.19. The fixed end moments are

$$M_{BC} = M_{CB} = \tfrac{2}{9} \times 90 \times 10 = 200 \text{ kN m}$$

The distribution of out-of-balance is

Joint B,
$$M_{BA} = \left(\frac{\frac{I}{15}}{\frac{I}{15} + \frac{I}{10}}\right) M_B = \tfrac{2}{5} M_B$$

$$M_{BC} = \left(\frac{\frac{I}{10}}{\frac{I}{15} + \frac{I}{10}}\right) M_B = \tfrac{3}{5} M_B$$

Joint C, $$M_{CB} = \left(\frac{\frac{I}{10}}{\frac{I}{10} + \frac{I}{6}}\right) M_C = \tfrac{3}{8} M_C$$

$$M_{CD} = \left(\frac{\frac{I}{6}}{\frac{I}{10} + \frac{I}{6}}\right) M_C = \tfrac{5}{8} M_C$$

The equilibrium equation for the structure is derived as previously explained, giving the following equations from which to eliminate H_A and H_D

$$M_{AB} + M_{BA} = -H_A \times 15$$
$$M_{DC} + M_{CD} = -H_D \times 6$$
$$H_A + H_D = 0$$

giving $$M_{AB} + M_{BA} + \tfrac{5}{2}(M_{CD} + M_{CD}) = 0$$

The final moments obtained must satisfy this relationship.

The distribution and balance with the structure held against sway are shown in the Table 9.7 (*a*) on the following page.

From this distribution

$$(M_{AB} + M_{BA}) + \tfrac{5}{2}(M_{CD} + M_{DC}) = -495$$

hence a sway correction is necessary.

Arbitrary column moments in the ratio $\frac{15^2}{6^2}$ are chosen for balance. If $M_{AB} = M_{BA} = 100$, and $M_{CD} = M_{DC} = 625$, the relationship is satisfied. The balance of joints from these moments is shown in Table 9.7 (*b*).

From this distribution

$$M_{AB} + M_{BA} + \tfrac{5}{2}(M_{CD} + M_{DC}) = +1{,}860$$

Hence to balance the structure the sway moments must be reduced in the ratio $\frac{495}{1{,}860}$, giving

	A		B		C	D
Non-sway . .	+ 50	+ 100	− 100	+ 172	− 172	− 86
Sway Correction .	+ 28	+ 29	− 29	− 62	+ 62	+ 114
FINAL . .	+ 78	+ 129	− 129	+ 110	− 110	+ 28

Hence the moments at the foot of the stanchions are

$$M_{AB} = 78 \text{ kN m}$$
$$M_{DC} = 28 \text{ kN m}$$

TABLE 9.7 (*a*)

A	B		C		D
	$\frac{2}{5}$	$\frac{3}{5}$	$\frac{3}{8}$	$\frac{5}{8}$	
		− 200	+ 200		
+ 40	+ 80	+ 120	− 75	− 125	− 63
		− 38	+ 60		
+ 8	+ 15	+ 23	− 22	− 38	− 19
		− 11	+ 11		
+ 2	+ 4	+ 7	− 4	− 7	− 3
		− 2	+ 3		
	+ 1	+ 1	− 1	− 2	
+ 50	+ 100	− 100	+ 172	− 172	− 86

TABLE 9.7 (*b*)

A	B		C		D
	$\frac{2}{5}$	$\frac{3}{5}$	$\frac{3}{8}$	$\frac{5}{8}$	
+ 100	+ 100			+ 625	+ 625
− 20	− 40	− 60	− 234	− 391	− 196
		− 117	− 30		
+ 24	+ 47	+ 70	+ 11	+ 19	+ 10
		+ 5	+ 35		
− 1	− 2	− 3	− 13	− 22	− 11
		− 7	− 1	+ 1	
+ 2	+ 3	+ 4	+ 2		
			− 1	− 1	
+ 105	+ 108	− 108	− 231	+ 231	+ 428

Example 9.8. Determine the bending moments at the joints and draw the bending moment diagram for the frame shown in Fig. 9.20 (*a*). The flexural rigidities of the members are as shown.

(*St. Andrews*)

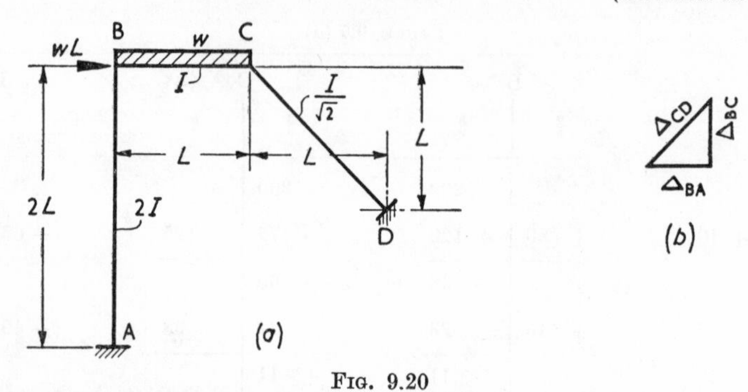

FIG. 9.20

Solution. This problem is more difficult than the previous one because the member CD is sloping instead of vertical. This brings in two main differences—

1. In the sum of the moments at the top and bottom of CD the vertical reaction V_D will occur. Hence a further equation must be obtained to eliminate it.

2. When the structure sways over, moments will be introduced into the horizontal member as well as the legs.

The question of the sway will be dealt with first. If the point B is displaced horizontally (i.e. at right angles to the direction of the member BA) by an amount Δ, then the triangle of displacements shown in Fig. 9.20 (*b*) gives the displacement of C at right angles to BC as Δ and its displacement perpendicular to CD as $\sqrt{2}\Delta$.

The moments induced at the ends of a built-in beam by a deflexion of one end by an amount Δ relative to the other are $\dfrac{6EI\Delta}{h^2}$, and therefore the moments introduced in the members of the frame when sway takes place are in the ratio

$$\text{AB} : \text{BC} : \text{CD} : : \frac{\Delta}{2L^2} : -\frac{\Delta}{L^2} : \frac{\Delta}{2L^2} : : \frac{1}{2} : -1 : \frac{1}{2}$$

The equation for $\Sigma M_B{}^T$ is derived as follows—

$$M_{AB} + M_{BA} = -H_A \times 2L$$

$$M_{DC} + M_{CD} = -(wL - H_A) \times L + V_D \times L$$

Moments about A give

$$V_D = \frac{M_{AB} + M_{DC} + \frac{3}{2}wL^2 + H_A L}{2L}$$

Eliminating V_D and H_A from these three equations gives

$$M_{AB} + 3M_{BA} + 2M_{DC} + 4M_{CD} = -wL^2$$

Having derived the sway and equilibrium equations the straightforward non-sway distribution can be carried out.

Fixed end moments—

$$M_{BC} = M_{CB} = \frac{wL^2}{12} = 1000 \text{ (say)}$$

Distribution of out-of-balance moments—

$$\text{Joint B,} \quad M_{BA} = \left(\frac{\frac{2I}{2L}}{\frac{I}{L} + \frac{2I}{2L}}\right) M_B = \frac{M_B}{2} = M_{BC}$$

TABLE 9.8 (*a*)

Non-sway Distribution

A	B		C		D
	$\frac{1}{2}$	$\frac{1}{2}$	$\frac{2}{3}$	$\frac{1}{3}$	
		− 1000	+ 1000		
+ 250	+ 500	+ 500	− 667	− 333	− 167
		− 333	+ 250		
+ 83	+ 167	+ 166	− 167	− 83	− 42
		− 84	+ 83		
+ 21	+ 42	+ 42	− 55	− 28	− 14
		− 28	+ 21		
+ 7	+ 14	+ 14	− 14	− 7	− 4
		− 7	+ 7		
+ 2	+ 3	+ 4	− 5	− 2	− 1
		− 2	+ 2		
	+ 1	+ 1	− 1	− 1	
+ 363	+ 727	− 727	+ 454	− 454	− 228

Joint C, $$M_{CB} = \left(\frac{\dfrac{I}{L}}{\dfrac{I}{L} + \dfrac{I}{2L}}\right) M_C = \tfrac{2}{3} M_C$$

$$M_{CD} = \left(\frac{\dfrac{I}{2L}}{\dfrac{I}{L} + \dfrac{I}{2L}}\right) M_C = \tfrac{1}{3} M_C$$

From this distribution

$$M_{AB} + 3M_{BA} + 2M_{DC} + 4M_{CD} = +272$$

But the $\Sigma M_B{}^T$ equation requires that this sum equals $-wL^4$, i.e. $-12{,}000$.

$\therefore$ Sway correction $= -12\,272$

TABLE 9.8 (*b*)

Sway Correction Distribution

A	B		C		D
	½	½	⅔	⅓	
− 1000	− 1000	+ 2000	+ 2000	− 1000	− 1000
− 250	− 500	− 500	− 667	− 333	− 167
		− 333	− 250		
+ 83	+ 167	+ 166	+ 167	+ 83	+ 42
		+ 84	+ 83		
− 21	− 42	− 42	− 55	− 28	− 14
		− 28	− 21		
+ 7	+ 14	+ 14	+ 14	+ 7	+ 4
		+ 7	+ 7		
− 2	− 3	− 4	− 5	− 2	− 1
		− 2	− 2		
	+ 1	+ 1	+ 1	+ 1	
− 1183	− 1363	+ 1363	+ 1272	− 1272	− 1136

The set of arbitrary moments in the required ratio is taken and balanced as in Table 9.8 (*b*).

From this distribution

$$M_{AB} + 3M_{BA} + 2M_{DC} + 4M_{CD} = -12\,360$$

Hence sway moments must be altered in the ratio $\frac{12\,272}{12\,360}$.

	A		B		C		D
Non-sway . .	+ 363	+ 727	− 727	+ 454	− 454	− 228	
Sway Correction .	− 1175	− 1353	+ 1353	+ 1263	− 1263	− 1128	
FINAL . .	− 812	− 626	+ 626	+ 1717	− 1717	− 1356	
	$0{\cdot}0676wL^2$		$0{\cdot}0522wL^2$		$0{\cdot}1436wL^2$		$0{\cdot}113wL^2$

The bending moment diagram is shown in Fig. 9.21.

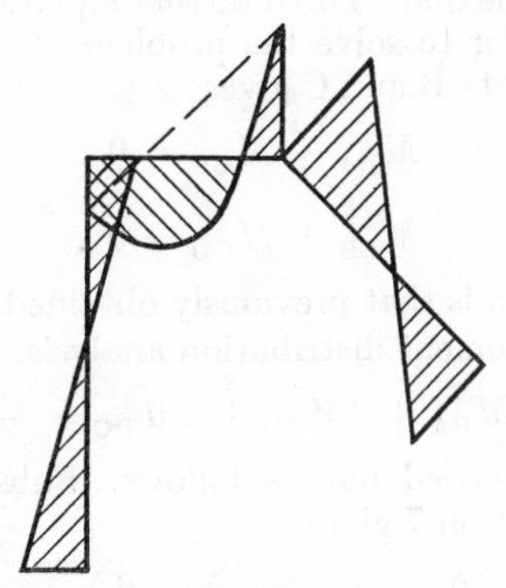

FIG. 9.21

This example will now be solved by the slope-deflexion method. The fundamental equations relating slope, deflexion, end-moments and areas of bending moment diagrams have already been given (pages 107, 118). These equations will be applied to each member in turn using the deflexion diagram given in Fig. 9.20 (*b*) and taking the convention that clockwise moments are positive. For the member AB, taking $\theta_A = 0$ (since end built-in)

$$M_{AB} = \frac{2EI}{L}\left(\theta_B - \frac{3\Delta}{2L}\right) \quad . \quad . \quad . \quad (1)$$

$$M_{BA} = \frac{2EI}{L}\left(2\theta_B - \frac{3\Delta}{2L}\right) \quad . \quad . \quad . \quad (2)$$

For the member BC, allowing for the distributed load on it,

$$M_{BC} = \frac{2EI}{L}\left(2\theta_B + \theta_C + \frac{3\Delta}{L}\right) - \frac{wL^2}{12} \quad . \quad . \quad (3)$$

$$M_{CB} = \frac{2EI}{L}\left(\theta_B + 2\theta_C + \frac{3\Delta}{L}\right) + \frac{wL^2}{12} \quad . \quad . \quad (4)$$

The positive sign in front of the Δ/L term should be noted. It is inserted because C moves upwards relative to B.

For the member CD, taking $\theta_D = 0$,

$$M_{CD} = \frac{2EI}{2L}\left(2\theta_C - \frac{3\sqrt{2}\Delta}{\sqrt{2}L}\right) \quad . \quad . \quad . \quad (5)$$

$$M_{DC} = \frac{2EI}{2L}\left(\theta_C - \frac{3\sqrt{2}\Delta}{\sqrt{2}L}\right) \quad . \quad . \quad . \quad . \quad (6)$$

These six equations contain nine unknowns: six moments, two slopes and one deflexion. Three further equations must therefore be built-up in order to solve the problem. Consideration of the equilibrium of joints B and C gives

$$M_{BA} + M_{BC} = 0 \quad . \quad . \quad . \quad . \quad . \quad (7)$$

and

$$M_{CB} + M_{CD} = 0 \quad . \quad . \quad . \quad . \quad . \quad (8)$$

The ninth equation is that previously obtained for the expression for ΣM_B^T in the moment distribution analysis, namely

$$M_{AB} + 3M_{BA} + 4M_{CD} + 2M_{DC} = -wL^2 \quad . \quad . \quad (9)$$

The solution is carried out as follows. Substituting for M_{BA} and M_{BC} in Equation 7 gives

$$4\theta_B - \frac{3\Delta}{L} + 4\theta_B + 2\theta_C + \frac{6\Delta}{L} - \frac{wL^3}{12EI} = 0$$

giving

$$\theta_B = -\frac{\theta_C}{4} - \frac{3\Delta}{8L} + \frac{wL^3}{96EI} \quad . \quad . \quad . \quad (10)$$

Substituting for M_{CB} and M_{CD} in Equation 8 gives

$$2\theta_B + 4\theta_C + \frac{6\Delta}{L} + \frac{wL^3}{12EI} + 2\theta_C - \frac{3\Delta}{L} = 0$$

giving

$$6\theta_C + \frac{3\Delta}{L} - \frac{\theta_C}{2} - \frac{3\Delta}{4L} + \frac{5wL^3}{48EI} = 0$$

$$\theta_C = -\frac{0{\cdot}01896wL^3}{EI} - 0{\cdot}409\frac{\Delta}{L} \quad . \quad . \quad (11)$$

Substituting this value for θ_C in Equation 10 gives

$$\theta_B = -0{\cdot}2725\frac{\Delta}{L} + 0{\cdot}015\,16\frac{wL^3}{EI}$$

Substituting for moments in terms of slopes and deflexions gives in Equation 9

$$2\theta_B - \frac{3\Delta}{L} + 12\theta_B - \frac{9\Delta}{L} + 8\theta_C - \frac{12\Delta}{L} + 2\theta_C - \frac{6\Delta}{L} = -\frac{wL^3}{EI}$$

$$14\theta_B + 10\theta_C - \frac{30\Delta}{L} = -\frac{wL^3}{EI}$$

$$1{\cdot}4\theta_B + \theta_C - \frac{3\Delta}{L} = 0{\cdot}1\frac{wL^3}{EI}$$

Substituting for θ_B and θ_C the expressions obtained in Equations 10 and 11 give

$$-0{\cdot}385\frac{\Delta}{L} + 0{\cdot}0212\frac{wL^3}{EI} - 0{\cdot}409\frac{\Delta}{L} - 0{\cdot}018\,96\frac{wL^3}{EI} - \frac{3\Delta}{L}$$

$$= 0{\cdot}1\frac{wL^3}{EI}$$

giving $$\frac{\Delta}{L} = 0{\cdot}026\,95\frac{wL^3}{EI}$$

Hence $$\theta_B = 0{\cdot}015\,16\frac{wL^3}{EI} - 0{\cdot}007\,34\frac{wL^3}{EI} = +0{\cdot}007\,82\frac{wL^3}{EI}$$

and $$\theta_C = -0{\cdot}018\,96\frac{wL^3}{EI} - 0{\cdot}0110\frac{wL^3}{EI} = -0{\cdot}029\,96\frac{wL^3}{EI}$$

Thus $$M_{DC} = \frac{EI}{L}\left(\theta_C - \frac{3\Delta}{L}\right) = -0{\cdot}111\,wL^2$$

$$M_{CD} = \frac{EI}{L}\left(2\theta_C - \frac{3\Delta}{L}\right) = -0{\cdot}141\,wL^2$$

$$M_{AB} = \frac{2EI}{L}\left(\theta_B - \frac{3\Delta}{2L}\right) = -0{\cdot}0652\,wL^2$$

$$M_{BA} = \frac{2EI}{L}\left(2\theta_B - \frac{3\Delta}{2L}\right) = -0{\cdot}0496\,wL^2$$

These values agree reasonably well with those obtained from the moment-distribution analysis. The work has been carried out on a slide rule and it is doubtful if sufficient accuracy can be obtained thereby.

This example is a useful one on which to employ slope-deflexion methods, since the feet of the stanchions are built-in. If they are pinned then the slopes are not zero and two further unknowns θ_A and θ_D come into the equations in place of M_{AB} and M_{DC}, which of course are zero if the ends are pinned.

Example 9.9. A rigid frame is loaded as shown in Fig. 9.22. Determine the fixing moments at A, D and F and sketch the deflected form of the frame.

$$I_{AB} = I_{CE} = 2I_{BC} = 2I_{CD} = 2I_{EF}$$ *(Aberdeen)*

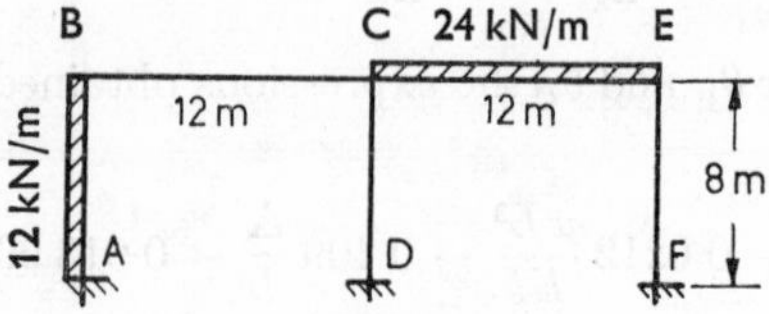

FIG. 9.22

Solution. The equilibrium condition is

$$\Sigma M_B{}^T = 8 \times 4 \times 12 = -384 \text{ kN m}$$

The fixed end moments are

$$M_{AB} = M_{BA} = 12 \times 8^2 \times \tfrac{1}{12} = 64 \text{ kN m}$$

$$M_{CE} = M_{EC} = 24 \times 12^2 \times \tfrac{1}{12} = 288 \text{ kN m}$$

The distribution of out-of-balance moments is

Joint B, $$M_{BA} = \left(\frac{\frac{2I}{8}}{\frac{2I}{8} + \frac{I}{12}}\right) M_B = \tfrac{3}{4} M_B$$

hence $$M_{BC} = \tfrac{1}{4} M_B$$

Joint C, $$M_{CB} = \left(\frac{\frac{I}{12}}{\frac{I}{12} + \frac{I}{8} + \frac{2I}{12}}\right) M_C = \tfrac{2}{9} M_C$$

$$M_{CD} = \left(\frac{\frac{I}{8}}{\frac{I}{12} + \frac{I}{8} + \frac{2I}{12}} \right) M_C = \tfrac{3}{9} M_C$$

$$M_{CE} = \left(\frac{\frac{2I}{12}}{\frac{I}{12} + \frac{I}{8} + \frac{2I}{12}} \right) M_C = \tfrac{4}{9} M_C$$

Joint E, $$M_{EC} = \left(\frac{\frac{2I}{12}}{\frac{I}{8} + \frac{2I}{12}} \right) M_E = \tfrac{4}{7} M_E$$

$$M_{EF} = \left(\frac{\frac{I}{8}}{\frac{I}{8} + \frac{2I}{12}} \right) M_E = \tfrac{3}{7} M_E$$

The distribution with the structure held against side sway, using initital moments multiplied by 10, is shown in Table 9.9 (*a*) on page 200.

From this distribution

$$\Sigma M_B{}^T = -1051 - 182 + 683 + 1366 - 813 - 1625 = -1622$$

But from the equilibrium equation

$$\Sigma M_B{}^T = -3840$$

Hence sway correction $= -2218$.

The distribution of an arbitary set of moments initially in the I/h^2 ratio of the stanchions is shown in Table 9.9 (*b*).

From this distribution $\Sigma M_B{}^T = -2498$.

Sway correction moments must be altered in ratio $\dfrac{2218}{2498}$ and added to non-sway moments to give the final moments in kN m as shown on page 202.

The deflected form of the structure is shown in Fig. 9.23.

TABLE 9.9 (*a*)

B			C			E
$\frac{3}{4}$	$\frac{1}{4}$	$\frac{2}{9}$	$\frac{3}{9}$	$\frac{4}{9}$	$\frac{4}{7}$	$\frac{3}{7}$
+ 640				− 2880	+ 2880	
	+ 320	+ 640	+ 960	+ 1280	+ 640	
− 720	− 240	− 120		− 1005	− 2011	− 1509
	+ 125	+ 250	+ 375	+ 500	+ 250	
− 94	− 31	− 15		− 72	− 143	− 107
	+ 10	+ 19	+ 29	+ 39	+ 20	
− 8	− 2	− 1		− 5	− 11	− 9
		+ 1	+ 2	+ 3		
− 182	+ 182	+ 774	+ 1366	− 2140	+ 1625	− 1625
A			D			F
− 640			+ 480			− 755
− 360			+ 187			− 53
− 47			+ 15			− 5
− 4			+ 1			
− 1051			+ 683			− 813

TABLE 9.9 (*b*)

B			C		E	
$\frac{3}{4}$	$\frac{1}{4}$	$\frac{2}{9}$	$\frac{3}{9}$	$\frac{4}{9}$	$\frac{4}{7}$	$\frac{3}{7}$
− 1000			− 500			− 500
+ 750	+ 250	+ 125		+ 143	+ 286	+ 214
	+ 26	+ 52	+ 77	+ 103	+ 52	
− 20	− 6	− 3		− 15	− 30	− 22
− 2	+ 2	+ 4	+ 6	+ 8	+ 4	
					− 2	− 2
− 272	+ 272	+ 178	− 417	+ 239	+ 310	− 310
A			D			F
− 1000			− 500			− 500
+ 375			+ 38			+ 107
− 10			+ 3			− 11
− 1						− 1
− 636			− 459			− 404

	AB	BA	BC	CB	CD
Non-sway . .	− 105·1	− 18·2	+ 18·2	+ 77·4	+ 136·6
Sway Correction .	− 56·5	− 24·2	+ 24·2	+ 15·8	− 37·0
FINAL . . .	− 161·6	− 42·4	+ 42·4	+ 93·2	+ 99·6

	CE	DC	EC	EF	FE
Non-sway . .	− 214·0	+ 68·3	+ 162·5	− 162·5	− 81·3
Sway Correction .	+ 21·3	− 40·7	+ 27·5	− 27·5	− 36·8
FINAL . . .	− 192·7	+ 27·6	+ 190·0	− 190·0	− 118·1

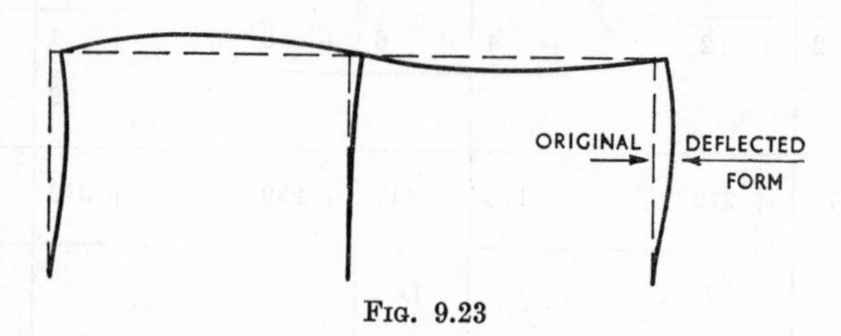

FIG. 9.23

Example 9.10. The culvert shown in Fig. 9.24 is of constant section throughout and the top beam is subjected to a central load of W. Assuming that this produces a uniformly distributed

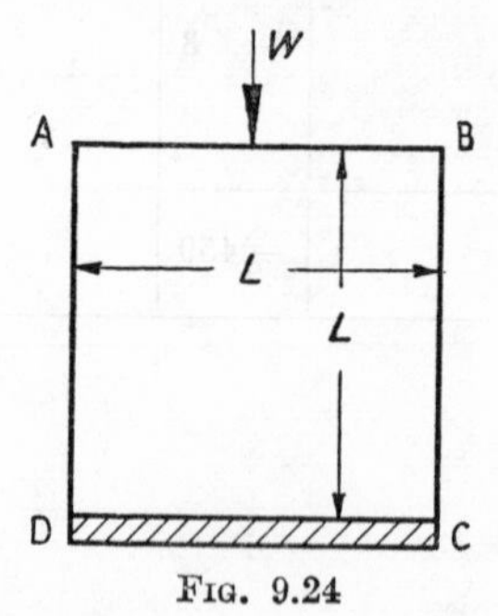

FIG. 9.24

reactive force under the base, determine the moments at the corners of the culvert and draw the bending moment diagram.

(*St. Andrews*)

Solution. This type of problem can be dealt with in a straightforward manner by moment distribution provided it is remembered that the frame is continuous. This means that balancing moments at one joint must have a carry-over to adjacent joints and that each joint must balance.
The fixed end moments are

$$M_{AB} = M_{BA} = \frac{WL}{8} = 1500 \text{ (say)}$$

$$M_{DC} = M_{CD} = \frac{WL}{12} = 1000$$

This out-of-balance at each joint is obviously distributed in the ratio of $\frac{1}{2} : \frac{1}{2}$. The balance is carried out as shown in Table 9.10 (*a*).

TABLE 9.10 (*a*)

D			A		B		C
½	½	½	½	½	½	½	½
+ 1000			− 1500	+ 1500			− 1000
− 500	− 500	+ 750	+ 750	− 750	− 750	+ 500	+ 500
+ 250	+ 375	− 250	− 375	+ 375	+ 250	− 375	− 250
− 312	− 312	+ 312	+ 312	− 312	− 312	+ 312	+ 312
+ 156	+ 156	− 156	− 156	+ 156	+ 156	− 156	− 156
− 156	− 156	+ 156	+ 156	− 156	− 156	+ 156	+ 156
+ 437	− 437	+ 812	− 812	+ 812	− 812	+ 437	− 437

The balancing has not been carried out to completion since it should be clear that any further balancing will only consist of adding equal positive and negative values to either side of the joint. These cannot affect the final summation. The joint moments are therefore

$$M_A = M_B = 0{\cdot}0677WL$$

$$M_C = M_D = 0{\cdot}0364WL$$

The bending moment diagram is given in Fig. 9.25.

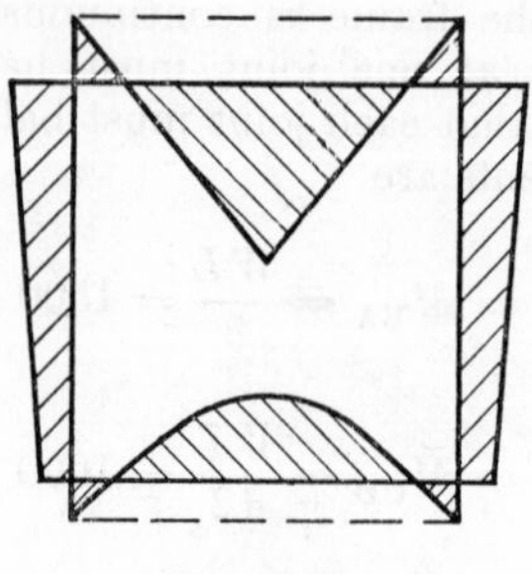

FIG. 9.25

EXAMPLES FOR PRACTICE ON PORTALS

1. The structure in Fig. 9.26 is hinged to the ground at E and fixed at A. The joint at D is rigid whilst B and C are pins. Under

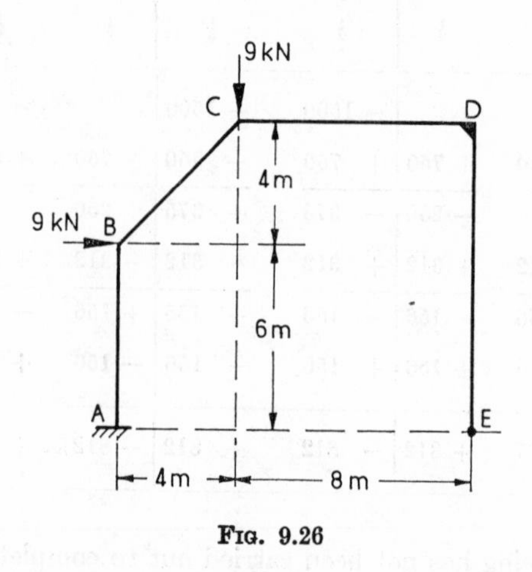

FIG. 9.26

the given loading determine the reactions at A and E, the bending moment at the rigid joint D, and draw the shearing force, bending moment and thrust diagrams for the member AB. *(Aberdeen)*

Answer. $H_A = 5$, $V_A = 4$; $M_A = 30$ kN m; $H_E = 4$, $V_E = 5$; $M_D = +40$ kN m.

2. A three-hinged portal is hinged to the springings at A and B and at the crown C (Fig. 9.27). The portal is subjected to a

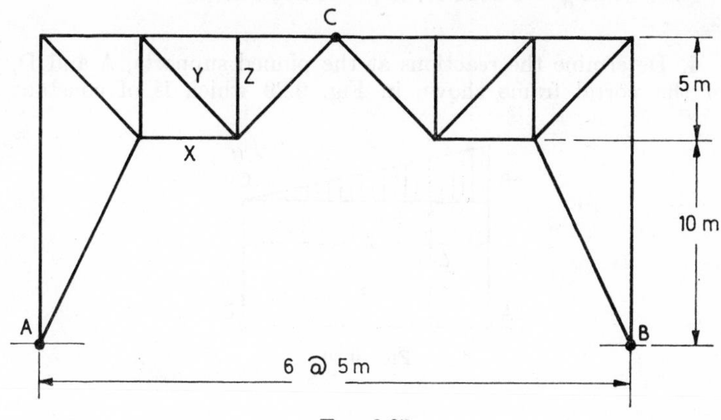

FIG. 9.27

uniform dead load of 25 kN/m covering the span and a live load longer than the span of 40 kN/m. Determine the range of force in each of the members X, Y and Z. *(Aberdeen)*

Answer.

Member	X	Y	Z
Max.	− 700	+ 719	− 325
Min.	− 200	+ 237	− 125

3. Draw the bending moment diagram for the portal frame shown in Fig. 9.28 which is pinned to the supports, A and D, and

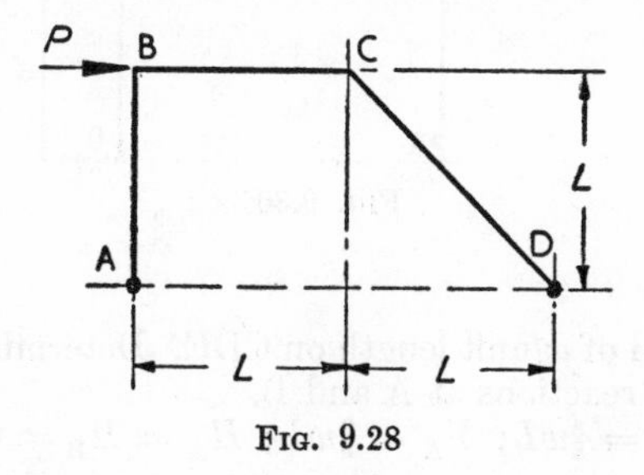

FIG. 9.28

carries a horizontal load P at B. The flexural rigidity of CD is twice that of AB and BC.

Answer. $M_B = 0{\cdot}234PL$; $M_C = 0{\cdot}266PL$.

4. Determine the reactions at the pinned supports, A and D, of the portal frame shown in Fig. 9.29 which is of constant

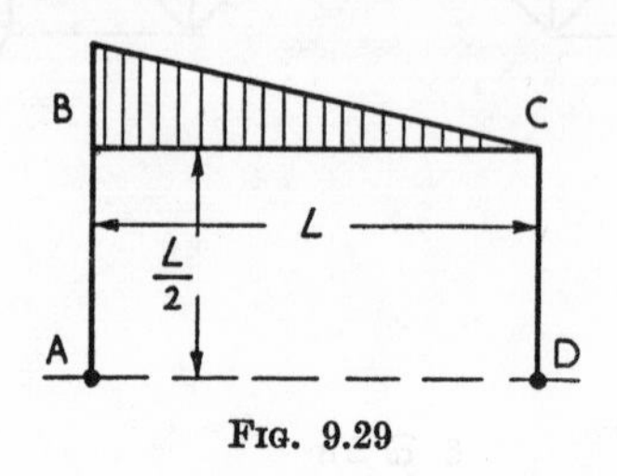

Fig. 9.29

flexural rigidity throughout and carries on BC a load which varies uniformly in intensity from wL at B to zero at C.

(*Oxford*)

Answer. $V_A = \dfrac{wL^2}{3}$; $H_A = \dfrac{wL^2}{16}$; $V_B = \dfrac{wL^2}{6}$; $H_D = \dfrac{wL^2}{16}$.

5. The portal frame shown in Fig. 9.30 is of constant section throughout. It is pinned at A and B and carries a uniformly

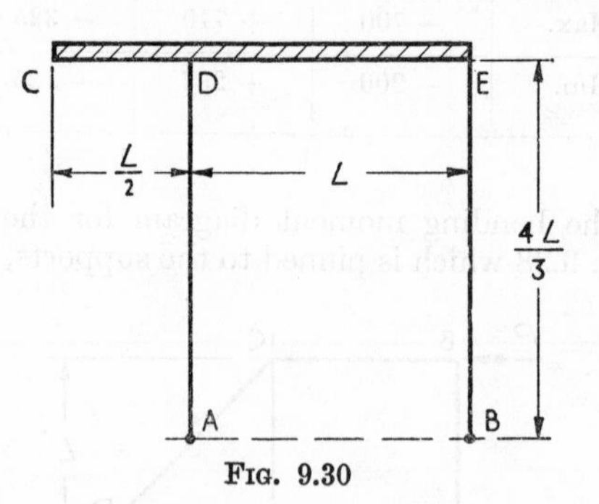

Fig. 9.30

distributed load of w/unit length on CDE. Determine the vertical and horizontal reactions at A and B.

Answer. $V_B = \frac{3}{8}wL$; $V_A = \frac{9}{8}wL$; $H_A = H_B = 0{\cdot}008\,25wL$.

6. Calculate the moments at the joints and draw the bending moment diagram for the frame shown in Fig. 9.31 which carries a

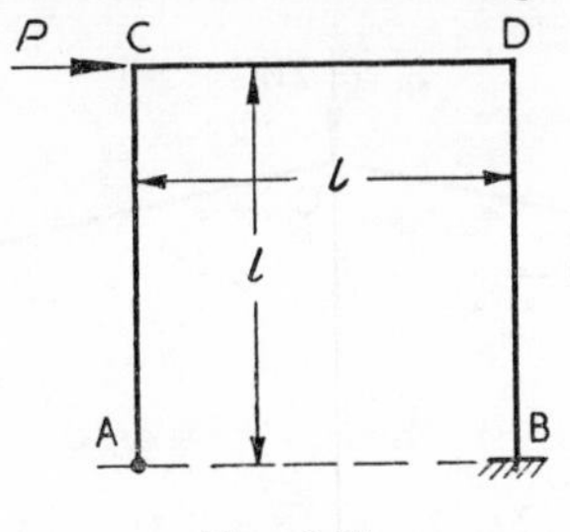

FIG. 9.31

horizontal load, P, as shown. The flexural rigidity is constant throughout, the support A is pinned and B is fully fixed.

(*St. Andrews*)

Answer. $M_B = 0{\cdot}454Pl$; $M_C = 0{\cdot}227Pl$; $M_D = 0{\cdot}319Pl$.

7. Calculate the horizontal thrust at the supports for the frame shown in Fig. 9.32 which consists of a quadrant AC of radius R and a vertical member CB of length R.

The member, BC, is subjected to a uniformly distributed load of w/unit length, the supports A and B are pin-jointed and C is a stiff joint. The flexural rigidity is constant throughout.

Answer. $H = 0{\cdot}41wR^2$.

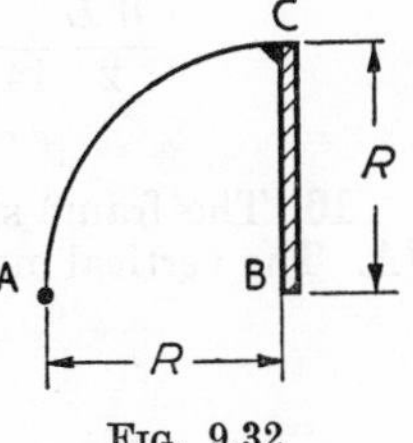

FIG. 9.32

8. If the joint at B of the frame shown in Fig. 9.33 remains rigid up to a moment of 100 kN m, what is the maximum value

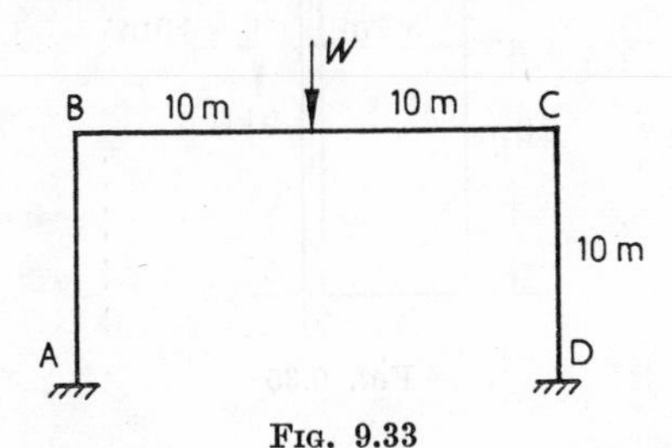

FIG. 9.33

of W in order that the frame shall behave as a stiff-jointed frame? EI is constant throughout.

Answer. 50 kN.

9. The structure shown (Fig. 9.34) is firmly fixed at the feet A and B of the stanchions and there are pins at E and F. The

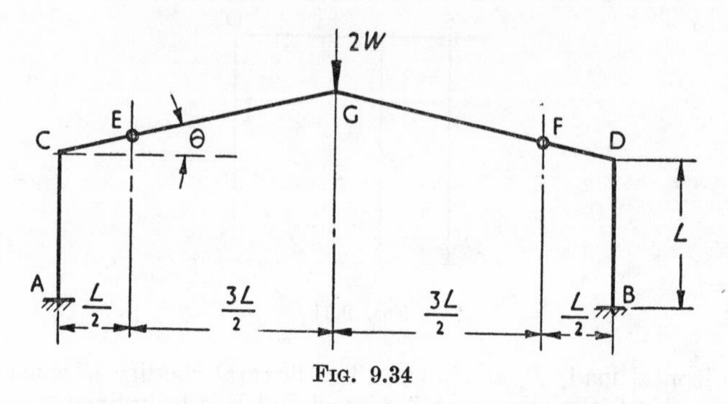

Fig. 9.34

flexural rigidity of the members is the same throughout. Show that the bending moment at the feet of the stanchions when a load $2W$ is carried at G is

$$\frac{WL}{2}\,\frac{28\tan\theta + 2\cos\theta + 3\sin\theta}{14\tan^2\theta + \cos\theta + 3\sec\theta + 6\sin\theta} \qquad \textit{(London)}$$

10. The frame shown in Fig. 9.35 supports a load of 9 kN at A. The vertical member is built-in at C and D and the horizontal

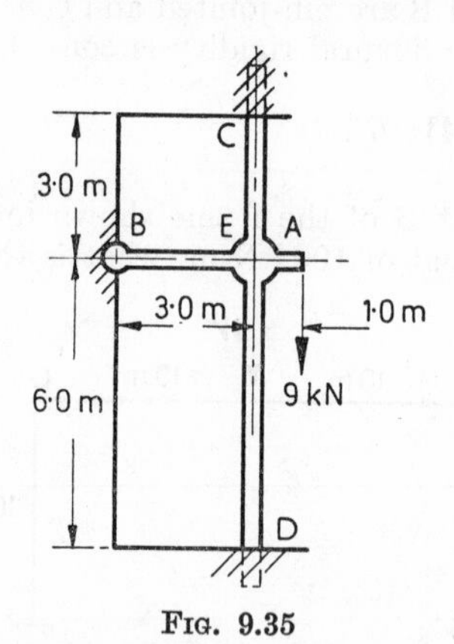

Fig. 9.35

member is pin-jointed at B. The members are of the same cross-section and are rigidly connected where they cross at E. Find the fixing moments at C and D. *(Oxford)*

Answer. $M_C = 2{\cdot}0$ kN m, $M_D = 1{\cdot}0$ kN m.

11. The stiff-jointed frame shown in Fig. 9.36 is hinged to the supports, A and B. The members are proportioned so that the

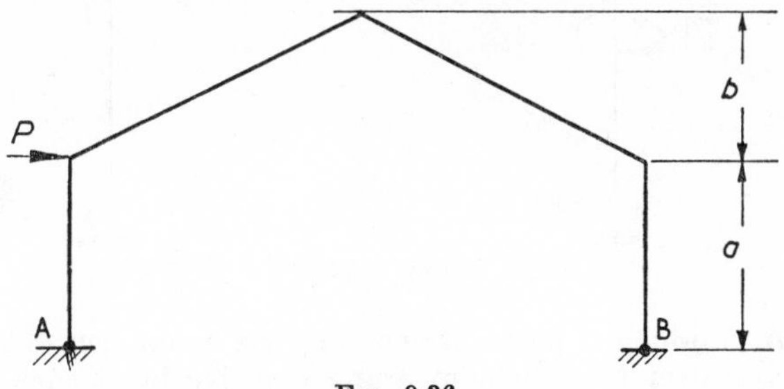

FIG. 9.36

ratio of the relevant moment of inertia to the length is the same for all members. Considering deflexions due to bending only and neglecting the bending moment due to axial forces, show that the force, P, causes a horizontal thrust at A which is independent of the span and is equal to

$$\frac{8a^2 + 9ab + 4b^2}{12a^2 + 12ab + 4b^2} P \qquad (Cambridge)$$

12. A main girder spans three 10 m bays and carries 48 kN/m loading (Fig. 9.37). Sketch the bending moment diagram for the

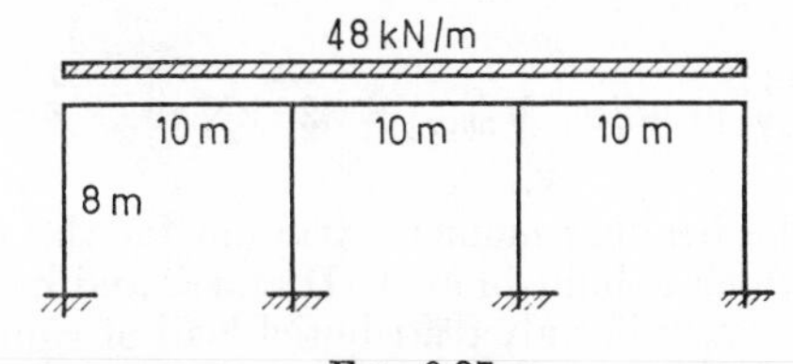

FIG. 9.37

girder and state the maximum value. All joints are considered rigid, M.I. of beam is $12{\cdot}5 \times 10^8$ mm^4 and of each column $5{\cdot}0 \times 10^8$ mm^4. *(Glasgow)*

Answer. $M_{\text{max.}} = 473$ kN m.

13. A portal frame (Fig. 9.38) is rigidly built-in at A and pinned at B and all other joints are rigid. Find by moment distribution or otherwise the bending moment diagram for the frame and the reactions at A and B. $I_{AC} = I_{BD} = \frac{1}{2}I_{CD}$. *(Glasgow)*

Answer. $H_A = 2{\cdot}0$, $H_B = 118{\cdot}0$; $V_A = 423{\cdot}9$, $V_B = 476{\cdot}1$ (all in kN).

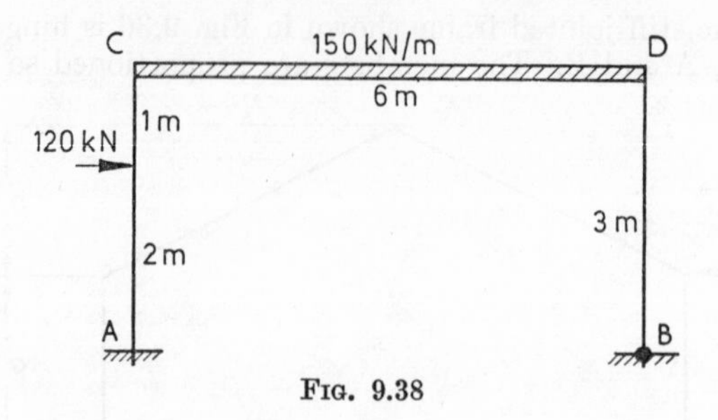

FIG. 9.38

14. A two-hinged portal frame supports a uniform dead load of 5 kN/m over the whole span and two live loads each 30 kN as shown in Fig. 9.39. The support D is adjusted so that the

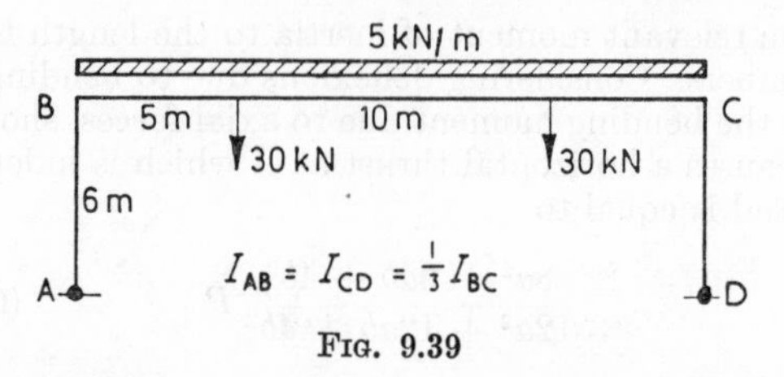

FIG. 9.39

bending moment at mid-span due to the dead load only is zero. Obtain the necessary adjustment and find the position and magnitude of the maximum bending moment in the structure due to all loads. *(Glasgow)*

Answer. $\dfrac{9333}{EI}$ m units, $M_{\text{max.}} = 320$ kN m.

15. Draw the bending moment diagram for the frame shown in Fig. 9.40, which is built-in at A, D and C and has a stiff joint at B. It carries a uniformly distributed load of w/unit length on BC and is of uniform section throughout.

Answer. $M_A = 0{\cdot}187wL^2$; $M_D = 0{\cdot}125wL^2$; $M_C = 0{\cdot}812wL^2$.

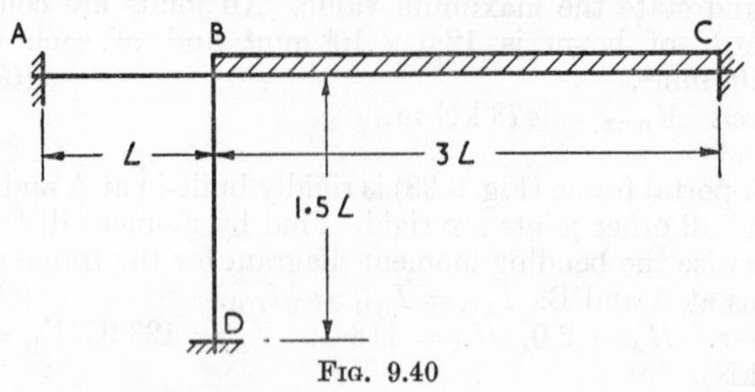

FIG. 9.40

16. The frame shown in Fig. 9.41 is of constant flexural rigidity throughout and carries a uniformly distributed pressure

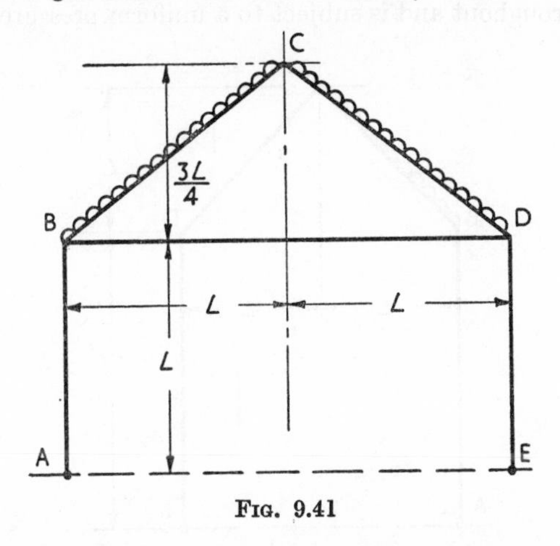

FIG. 9.41

loading on the sloping members. The eaves are tied so that under load there is no horizontal movement of B and D. Calculate the moments at B, C and D and the horizontal force at B and D. (*Glasgow*)

Answer. $M_D = M_B = 0{\cdot}063wL^2$; $M_C = 0{\cdot}163wL^2$; $R = 0{\cdot}37wL$.

17. The portal frame, ABCD, as shown in Fig. 9.42 is of constant section throughout. A crane girder applies loads of

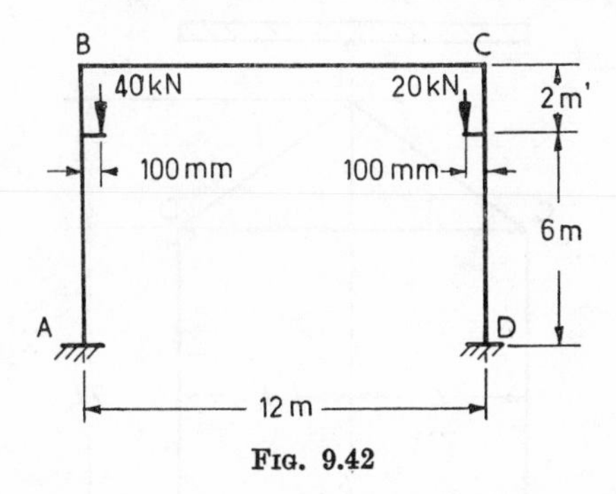

FIG. 9.42

40 kN and 20 kN to AB and CD respectively, each load being applied at 100 mm from the member. Calculate the fixing moments at A and D. (*Glasgow*)

Answer. $M_A = 0{\cdot}74$ kN m; $M_D = -1{\cdot}55$ kN m

18. Calculate the bending moments at B, C and D in the stiff-jointed frame shown in Fig. 9.43 which is of constant section throughout and is subject to a uniform pressure of w/unit

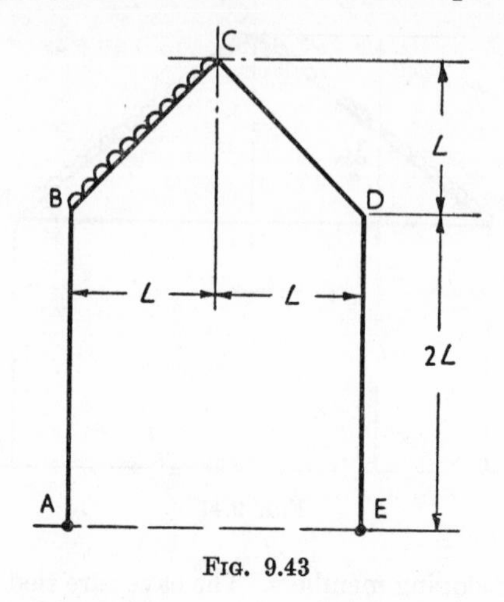

FIG. 9.43

length normal to BC. Sketch the bending moment diagram for the frame. EI is constant throughout.

Answer. $M_B = 0{\cdot}95wL^2$; $M_D = 1{\cdot}05wL^2$; $M_C = 0{\cdot}075wL^2$.

19. The structure shown in Fig. 9.44 consists of a stiff-jointed frame, ACEDB, of constant flexural rigidity EI with a tie CD

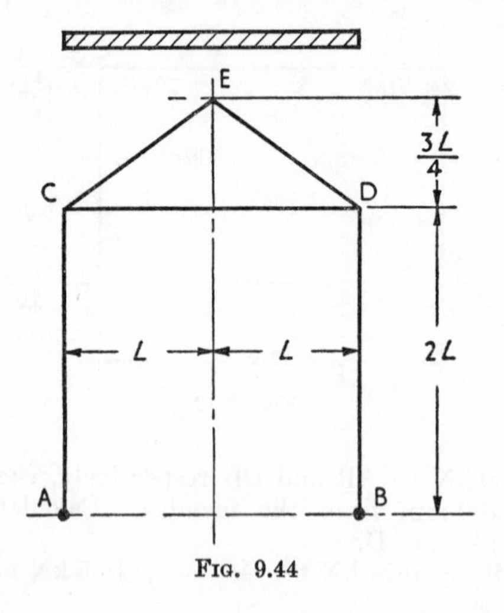

FIG. 9.44

whose extensibility $= \dfrac{48EI}{L^2}$. Calculate the load in the tie and the horizontal thrust on the pins A and B when it carries a uniformly distributed load as shown. (*St. Andrews*)

Answer. Load in tie $= 0{\cdot}645\,wL$; $H = 0{\cdot}027wL$.

20. Draw the bending moment diagram for the frame shown in Fig. 9.45 indicating the critical values. The flexural rigidity of AB and BC is twice that of CD. (*Glasgow*)

Answer. $M_A = 0{\cdot}072WL$; $M_B = 0{\cdot}127WL$; $M_C = 0{\cdot}081WL$; $M_D = 0{\cdot}019WL$.

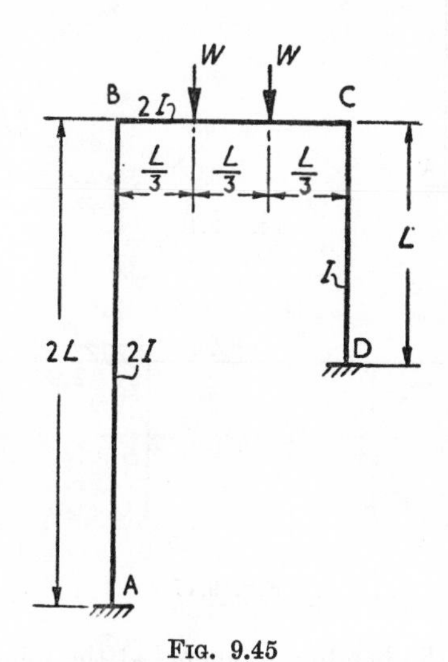

FIG. 9.45

21. Calculate the moments at the joints and draw the bending moment diagram for the frame shown in Fig. 9.46 which is of constant section throughout.

Answer. $M_A = 78$ kN m; $M_B = 113{\cdot}5$ kN m; $M_C = 19{\cdot}6$ kN m; $M_D = 69{\cdot}1$ kN m.

22. The portal frame shown in Fig. 9.47 has rigid joints and is built-in at A, B and C. It carries a concentrated load, W, at the mid-point of DE and a uniformly distributed load of intensity w on EF. If the flexural rigidities are as shown, determine the relationship between W and w in order that EB shall be axially loaded. (*Oxford*)

Answer. $W = \dfrac{wL}{3}$.

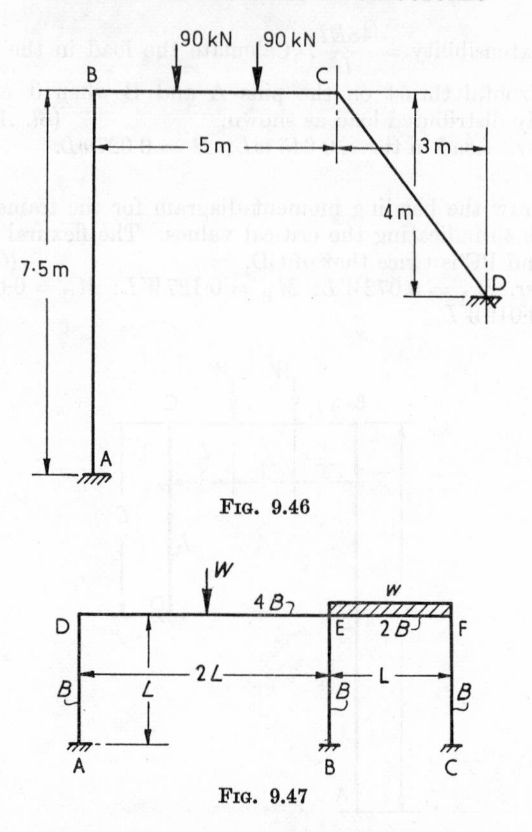

Fig. 9.46

Fig. 9.47

23. Calculate the bending moments at the joints and draw the bending moment diagram for the portal frame shown in Fig. 9.48 which carries the loads shown, if there is a point of contraflexure

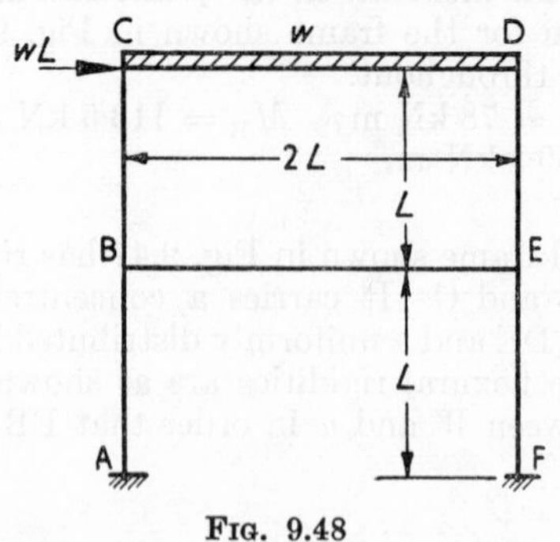

Fig. 9.48

at mid-height in each of the stanchions, BC and DE. The flexural rigidity of the members is constant throughout and the feet of the stanchions are built-in. (*St. Andrews*)

Answer. $M_C = 0{\cdot}036wL^2$; $M_D = 0{\cdot}536wL^2$; $M_A = 0{\cdot}490wL^2$; $M_F = 0{\cdot}262wL^2$.

24. Draw the bending moment diagram for the frame shown in Fig. 9.49. The members are all of equal flexural rigidity and the

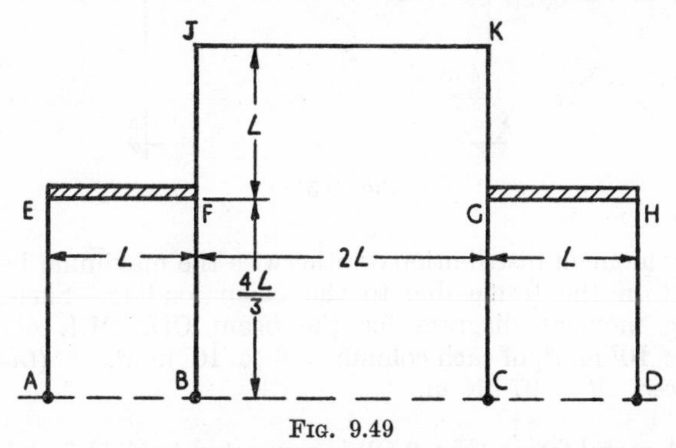

Fig. 9.49

stanchions are pinned at their bases. The loading is a uniformly distributed load of w/unit length of EF and GH. (*St. Andrews*)

Answer. $M_E = M_H = 0{\cdot}0310\,wL^2$; $M_{FE} = 0{\cdot}0650wL^2$; $M_{FB} = 0{\cdot}0477wL^2$; $M_{FJ} = 0{\cdot}0173wL^2$; $M_{JF} = 0{\cdot}0045wL^2$.

25. In the two-storeyed frame of Fig. 9.50 each beam carries 7·5 kN/m. Draw the bending moment diagram for one column

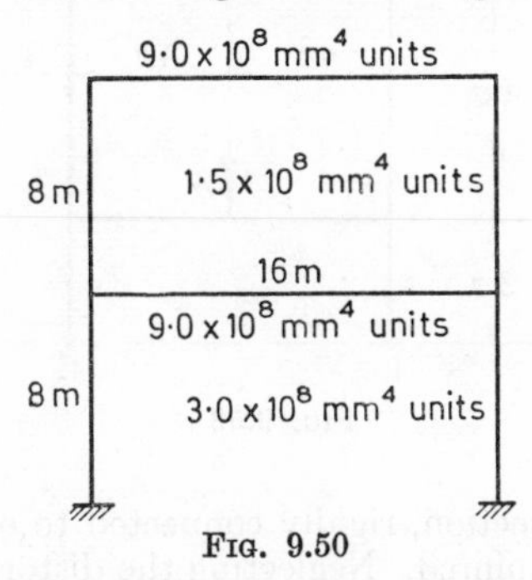

Fig. 9.50

and find the direct forces in the beams. All joints are rigid. The sectional moments of inertia are given in the figure. (*Glasgow*)

Answer. $M_A = 30$; $M_{BA} = 59$; $M_{BC} = 59$; $M_{CB} = 73$ (all in kN m). Upper beam, 16·4 kN; Lower beam, 5·10 kN.

26. A portal frame has rigid joints and is rigidly fixed at the base. A restraint at C as shown in Fig. 9.51 prevents side sway.

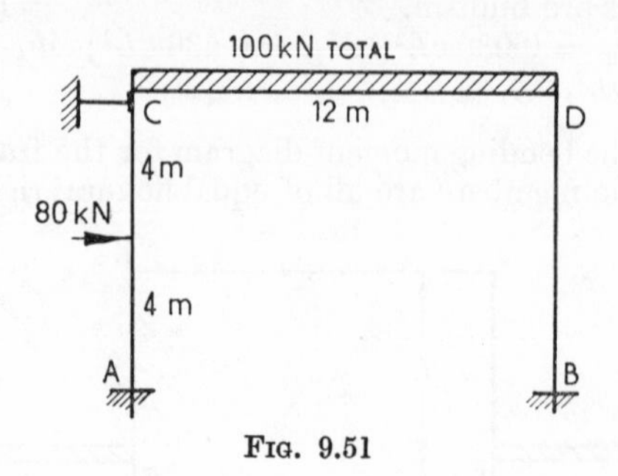

FIG. 9.51

Find by moment distribution or otherwise the maximum bending moment in the frame due to the given loading. Sketch the bending moment diagram for the beam CD. M.I. of beam $= 32 \times 10^7$ mm^4, of each column $= 8 \times 10^7$ mm^4. (*Glasgow*)

Answer. $M = 97$ kN m.

27. A metal frame (Fig. 9.52) is connected to rigid foundations by pin connexions at B and C. The frame is composed of members

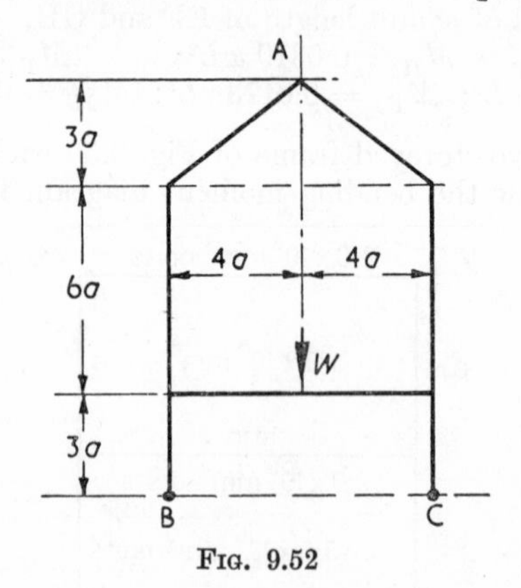

FIG. 9.52

of identical cross-section, rigidly connected to each other except at A which is pin-jointed. Neglecting the distortion of the structure due to direct stress and shear, show that the horizontal thrust at B due to a test load W as shown is about $0{\cdot}21W$. (*Cambridge*)

Answer. Given in question.

28. Fig. 9.53 shows a framework consisting of two vertical stanchions, AD and BC, fixed vertically at their lower ends,

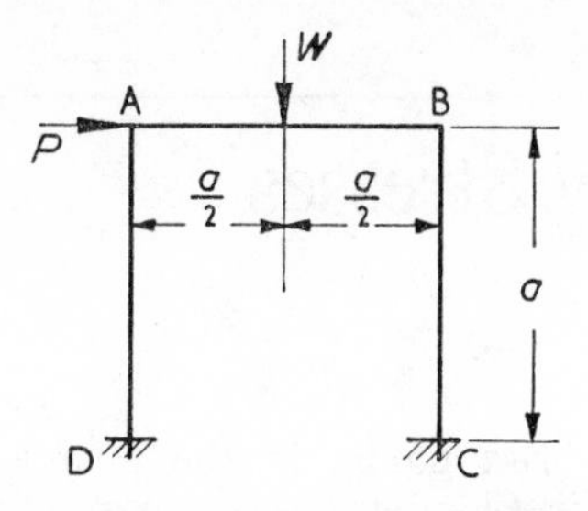

FIG. 9.53

their upper ends being riveted to a cross-beam, AB. The three members are of the same section and equal lengths a. If the framework is loaded in the manner indicated, show that for the member AB the terminal bending moments are

$$\frac{Wa}{12} \pm \frac{3}{14} Pa$$

and that the thrust in the member is

$$\frac{W}{8} \pm \frac{P}{2}$$

(*Cambridge*)

Chapter 10

Secondary Stresses

In determining the loads in the members of the various frames analysed in Chapter 1 the joints between the members have in all cases been assumed to be pinned. Whilst in the early days of truss construction pinned connexions of one type or another were actually used, the members have been for many years now riveted together—and in more recent years welded—with the result that the joints between the members are much closer in practice to the rigid joint, which occurred in the frames analysed in the previous chapter, than to the pinned joint which was assumed in Chapter 1.

Thus if a true picture is required of the forces in the truss shown in Fig. 10.1 it must be analysed not as a pin-jointed

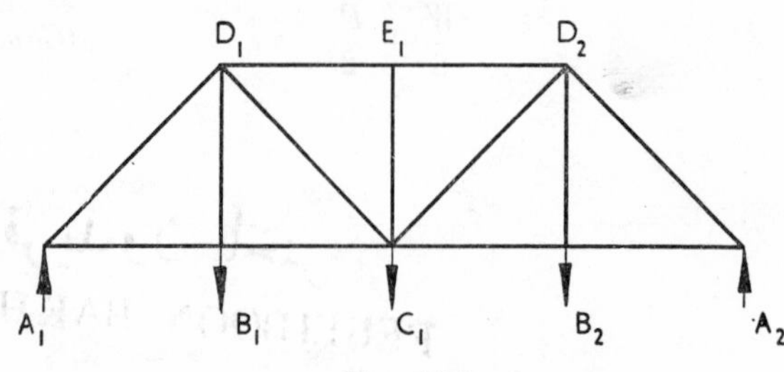

Fig. 10.1

structure as in Chapter 1 but as a rigid-jointed frame using the methods of Chapter 9. An accurate analysis will give two sets of forces—(i) direct compressions and tensions in the members, and (ii) moments in each member due to the stiffness of the joints. Such an analysis is a very lengthy process.

Where these accurate analyses have been carried out, however, the results show that the first set of forces is practically the same as that obtained by the analysis assuming pinned joints. If this is so, then the deflected form of the truss under load is for all practical purposes similar to that obtained for a pinned-jointed truss. The deflected form of such a truss can be drawn by means of a Williot-Mohr diagram. From this diagram the relative deflexions of any two panel points say B_1 and C_1 can be found. But any deflexion of C_1 relative to B_1 will cause end moments at both B_1 and C_1. If δ_B and δ_C are the deflexions measured perpendicular to B_1C_1 of points B_1 and C_1 respectively, then

$$M_{BC} = M_{CB} = -\frac{6EI_{BC}(\delta_C - \delta_B)}{L^2_{BC}}$$

where EI_{BC} = flexural rigidity of BC

L_{BC} = length of BC

Similar moments will occur at the ends of any member due to a movement, measured perpendicular to the length of the member, of these ends relative to one another. Thus from the deflexion diagram a series of moments at the ends of the members is obtained. The laws of static equilibrium, however, demand that each joint must be balanced. If the moments obtained from the deflexions do not give a balanced joint, any out-of-balance moment must be divided between the members in the ratio of their stiffnesses and the joint balanced in the usual way, moments being also carried over to the other end of a member as in the standard moment-distribution method. This process is continued until all the joints are balanced thus giving the final analysis of the structure.

Even with this simplification the problem is not a short one and most examples that the student meets deal with very simple trusses.

In the primary analysis with pin joints the weight of the member is neglected. This weight will also cause a secondary stress. Its effect can be allowed for by adding moments of $\frac{\pm wL^2}{12}$ to the ends of the member before proceeding with the balancing of the joints.

A further cause of secondary stress is the non-concurrency of the forces at a joint. Consider for example a support of a roof truss as shown in Fig. 10.2. The actual line of the support is A but the members meet at B, a distance d from A. The theoretical joint is therefore subjected to an anti-clockwise moment $R_A \times d$. When balancing up the moments meeting at the joint there must therefore be a residual moment of $- R_A \times d$.

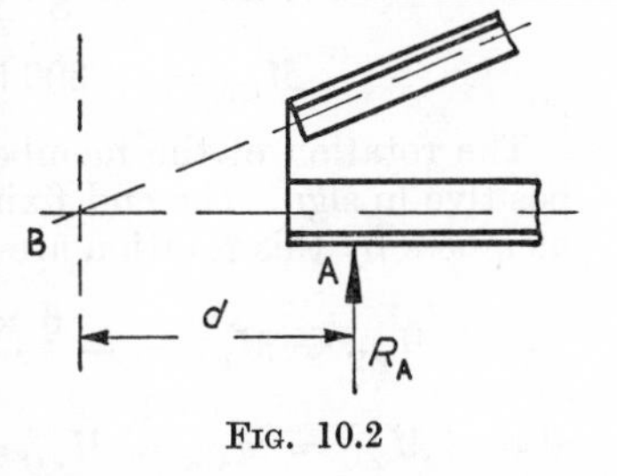

FIG. 10.2

Examples on Secondary Stresses

Example 10.1. The framework in Fig. 10.3 carries the loads as shown. The rotation of the members as found from a Williot–Mohr diagram are each $+\frac{1}{1000}$ radians. The $\frac{I}{L}$ values are

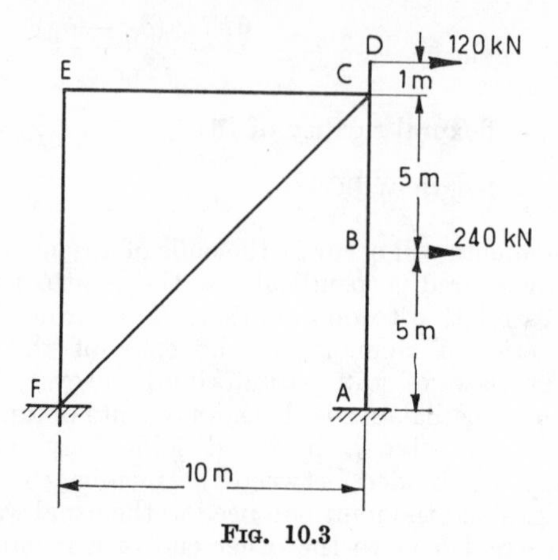

FIG. 10.3

$240 \times 10^3\ \text{mm}^3$ units for CE, CF and FE, and $480 \times 10^3\ \text{mm}^3$ units for AC. Determine the maximum bending moment in the member ABCD. The joints are all rigid and $E = 200\ \text{kN/mm}^2$.

(*Aberdeen*)

Solution. This example will be solved by the moment distribution method. Due to the external loading the fixed end moments are

$$M_{CD} = 120 \times 1 = -120\ \text{kN m}$$

$$M_{AC} = \frac{240 \times 10}{8} = -300\ \text{kN m}$$

$$M_{CA} = +300\ \text{kN m}$$

The rotation of the members is taken as clockwise since it is positive in sign. The end fixing moments induced in the various members by this rotation are

$$M_{EF} = M_{FE} = -\frac{6 \times 200 \times 240}{1000} = -288\ \text{kN m}$$

also $$M_{EC} = M_{CE} = M_{FC} = M_{CF} = -288\ \text{kN m}$$

$$M_{AC} = M_{CA} = -\frac{6 \times 200 \times 480}{1000} = -576\ \text{kN m}$$

The distribution of the out-of-balance moment is as follows—

Joint E, $M_{EF} = M_{EC} = \frac{1}{2}M_E$

Joint C, $M_{CE} = \left(\dfrac{240}{240 + 240 + 480}\right) M_C = \dfrac{M_C}{4} = M_{CF}$

$$M_{CA} = \left(\frac{480}{240 + 240 + 480}\right) = \frac{M_C}{2}$$

The moments are balanced as in Table 10.1 where the initial moments due to loading and rotation have been kept separate.

TABLE 10.1

E		C			A
$\frac{1}{2}$	$\frac{1}{2}$	$\frac{1}{4}$	$\frac{1}{4}$	$\frac{1}{2}$	
			− 120		
− 288	− 288	− 288	− 288	− 576	− 576
+ 288	+ 288	+ 144		+ 300	− 300
	+ 104	+ 207	+ 207	+ 414	+ 207
− 52	− 52	− 26			
	+ 3	+ 6	+ 7	+ 13	+ 7
− 1	− 2	− 1		+ 1	
− 53	+ 53	+ 42	− 74	+ 152	− 662

The bending moment diagram for the member ABCD is shown in Fig. 10.4 from which it is clear that the maximum moment is at A and has a value of 662 kN m.

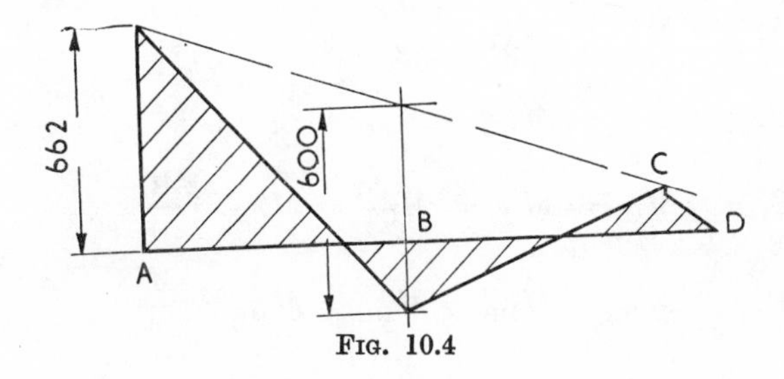

FIG. 10.4

Example 10.2. The Warren girder shown in Fig. 10.5 has a span of 40 m and is loaded uniformly. The rotations of the

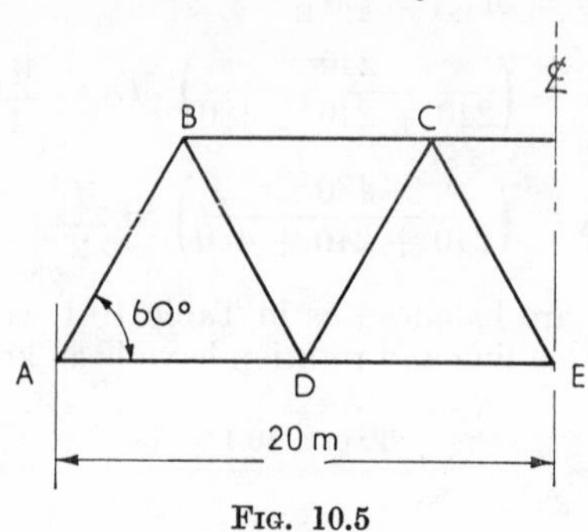

FIG. 10.5

members in the left-half of the girder are $\frac{1}{1000}$ clockwise. Allowing for the dead weight of the chords as 13 kN/m find the secondary moments in the members assuming rigid joints. The length and moment of inertia of the members is constant at 10 m and 5×10^9 mm^4 units respectively. Take $E = 200$ kN/mm^2.

(*Glasgow*)

Solution. The fixed end moments due to dead load in the horizontal members AD, DE and BC are

$$13 \times 10^2 \times \frac{1}{12} = 108 \text{ kN m}$$

Those in the sloping members are 54 kN m since the horizontal span is halved.

The fixed end moments due to the clockwise rotation of each member are

$$\frac{6 \times 200 \times 5 \times 10^9}{1000 \times 10 \times 1000} = 600 \text{ kN m}$$

The distribution of out-of-balance moment is as follows—

Joint A, $\quad M_{AB} = M_{AD} = \frac{M_A}{2}$

Joint B, $\quad M_{BA} = M_{BD} = M_{BC} = \frac{M_B}{3}$

Joint C, $\quad M_{CD} = M_{CE} = M_{CC'} = M_{CE} = \frac{M_C}{4}$

Joint D, $\quad M_{DA} = M_{DB} = M_{DC} = M_{DE} = \frac{M_D}{4}$

The balancing up of the joints is shown in Table 10.2. This is set out with the upper chord points above and the lower chord points below. The student must, however, remember to carry over the moments from upper to lower chord and vice versa. It is possible in dealing with the secondary stress problem that it is advantageous to write down the moments at the sides of the joint. The first few steps of this solution has been given in Fig. 10.6 and the student can adopt whichever method appeals more to him.

TABLE 10.2

B			C			
A ⅓	D ⅓	C ⅓	B ¼	D ¼	E ¼	C′ ¼
− 546 + 636	− 654 + 636	− 708 + 636	− 492 + 450	− 546 + 450	− 654 + 450	− 108 + 450
+ 340 − 288	+ 300 − 288	+ 225 − 289	+ 318 − 98	+ 300 − 98	0 − 98	− 225 − 99
− 155 + 105	− 110 + 105	− 49 + 104	− 145 + 52	− 110 + 53	0 + 53	+ 45 + 52
+ 63 − 44	+ 43 − 44	+ 26 − 44	+ 52 − 17	+ 43 − 17	0 − 17	− 26 − 18
− 24 + 17	− 18 + 17	− 9 + 17	− 22 + 8	− 18 + 8	0 + 8	+ 9 + 7
+ 10 − 7	+ 7 − 7	+ 4 − 7	+ 8 − 3	+ 7 − 3	0 − 3	− 4 − 2
− 4 + 3	− 3 + 3	− 1 + 2	− 3 + 1	− 3 + 1	0 + 1	+ 1 + 2
+ 106	− 13	− 93	+ 109	+ 67	− 260	+ 84

A		D				E	
B ½	D ½	A ¼	B ¼	C ¼	¼ E	D	C
− 654 + 681	− 708 + 681	− 492 + 600	− 546 + 600	− 654 + 600	− 708 + 600	− 492 0	− 546 0
+ 318 − 309	+ 300 − 309	+ 340 − 220	+ 318 − 221	+ 225 − 221	0 − 221	+ 300 0	+ 225 0
− 144 + 127	− 110 + 127	− 155 + 87	− 144 + 87	− 49 + 87	0 + 87	− 110 0	− 49 0
+ 53 − 48	+ 43 − 48	+ 63 − 35	+ 53 − 36	+ 26 − 36	0 − 35	+ 43 0	+ 26 0
− 22 + 20	− 18 + 20	− 24 + 14	− 22 + 14	− 9 + 14	0 + 13	− 18 0	− 9 0
+ 8 − 8	+ 7 − 7	+ 10 − 5	+ 8 − 6	+ 4 − 6	0 − 5	+ 7 0	+ 4 0
− 3 + 3	− 3 + 3	− 4 + 2	− 3 + 2	− 1 + 2	0 + 2	− 3 0	− 1 0
+ 31	− 31	+ 181	+ 104	− 18	− 267	− 273	− 250

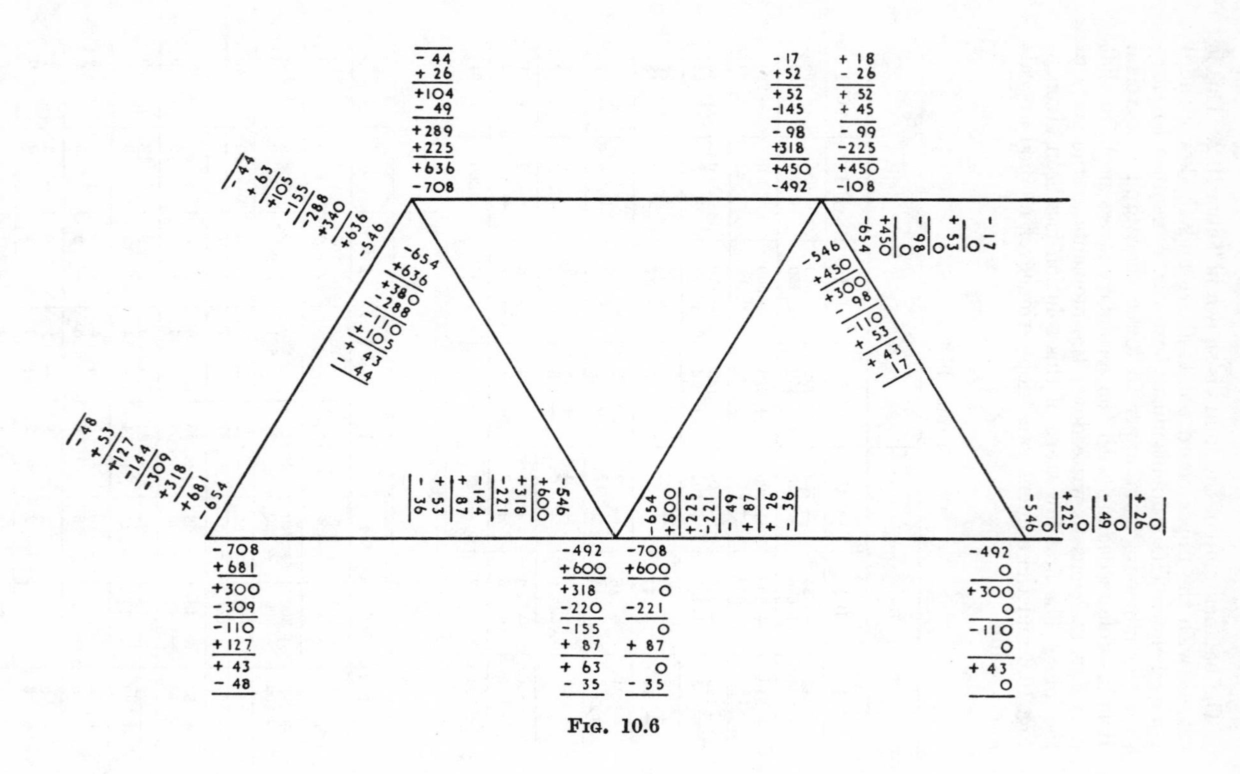

-44 +26 +104 -49 +289 +225 +636 -708
-44 +63 +105 -155 -288 +340 +636 -546
-654 +636 +380 -288 -110 +105 +43 -44
-48 +53 +127 -144 -309 +318 +681 -654
-708 +681 +300 -309 -110 +127 +43 -48
-546 +600 +318 -221 -144 +87 +53 -36
-492 +600 +318 -220 -155 +87 +63 -35
-708 +600 0 -221 0 +87 0 -35
-654 +600 +225 -221 -49 +87 +26 -36
-17 +52 +52 -145 -98 +318 +450 -492
+18 -26 +52 +45 -99 -225 +450 -108
-654 +450 0 -98 0 +53 0 -17
-546 +450 +300 -98 -110 +53 +43 -17
-546 0 +225 0 -49 0 +26 0
-492 0 +300 0 -110 0 +43 0

Fig. 10.6

Joint E will by symmetry always remain a balanced joint.

There will also be no rotation of the member CC′ due to symmetry, consequently the fixed end moments in this member are just those due to dead loading only, i.e. ± 108 kN m.

The balancing process is rather longer when dealing with secondary stresses than with the average stiff-jointed frame. In Table 10.2 the first heading in the column represents the joint under consideration, e.g. B, the letters on the line below give the members meeting at that joint together with a fraction indicating the division of the out-of-balance moment. The first line of figures gives the fixed end moments due to rotation and dead loading; the second gives the moments required to balance the joint. A line is then ruled across indicating that the joint is balanced. The next line of figures is the carry-over from previously inserted balancing moments. Particular attention should be paid to this line in the case of joint C, member CC′. The balancing moment for this member at joint C is + 450. By symmetry the balancing moment at the joint at the other end of the member will be − 450. It is a half of this latter moment which is transferred to the CC′ member at joint C. This explains the figure of − 225 in the table for this joint.

The final moments are summed in the table and have the following values.

Joint A, $M_{AB} = +31$ kN m
$M_{AD} = -31$ „

Joint B, $M_{BA} = +106$ „
$M_{BD} = -13$ „
$M_{BC} = -93$ „

Joint C, $M_{CB} = +109$ „
$M_{CD} = +67$ „
$M_{CE} = -260$ „
$M_{CC'} = +84$ „

Joint D, $M_{DA} = +181$ „
$M_{DB} = +104$ „
$M_{DC} = -18$ „
$M_{DE} = -267$ „

Joint E, $M_{ED} = -273$ „
$M_{EC} = -250$ „

Example 10.3. A trussed beam ABCD supports a vertical load of 100 kN at B. The clear span is 10 m between supports which have an eccentricity of 200 mm at the joints A and D (Fig. 10.7).

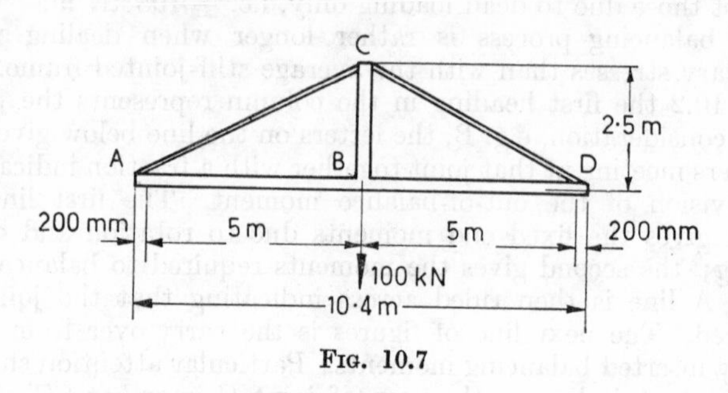

FIG. 10.7

The joints A, B and D are rigid and the joint at C is assumed to be pinned. Compare the maximum bending moments in the beam ABC for—

(i) eccentricity of 200 mm at supports,
(ii) assuming no eccentricity at supports.

The $E\delta/L$ values for the members are

AC and DC, 650 N/mm².

AB and BD, 800 N/mm².

$I_{ABD} = I_{BC} = 1{\cdot}5 \times 10^8$ mm⁴ units.

$I_{AC} = I_{CD} = 6{\cdot}0 \times 10^7$ mm⁴ units.

(*Glasgow*)

Solution. No data is given in the question with regard to the direction of the rotation but a reasonable assumption is that those to the left of the central member BC rotate in a clockwise direction whilst those to the right rotate anticlockwise. On this assumption the moments introduced at the joints are

$$M_{AB} = M_{BA} = -\frac{6EI\delta}{l^2} = -\frac{6 \times 1{\cdot}5 \times 10^8 \times 800}{5 \times 10^3 \times 10^6}$$

$$= -144 \text{ kN m}$$

$$M_{BD} = M_{DB} = +144 \text{ kN m}$$

$$M_{AC} = M_{CA} = -\frac{6 \times 6{\cdot}0 \times 10^7 \times 650}{\sqrt{5{\cdot}0^2 + 2{\cdot}5^2} \times 10^3 \times 10^6} = -41{\cdot}8 \text{ kN m}$$

$$M_{DC} = M_{CD} = +41{\cdot}8 \text{ kN m}$$

The reaction at A, R_A, is 50 kN, and the eccentricity is 0·2 m consequently the residual moment at A is

$$-50 \times 0{\cdot}2 = -10 \text{ kN m}$$

The distribution of the out-of-balance moment at the joints (C being taken as pinned) is as follows—

$$\text{Joint A,} \quad M_{AC} = \frac{\frac{3}{4} \times \frac{6{\cdot}0 \times 10^7}{5{\cdot}6 \times 10^3}}{\frac{3}{4} \times \frac{6{\cdot}0 \times 10^7}{5{\cdot}6 \times 10^3} + \frac{1{\cdot}5 \times 10^8}{5 \times 10^3}} = 0{\cdot}21 M_A$$

$$M_{AB} = \frac{\frac{1{\cdot}5 \times 10^8}{5 \times 10^3}}{\frac{3}{4} \times \frac{6{\cdot}0 \times 10^7}{5{\cdot}6 \times 10^3} + \frac{1{\cdot}5 \times 10^8}{5 \times 10^3}} = 0{\cdot}79 M_A$$

The distribution at D will be obtained in a similar manner.

$$M_{DB} = 0{\cdot}79 M_D \quad \text{and} \quad M_{CD} = 0{\cdot}21 M_D$$

If joints A and D are balanced simultaneously there will, by symmetry, never be an out-of-balance moment at B, hence there is no need to determine the distribution of out-of-balance moment at that joint.

Although the question asks for the eccentricity case to be dealt with first it is simpler to deal with the case of no eccentricity. The distribution for this is shown in Table 10.3 (*a*).

TABLE 10.3 (*a*)

C	A			B		D		C
	0·21	0·79				0·79	0·21	
−41·8	−41·8	−144·0	−144·0	0	+144·0	+144·0	+41·8	+41·8
+41·8	+20·9						−20·9	−41·8
	+34·6	+130·3	+65·2		− 65·2	−130·3	−34·6	
0	+13·7	− 13·7	− 78·8	0	+ 78·8	+ 13·7	−13·7	0

The eccentric loading involves residual moments of − 10 at A and + 10 at C. These are divided in the ratio of the stiffnesses of the members and moments carried over as shown in the first line of Table 10.3 (*b*). The second line gives the no-eccentricity moments obtained from Table 10.3 (*a*).

TABLE 10.3 (*b*)

D	A		B	C	D
0·21	0·79			0·79	0·21
− 2·1 + 13·7	− 7·9 − 13·7	− 4·0 − 78·8	+ 4·0 + 78·8	+ 7·9 + 13·7	+ 2·1 − 13·7
− 11·6	− 21·6	− 82·8	+ 82·8	+ 21·6	+ 11·6

The maximum moment will in each case occur at B and has the following values—

(i) with eccentricity at joints, 82·8 kN m
(ii) with no eccentricity, 78·8 kN m

EXAMPLES FOR PRACTICE ON SECONDARY STRESSES

1. The truss ABC shown in Fig. 10.8 has rigid joints and is tied to a wall by a member AD which is pinned at both ends. The beam BC carries a load of 100 kN at its mid-point E.

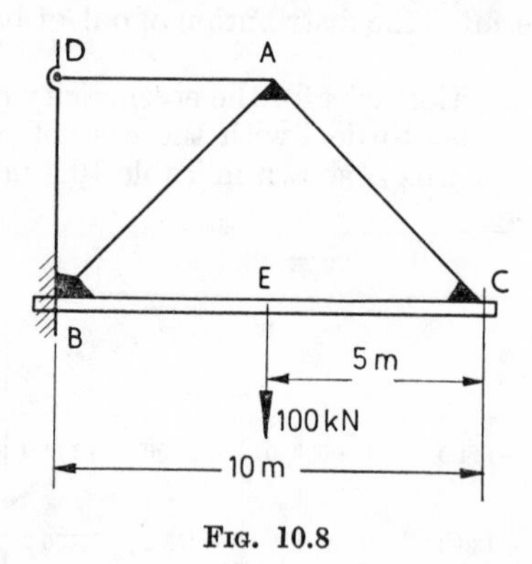

FIG. 10.8

The rotations of the members under this load are $\frac{10}{E}$, $\frac{20}{E}$ and $\frac{30}{E}$ for members AB, AC and BC respectively, all being in a clockwise direction and E in kN/mm² units. The I/L values for each member are $1{\cdot}8 \times 10^3$ mm³ units. Determine the bending moment under the load. (*Glasgow*)

Answer. 1355 kN m.

2. The horizontal and vertical components of the displacement of each joint of a Warren truss under symmetrical loading are given in the Table 10.4. The second moments of area, I in

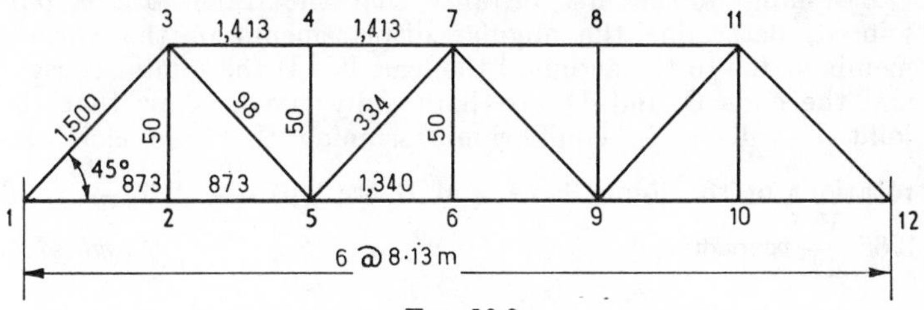

FIG. 10.9

$mm^4 \times 10^6$ units, of all the members are given in the diagram (Fig. 10.9).

Assuming the joints to be rigid, sketch the deflected form of the truss. Calculate, to the nearest kN m, the maximum bending moment caused by the joint rigidity in the member 1–3. $E = 200$ kN/mm².

TABLE 10.4

Joint	$x + \rightarrow$ mm $\times 10^{-1}$	$y + \downarrow$ mm $\times 10^{-1}$
1	0	0
2	0·19	3·02
3	1·04	2·74
4	0·84	4·69
5	0·38	4·69
6	0·58	5·59
7	0·58	5·29

(*London*)

Answer. 254 kN mm

3. In the cantilever framework illustrated by Fig. 10.10 the points of attachment C and D are fixed in position.

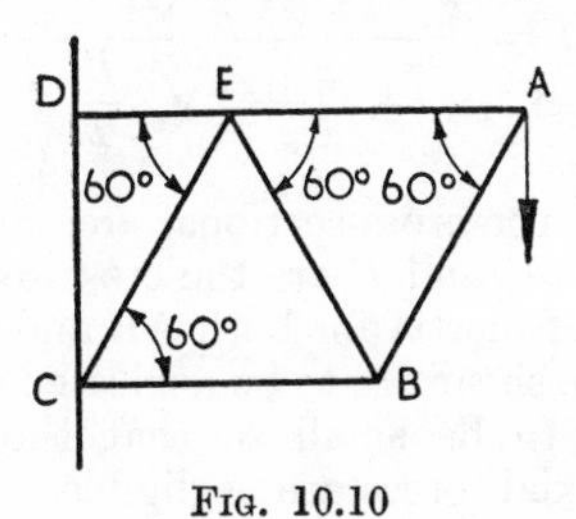

FIG. 10.10

For the tension members the sectional area is A and the moment of inertia of the cross-section is I. For the compression members the corresponding values are $2A$ and $4I$.

Assuming in the first instance that the framework is pin-jointed, determine the angular displacements of the various members due to the action of the load W. If the joints are rigid and the ends C and D are both fully fixed, show that the joint E will be in equilibrium assuming that the clockwise rotations of the joints E, A and B are $2{\cdot}67\,\dfrac{W}{AE}$, $2{\cdot}67\,\dfrac{W}{AE}$ and $3{\cdot}65\,\dfrac{W}{AE}$ respectively. (*Cambridge*)

4. For the frame shown in Fig. 10.11 the joints are welded and the vertical loads $\dfrac{W}{2}$ are applied to extensions of the member BC. Show that the bending moment M induced in the member BC

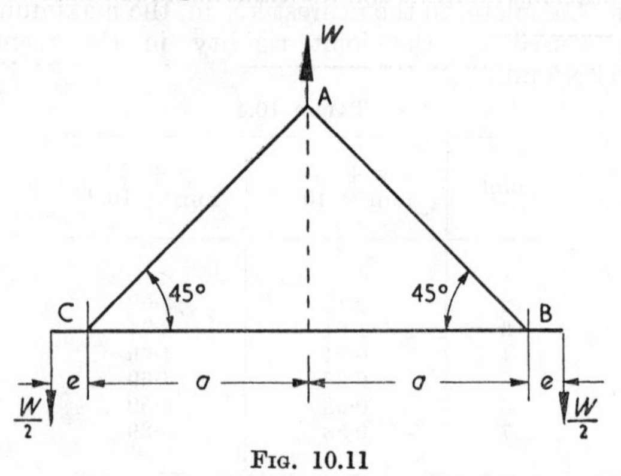

FIG. 10.11

by the eccentricity of the loading and the rigidity of the joints is given by

$$M = \frac{\dfrac{W}{2}e + \dfrac{3WI'}{a\sqrt{2}}\left(\dfrac{1}{A} + \dfrac{\sqrt{2}}{A}\right)}{1 + 2\sqrt{2}\,\dfrac{I'}{I}}$$

where A and I are the cross-sectional area and relevant moment of inertia for BC; A' and I' are the cross-sectional area and the relevant moment of inertia for both AB and AC.

Deflexions due to shear are to be neglected and all deformations may be assumed to be small so that the effects of bending moments due to axial forces are negligible. (*Cambridge*)

5. In Fig. 10.12 the member ABC is continuous, built-in at A and rigidly connected to the sloping member at C. All other joints are hinged. Sketch the bending moment diagram for ABC and find the greatest stress in the portion AB.

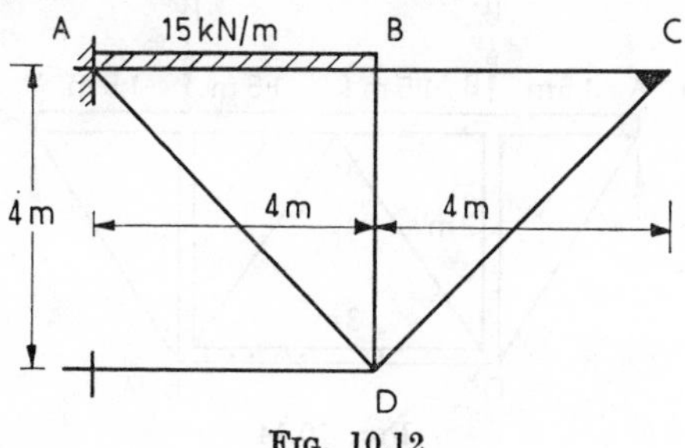

Fig. 10.12

The rotations (δ/L) of the relevant members may be taken as 1/1,000 clockwise.

$I_{ABC} = 12 \times 10^6$ mm^4 units $\qquad I_{CD} = 5 \times 10^6$ mm^4

Depth ABC = 130 mm $\qquad E = 200$ kN/mm^2.

(*Glasgow*)

Answer. 150 N/mm^2.

6. The frame shown in Fig. 10.13 is rigidly fixed at A and B and all joints are rigid. Find the bending moment values for member ADE due to the given load. Assume that the I/l ratio

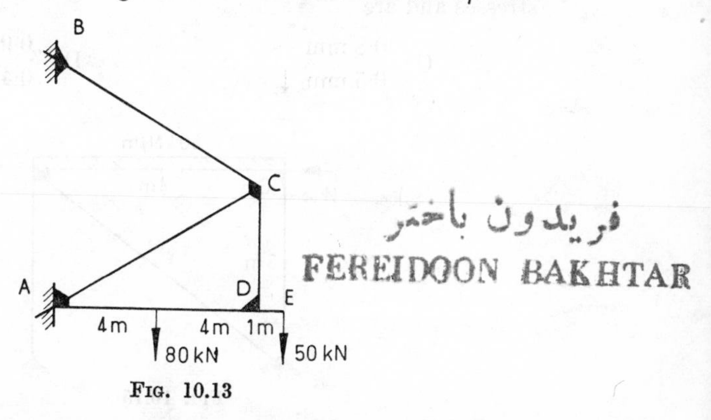

Fig. 10.13

for each member is $13{\cdot}0 \times 10^3$ mm^3 and that the rotation of each from a Williot diagram is 1/1000 radians. Take $E =$ 200 kN/mm^2. (*Glasgow*)

Answer. $M_{AD} = 97{\cdot}3$; $M_{DA} = 61{\cdot}0$; $M_{load} = 80{\cdot}8$ (all in kN m).

7. In a small girder frame, Fig. 10.14, the chords and verticals are rigidly connected, and have constant moment of inertia equal to $16 \cdot 6 \times 10^8$ mm^4 units. The diagonals are light and of negligible

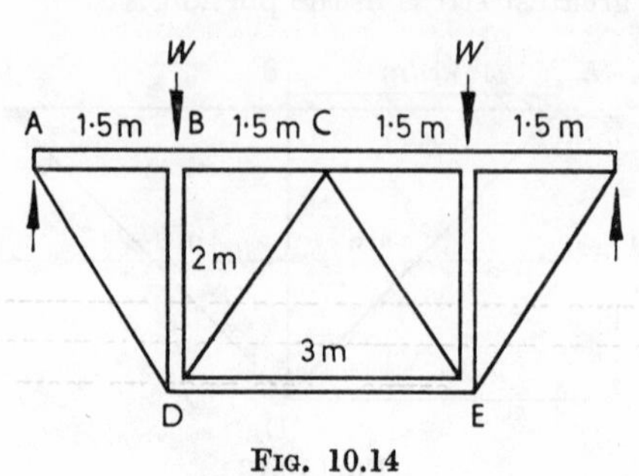

Fig. 10.14

inertia. Under the symmetrical loading shown the vertical deflexions of B and C and D relative to A are 1·5, 2·0 and 2·5 mm and the horizontal 0·5, 1·0 and 1·5 mm to the left.

Draw the bending moment diagram for the top chord member.

(*Glasgow*)

Answer. $M_{BA} = 214$; $M_{BC} = 156$; $M_{CB} = 145$ kN m.

8. A frame ABCD with rigid joints throughout is shown in Fig. 10.15. Deflexions of C and D are obtained from the primary stresses and are

C 0·8 mm → 0·5 mm ↓ D 0·9 mm → 0·4 mm ↓

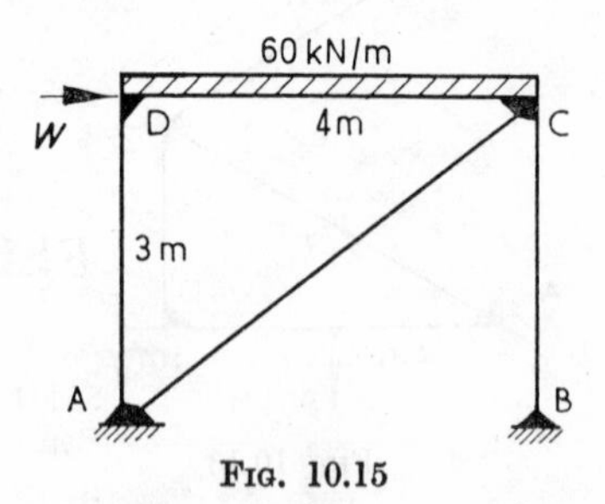

Fig. 10.15

If $K = 5 \times 10^4$ mm^3 for each member find the secondary moments in DC and AC stating maximum values.

Answer. $M_{DC} = -38 \cdot 4$; $M_{CD} = +79 \cdot 9$; $M_{AC} = -23 \cdot 8$; $M_{CA} = -37 \cdot 2$. (All in kN m.)

Chapter 11

Models

THERE ARE TWO MAIN USES of models in structural analysis and design. One is to use experiments on a scale model to determine the stresses in the prototype. This finds its main application in such structures as arch dams where the analysis is very difficult and where there is no standard method of design. The conversion of the model stresses to those in the prototype depends on a

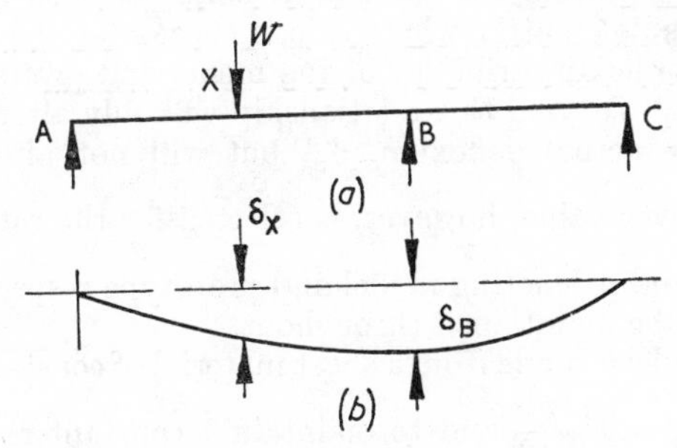

FIG. 11.1

number of factors amongst them being the linear scale, the scale of elastic moduli of model and prototype, and the load scale. The choice of a material in which to make the model is limited since Poisson's ratio for both the material of model and prototype must be the same. It is not proposed in this book to work through the dimensional analysis of this type of model for it does not lend itself to tutorial questions.

The main use of structural models by students is in the drawing of influence lines for redundant structures. The principle on which this method is based is the theorem of Muller–Breslau. This states that the influence line for the force in a redundant member is the same as the deflected form of the structure when unit load replaces the redundancy, the scale being so adjusted that the deflexion at the load point is unity.

Consider the continuous beam ABC shown in Fig. 11.1 (*a*). This is a redundant structure and R_B will be taken as a redundant element. Assume that a load W acts at X and it is required to find the reaction at B due to this load.

If the support B is removed and replaced by a unit load, the beam will deflect as shown in Fig. 11.1 (*b*), the deflexions at B and X being δ_B and δ_X respectively. But if δ_X is the deflexion at X when unit load acts at B it is also, by Clerk-Maxwell's theorem, the deflexion at B when unit load acts at X. Thus if the beam were propped at B to prevent deflexion when a load W was applied at X the force on the prop would be $W\frac{\delta_X}{\delta_B}$. In other words, if the scale of the deflected form of the beam could be adjusted so that δ_B was unity, this deflected form would be the influence line for the redundant reaction.

Model analysis of this type is particularly useful when the structure is not easily analysed, especially in cases where the section is not uniform.

The fact that the deflexion at B is adjusted to unity means that it does not matter what scale is chosen for the relationship between the flexural rigidity of the model and prototype at any given point. Varying this relationship will only alter the magnitude of the actual deflexion, δ_B, but will not alter the ratio $\frac{\delta_X}{\delta_B}$. Whatever value, however, is selected for the ratio $\frac{EI_m}{EI_p}$ (the suffixes m and p denoting model and prototype respectively), this value must be maintained throughout.

If the model is made from a sheet material of constant thickness t, then $EI_m = E_m\frac{td^3}{12}$ and to maintain a constant ratio between the flexural rigidities the depth of the model, d, must at all points be proportioned to $\sqrt[3]{EI_p}$.

The above remarks apply only to redundant structures where in normal analysis bending energy only would be considered. If the structure is such that in analysis the direct energy only would be taken into account, then the depth of the model, d, must at all points be proportional to AE_p. Such a structure is shown in Example 11.2.

Worked Examples on Model Analysis

Example 11.1. A test on a xylonite model fixed-end arch of 250 mm span gave the following results—

Anticlockwise rotation of left-hand support 0·5°.

Movement of a point X on the arch 0·52 mm to the left; 0·28 mm upwards.

If a similar arch was constructed of 40 m span what would be the bending moment at the left-hand support due to a load of

500 kN, acting at 30° to the vertical (towards the right) applied at a point corresponding to X.

Solution. The arch is shown diagrammatically in Fig. 11.2. The redundant reaction is the moment at the left-hand support.

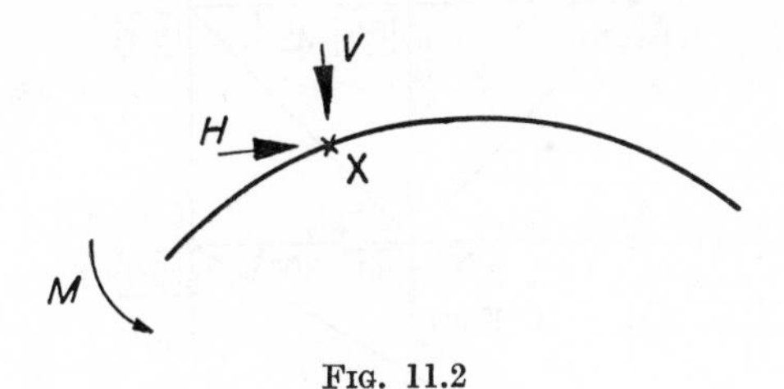

FIG. 11.2

If this is turned anticlockwise through θ let X move δ_H horizontally and δ_V vertically.

Applying the generalized form of Clerk-Maxwell's theorem gives

$$M \times \theta = H \times \delta_H = V \times \delta_V$$

$$\theta = 0{\cdot}0087 \text{ radians}$$

$$M = +\frac{0{\cdot}52}{0{\cdot}0087} H = 59{\cdot}0H \text{ mm units}$$

$$\text{Due to } V,\ M = +\frac{0{\cdot}28}{0{\cdot}0087} V = 31{\cdot}8V \text{ mm units}$$

In actual arch, $V = 250\sqrt{3}$ and $H = 250$ kN and the scale is 1 to 160, hence

$$M \text{ in actual arch} = \begin{cases} +\,59{\cdot}0 \times 250 \times 160 \times 10^{-3} \\ \qquad = 2360 \text{ kN m} \\ +\,31{\cdot}8 \times 250\sqrt{3} \times 160 \times 10^{-3} \\ \qquad = 2205 \text{ kN m} \end{cases}$$

giv ng $M = 4565$ kN m

Example 11.2. Write a short note on the large displacement method of model analysis.

A scale model is made of the frame shown in Fig. 11.3. The values indicated in boxes give the vertical displacement in millimetres of the point concerned when the points A and G are made to approach each other by 15 mm. Calculate the forces in the members when it carries the loads shown. (*Glasgow*)

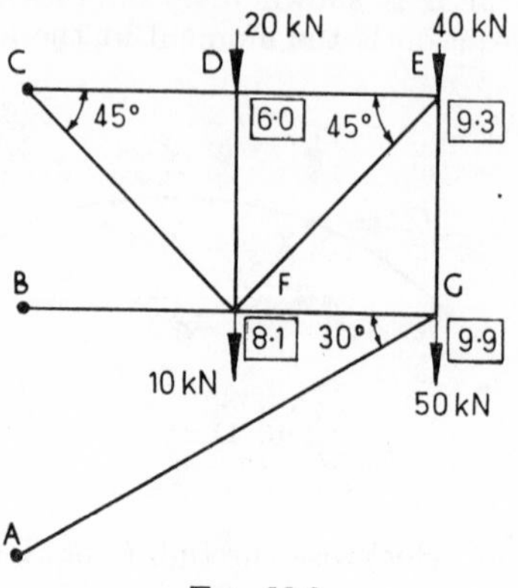

FIG. 11.3

Solution. The force in the member AG for a load W acting at any given point, X, is equal to $\dfrac{\delta_X}{\delta_{AG}} \times W$, where δ_X is the deflexion at X in the direction of W when points A and G approach one another by δ_{AG}.

The force in AG due to the load system shown with the displacements as given can therefore be obtained as follows—

Due to 2 tons at D, $F_{AG} = \dfrac{6 \cdot 0}{15 \cdot 0} \times 20 = 8 \cdot 0$ kN

Due to 4 tons at E, $F_{AG} = \dfrac{9 \cdot 3}{15 \cdot 0} \times 40 = 24 \cdot 8$ kN

Due to 1 ton at F, $F_{AG} = \dfrac{8 \cdot 1}{15 \cdot 0} \times 10 = 5 \cdot 4$ kN

Due to 5 tons at G, $F_{AG} = \dfrac{9 \cdot 9}{15 \cdot 0} \times 5 = 33 \cdot 0$ kN

Hence the total force in AG $= 71 \cdot 2$ kN

This force is then placed to act at G in the direction GA and the frame can now be solved by inspection giving the forces in the other members as follows.

CD = 54·3 kN tension

DE = 54·3 kN tension

EG = 14·4 kN tension

EF = 61·7 kN compression

FB = 82·1 kN compression

CF = 119·3 kN tension

DF = 20·0 kN compression

FE = 76·8 kN compression

Example 11.3. The force analysis of a two-bay rigid frame AD shown in Fig. 11.4 is to be checked by means of a spline model (Scale Ratio = 20).

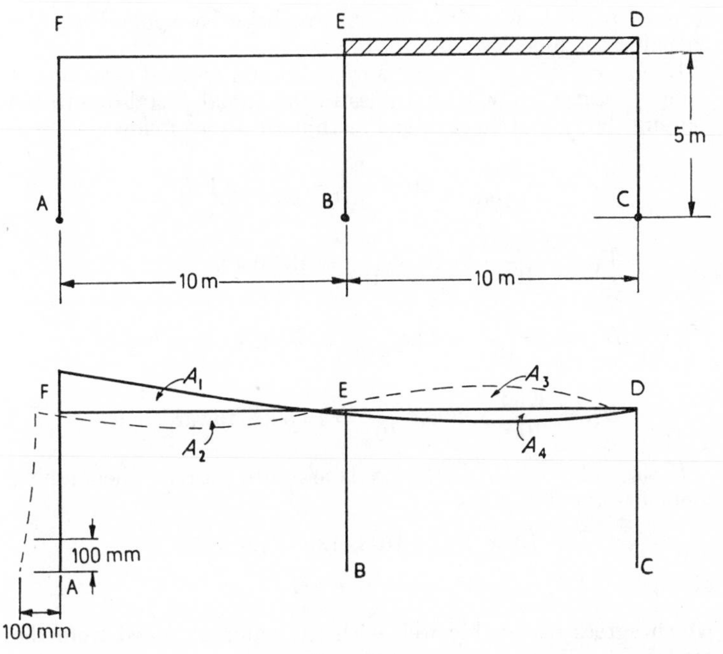

Fig. 11.4

The necessary movements and areas are given in the table. Complete the force analysis and draw the bending moment diagram for the frame.

Indicate the direction of sway and state briefly how its magnitude may be obtained.

Movement at A	A_1	A_2	A_3	A_4
100 mm ↑	22 500			2580
100 mm ←		3250	245	

Areas are given in square millimetres. It carries a load of 30 kN/m run on DE. *(Glasgow)*

Solution. The portal as shown has three redundancies. Sufficient information is given in the question to enable four reactions to be found. This is an advantage in model analysis as the fourth reaction can also be obtained from the static equilibrium equations when the other three have been found thereby checking the model work.

If H_A, V_A, H_C, V_C are the horizontal and vertical reactions at A and C respectively, their values in the actual portal due to the loading shown can be obtained arithmetically as follows—

$$V_A = -\frac{2580}{100} \times 20 \times \frac{30}{10^3} = -15{\cdot}5 \text{ kN}$$

$$V_C = \frac{22\,500}{100} \times 20 \times \frac{30}{10^3} = 135{\cdot}0 \text{ kN}$$

$$H_A = -\frac{245}{100} \times 20 \times \frac{30}{10^3} = 1{\cdot}5 \text{ kN towards left}$$

$$H_C = \frac{3250}{100} \times 20 \times \frac{30}{10^3} = 19{\cdot}5 \text{ kN towards left}$$

If the value of $V_C = 135$ kN is assumed correct, then taking moments about B gives

$$10 \times V_A = 10 \times 135 - 30 \times 5$$

giving

$$V_A = -15{\cdot}0 \text{ kN}$$

which agrees reasonably well with the value obtained from the model.

Before proceeding with the remaining parts of the question it is felt that an explanation of the method of obtaining the forces would be helpful. The deflected form when A is displaced vertically gives the influence line for V_A when the displacements are divided by 100 and measured in millimetres. The shape of this line will not be affected by the scale adopted. In dealing with a distributed load, however, it is the area under the influence line which is required and the ratio of the area to the displacement at A *is* affected by the scale. If the linear scale is $n:1$, then the ratio of the area under the influence line to the displacement is in the actual structure n times its value in the model. Hence the insertion of 20 in the numerator. But since all the units have been measured in millimetres the intensity of loading must be given in kN/mm run. This is the reason for the 10^3 in the denominator.

Having obtained V_A, H_A, V_C and H_C from the model, the reactions at B are obtained by considering the vertical and horizontal equilibrium of the structure. This gives

$$V_B = 180 \text{ kN} \quad \text{and} \quad H_B = 21{\cdot}0 \text{ kN}$$

The moments at the joints are as follows—

$$M_{FA} = -1{\cdot}5 \times 5 = -7{\cdot}5 \text{ kN m}$$
$$M_{DC} = +19{\cdot}5 \times 5 = +97{\cdot}5 \text{ kN m}$$
$$M_{EB} = +21{\cdot}0 \times 5 = +105 \text{ kN m}$$
$$M_{EF} = -1{\cdot}5 \times 5 + 15{\cdot}5 \times 10 = 147{\cdot}5 \text{ kN m}$$
$$M_{ED} = +19{\cdot}5 \times 5 - 135 \times 10 + 300 \times 5$$
$$= +247{\cdot}5 \text{ kN m}$$

The bending moment diagram for the structure is shown in Fig. 11.5.

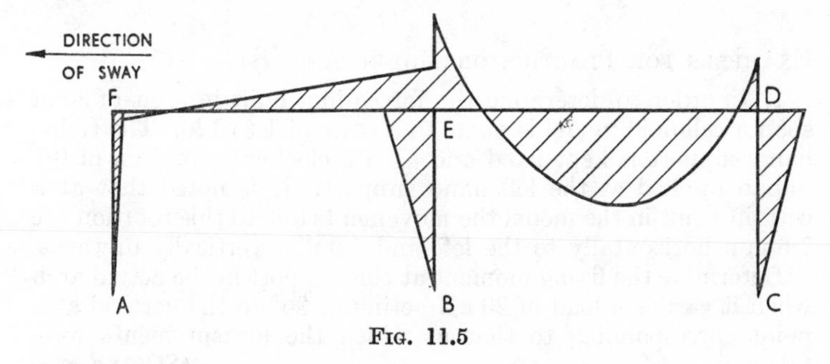

FIG. 11.5

The direction of the sway is also shown in Fig. 11.5.

The magnitude can be obtained from the application of the slope deflexion equations.

The previous sentence would probably suffice in answer to the question given but since the problem of the calculation of sway has not been dealt with elsewhere in this book the last portion of the question will be amplified to assist the reader in problems of this nature.

The standard slope-deflexion equations for the ends A and B of the beam are

$$\theta_A = -\frac{l}{6EI}\left(2M_A + M_B + \frac{6A\bar{x}'}{l^2}\right) + \frac{\delta}{l}$$

$$\theta_B = \frac{l}{6EI}\left(M_A + 2M_B + \frac{6A\bar{x}}{l^2}\right) + \frac{\delta}{l}$$

This standard equation will be applied firstly to beam EF and then to AF.

There is no relative vertical deflexion of points E and F, hence $\delta = 0$ in the equations previously given. There is also no lateral load and

$$\theta_{FE} = -\frac{10}{6EI}(-2 \times 7{\cdot}5 + 147{\cdot}5) = -\frac{1625}{6EI}$$

But $\quad \theta_{FE} = \theta_{FA} = \dfrac{5}{6EI}(-2 \times 7{\cdot}5) + \dfrac{\delta_{FA}}{5} = -\dfrac{75}{6EI} + \dfrac{\delta_{FA}}{5}$

This gives $\delta_{FA} = -\dfrac{1290}{EI}$

showing that the sway is as indicated and that F moves to the left of A.

Examples for Practice on Model Analysis

1. In order to determine the forces in a built-in arch of 30 m span a celluloid model is made to a scale of 1 to 120. The right-hand support is kept fixed and an anticlockwise rotation of 0·1 radian applied at the left-hand support. It is noted that at a certain point in the model the movements due to this rotation are 7·6 mm horizontally to the left and 4·0 mm vertically upwards.

Determine the fixing moment at this support in the actual arch when it carries a load of 20 kN acting at 30° to the vertical at a point corresponding to that at which the measurements were taken. *(St. Andrews)*

Answer. $M = 174{\cdot}5$ kN m

2. Explain what is meant by a displacement corresponding with a specified force.

In order to determine the actions at one of the abutments of an arch with built-in ends due to a single load acting in a given direction at a certain point, A, a model was constructed to a scale of 1 : 100. The following readings were taken during two tests on the model in each of which displacements were applied to the same end while the other end was held fixed.

Test I

Upward vertical displacement of end	1·3 mm
Horizontal displacement of end	0·00 mm
Clockwise rotation of end	0·02 rad
Corresponding displacement at A	— 0·8 mm

Test II

Clockwise rotation of end	— 0·06 rad
Vertical and horizontal displacement of end	0·00 mm
Corresponding displacement of A	0·6 mm

Calculate the values of the fixing moment and of the vertical component of the reaction at the end under consideration due to a load of 15 kN applied as above, for the full scale arch.

Answer. 7 kN; 15 kN m *(Oxford)*

3. It is required to find the horizontal thrust and end fixing moments in a built-in bridge of 100 m span due to a concentrated load of 50 kN acting at 60° to the horizontal at some point P. A model having a 0·60 m span is constructed and the following data are collected. One abutment, A, is fixed, the other, B, is free to move.

Movement of Abutment B	Movement of P	
Horizontal, in direction AB, 12 mm	Horizontal Vertical	8 mm 4 mm
Vertical, downwards, 12 mm	Horizontal Vertical	– 3 mm 4·3 mm
Angular, clockwise, 0·05 radian	Horizontal Vertical	3·5 mm – 7·0 mm

(London)

Answer. $H_B = 31{\cdot}2$ kN; $M_B = 719$ kN m

4. A small scale model of a parabolic arch rib is fixed rigidly in position and direction at the right-hand abutment but the left-hand abutment may be released or displaced linearly without any rotational movement. In both cases the vertical displacement of equally spaced points on the arch centre line may be measured and recorded.

Dimensions and Data—

Span 600 mm
Central Rise 100 mm

Horizontal Distance from L.H. Abutment	75	150	225	300	375	450	525
Vertical displacement due to rotation of 15° applied to the left-hand abutment	−24·4 mm	−20·0 mm	−5·1 mm	13·0 mm	18·8 mm	16·0 mm	4·8 mm
Vertical displacements due to a horizontal displacement of 12 mm	3·0 mm	10·0 mm	15·5 mm	18·3 mm	15·5 mm	10·0 mm	3·0 mm

Answers required for an arch of 90 m span

(*a*) I.L. for horizontal thrust at L.H. Abutment

(*b*) I.L. for B.M. at L.H. Abutment

(*c*) I.L. for B.M. at crown

(*d*) Maximum B.M. at crown due to a point load of 10 kN.

Answer. (*d*) 70·5 kN m (*Durham*)

5. A slender spline of elastic material of uniform cross-section was pinned through its centre line to a drawing board by four pins A, B, C and D so that rotational movement about each pin was possible. The pins were in a straight line with spacings AB 450 mm, BC 500 mm and CD 450 mm. The pin at A was then withdrawn and the free end of the spline was moved up 100 mm. The resulting vertical displacements of points along the spline were as given in the table below.

Deduce from the above information the force on one of the outer supports of a three-span continuous beam, of constant section, covering spans of 9 m, 10 m and 9 m (in order) and carrying a uniformly distributed load of 40 kN/m run over the full length of the outer spans, together with concentrated loads of 100 kN at the third points of the middle span. Check your result by a separate method. (*London*)

Distance along span (as fraction of span)	Vertical Deflexion of Point (in.)		
	Span AB	Span BC	Span CD
$\frac{1}{6}$	78·5	—	1·6
$\frac{1}{3}$	58·0	− 9·4	2·0
$\frac{1}{2}$	40·5	—	2·2
$\frac{2}{3}$	24·6	− 6·1	2·0
$\frac{5}{6}$	11·7	—	1·3

Answer. Model 127·7 kN; Analysis 147·5 kN.

6. A two-hinged parabolic arch is of 40 m span and 8 m rise and the moment of inertia of its cross-section varies as the secant of the angle of slope of the centre line.

A suitably proportioned celluloid model was tested by the indirect method, the vertical deflexions of points along the axis of the arch due to a relative horizontal displacement of 25 mm of the pins being measured. The results were as follows—

Distance from left-hand support (as fraction of span)	Vertical Deflexion mm
0	0
0·1	6·5
0·2	13·0
0·3	18·0
0·4	21·2
0·5	22·6
0·6	21·0
0·7	18·4
0·8	13·2
0·9	6·8
1·0	0

Deduce, from the model test, the horizontal thrust at the abutments of the arch due to a total uniformly distributed load of 60 kN/m run over the full span length and compare this value with the calculated thrust.

Answer. 1355 kN (model); 1500 kN (calculation).

7. Show how the scale factor operates in the analysis of redundant structures when spline models are used.

A spline model was used for the analysis of the frame loaded as shown in Fig. 11.6 The following data were obtained from the model.

Movement A	Corresponding Movement at G	Corresponding Area Under Load BC	Corresponding Area Under Load DE
Horizontal 50 mm to right	– 38 mm	2000 mm²	700 mm²
Rotational $\pi/6$ clockwise	– 75 mm	550 mm²	200 mm²

Full scale/model = 16.

Sketch the Bending Moment Diagram for column AGB giving the critical values. *(Glasgow)*

Answer. $M_A = 44{\cdot}4$; $M_G = -14{\cdot}3$; $M_B = +26{\cdot}9$ kN m.

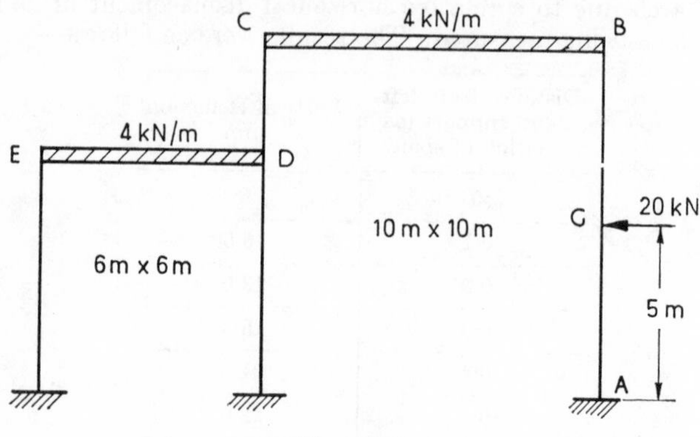

FIG. 11.6

8. The analysis of a rigid frame of the type shown in Fig. 11.7 was carried out by the use of a spline model (scale ratio 15). For the particular case of horizontal loading, a movement was given to E such that EE′ = 75 mm. The deflexion area covered

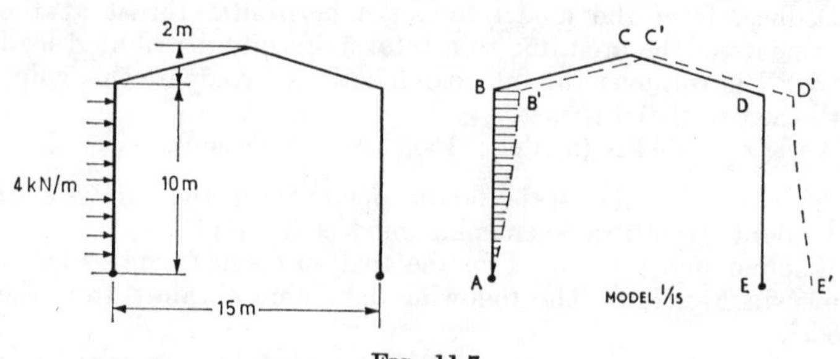

FIG. 11.7

by the load was measured and found to be 12 000 mm². Determine the values of the horizontal thrusts at A and E and draw neatly the bending moment diagram for the whole frame. Deduce any formulae used. *(Glasgow)*

Answer. $H_A = 30{\cdot}4$ kN; $H_E = 9{\cdot}6$ kN; $M_B = 104$ kN m; $M_D = 96$ kN m.

9. In a 1/50 scale model of a fixed frame 16 m high and of 20 m span, the deflexions obtained by giving B (i) a horizontal movement of 5 mm, (ii) a rotation of $\frac{1}{20}$ of a radian, are as shown

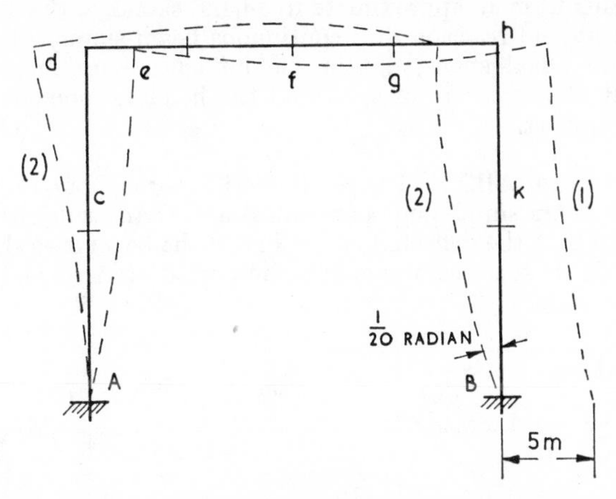

FIG. 11.8

in Fig. 11.8 and as detailed in the following table which gives the horizontal movements of the columns and the vertical movements of the beam in millimetres.

Point	c	d	e	f	g	h	k
Curve 1 .	1·0	2·5	0·4	0·6	0·4	2·5	4·0
Curve 2 .	1·6	4·3	0·5	0·8	0·7	4·3	4·6

Assuming the Reciprocal Theorem prove that these curves give the forms of the influence lines for two of the reactions. Sketch the influence line for the bending moment at the centre of the beam for the left half of the frame. (*Glasgow*)

Answer—

Load at	c	d	e	f
Central M	+ 0·70	0	− 1·80	− 3·90

10. Discuss briefly the model method of obtaining influence lines in redundant structures on the principle of the reciprocal theorem. Obtain the necessary relations for finding the value of a redundant fixing moment, including the scale of the model.

Starting from an approximate free-hand sketch of the influence line for an end reaction of a continuous beam simply supported over two equal spans, obtain the influence lines for bending moment at the centre of span and the bending moment at the centre support. *(Glasgow)*

11. A beam ABC is of variable section throughout its length. For the beam simply supported at B and C and subjected to a unit load at A the deflected centre line of the beam is as shown in Fig. 11.9. If this beam is simply supported at A, B and C and

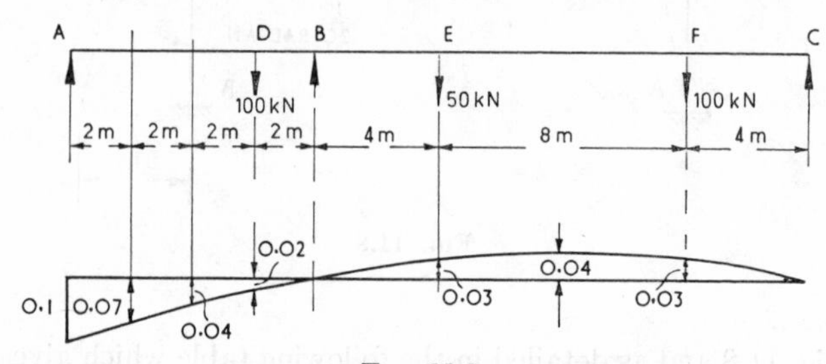

Fig. 11.9

subjected to the given vertical loads at points D, E and F, determine the reaction at B and draw the shear force and bending moment diagram for this loading. *(Aberdeen)*

Answer. 212·5 kN.

Chapter 12

Struts

THE SUBJECT OF STRUTS and columns covers a very wide field and it is not the object of this book to deal with it all. For tutorial purposes it may be divided into three main groups—

1. Examples dealing with instability problems.
2. Examples dealing with the application of the standard strut formulae.
3. Examples on laterally loaded struts solved analytically or geometrically.

This chapter will be divided into sections to follow each of the above groups.

Worked Examples on Instability Problems

Example 12.1. Calculate the Euler critical load for a strut which is pin-jointed at one end and fixed at the other.

Solution. The strut is shown in Fig. 12.1 the upper end being pinned whilst the lower is fixed, M being the fixing moment.

If y is the horizontal deflexion at a point distant x from the lower end of the strut, the bending moment M_x at that point is given by

$$M_x = -Py + M\left(1 - \frac{x}{L}\right)$$

where P = vertical load and L = strut height.

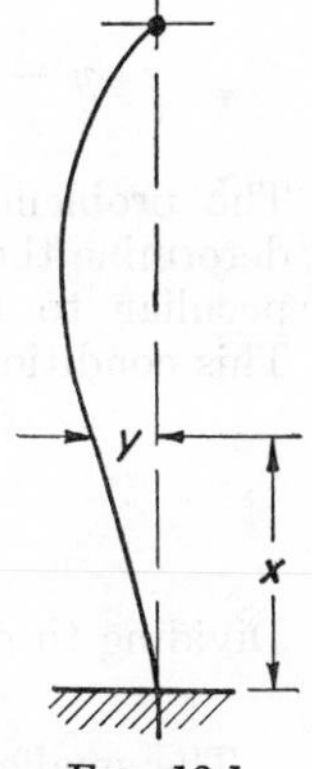

FIG. 12.1

In the solution of any instability problem on struts the expression for the bending moment is equated to $EI\dfrac{d^2y}{dx^2}$ giving a second order differential equation which is solved for y. The arbitrary constants are obtained from the two given end conditions.

Applying this method to the example in question gives

$$EI\frac{d^2y}{dx^2} = -Py + M\left(1 - \frac{x}{L}\right)$$

Writing $$\frac{P}{EI} = \mu^2$$

gives $$\frac{d^2y}{dx^2} + \mu^2 y - \frac{M\mu^2}{P}\left(1 - \frac{x}{L}\right) = 0$$

the solution of which is

$$y - \frac{M}{P} = -\frac{M}{P}\cdot\frac{x}{L} + A\sin\mu x + B\cos\mu x$$

where A and B are the constants of integration.

Differentiating the expression for y gives

$$\frac{dy}{dx} = -\frac{M}{PL} + \mu A\cos\mu x - \mu B\sin\mu x$$

When $x = 0$, $\frac{dy}{dx} = 0$, hence $A = \frac{M}{\mu PL}$

Also when $x = 0$, $y = 0$, hence $B = -\frac{M}{P}$

The complete equation therefore is

$$y - \frac{M}{P} = -\frac{M}{P}\cdot\frac{x}{L} + \frac{M}{\mu PL}\sin\mu x - \frac{M}{P}\cos\mu x$$

The problem now is to eliminate M from this equation and determine the value of P to satisfy the third condition, which is peculiar to this problem, namely that when $x = L$, $y = 0$. This condition gives, for $x = L$ in the above equations,

$$0 = \frac{M}{\mu PL}\sin\mu L - \frac{M}{P}\cos\mu L$$

Dividing throughout by $\frac{M}{P}$ gives $\tan\mu L = \mu L$

The smallest solution (apart from $\mu L = 0$ which is invalid) is

$$\mu L = 4{\cdot}493$$

giving $$P = \frac{20{\cdot}1EI}{L^2}$$

Example 12.2. In the portal structure shown in Fig. 12.2 the uprights are hinged to the horizontal member at B and C. The moment of inertia I is uniform throughout.

Assuming that the structure buckles into the form indicated,

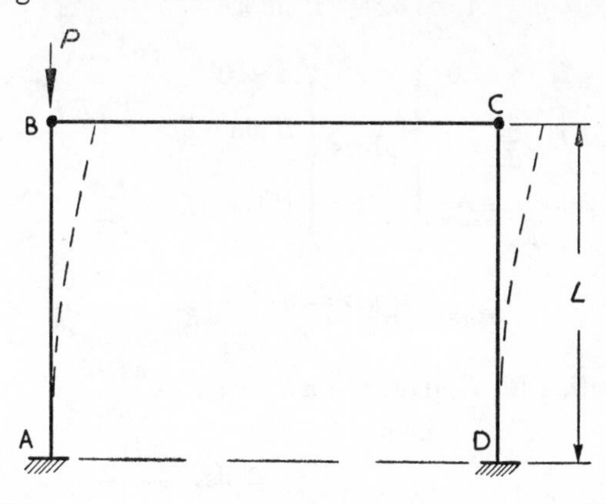

FIG. 12.2

show that the value of the load P corresponding to the onset of elastic stability is given by

$$\tan nL = nL - \frac{n^3L^3}{3}$$

where $$n = \sqrt{\frac{P}{EI}}$$

Consider bending effects only. (*Glasgow*)

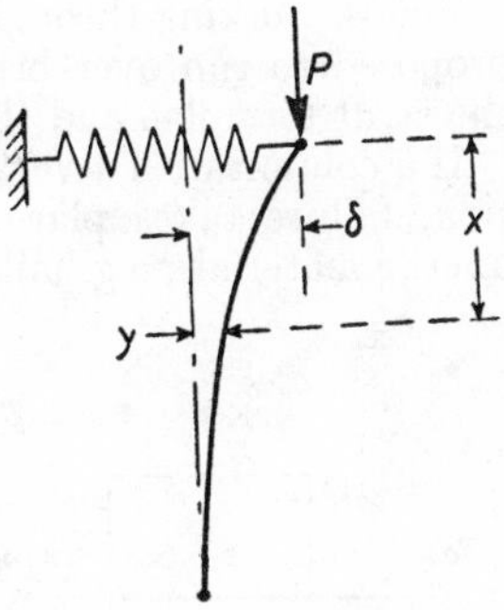

FIG. 12.3

Solution. The Problem is equivalent to a strut with spring support. Stiffness of spring $K = \frac{3EI}{L^3}$ = Force required to produce unit deflexion at the end of a cantilever.

Applying the relationship

$$M = EI\frac{d^2y}{dx^2}$$

gives in this case $EI\frac{d^2y}{dx^2} = P(\delta - y) - K\delta x$

i.e. $$\frac{d^2y}{dx^2} + n^2y = n^2\delta - \frac{K}{EI}\delta x \quad \text{with } n^2 = \frac{P}{EI}$$

giving $$y = A \cos nx + B \sin nx + \delta - \frac{K}{P}\delta x$$

$$\left.\begin{aligned} x &= L \quad y = 0 \\ x &= L \quad \frac{dy}{dx} = 0 \\ x &= 0 \quad y = \delta \end{aligned}\right\} \text{gives} \left\{\begin{aligned} &A = 0 \\ &B \sin nL = \delta\left(\frac{KL}{P} - 1\right) \\ &B \cos nL = \frac{K\delta}{Pn} \end{aligned}\right.$$

or $$\tan nL = nL - \frac{nP}{K}$$

Substituting for K gives $\tan nL = nL - \dfrac{nPL^3}{3EI}$

$$= nL - \frac{n^3L^3}{3}$$

Worked Examples on the Strut Formulae

Before working through some examples on strut formulae it is proposed to run over briefly the fundamental principles behind the strut formulae and their method of application.

If a column with a very low l/k ratio is compressed to destruction, failure takes place as shown in Fig. 12.4 in the case of ductile materials, e.g. mild steel, and as in Fig. 12.5 in the case of

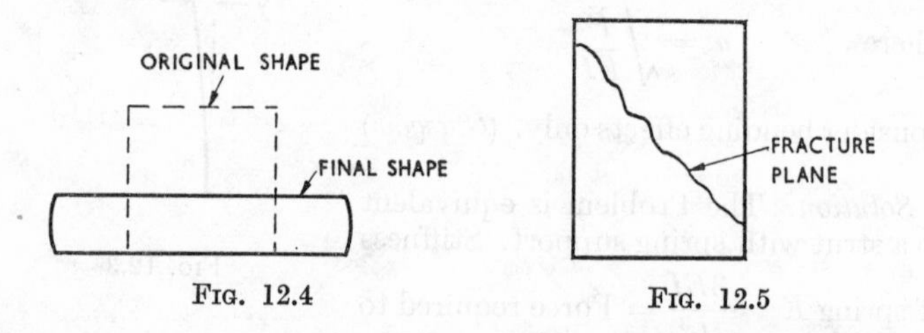

FIG. 12.4 FIG. 12.5

brittle materials such as a concrete cube. In the case of columns with a high l/k ratio failure takes place by buckling, the Euler formula giving a reasonable value for the ultimate load carried.

There is, however, a very wide range of l/k values for which both the simple compression and the Euler values do not apply. In this range the failure is generally by buckling but the Euler expression gives too high a value for the failing load. The buckling failure is caused by initial imperfections in manufacture and slight eccentricities of loading which together result in a measure of bending on struts, even those with a low l/k ratio. The various strut formulae have been evolved in order to give an expression for the failing load (and from this the design stress) which allows for these eccentricities.

The transformation from failing load to design stress is carried out in one of two ways either by (i) the factor of safety or (ii) the load factor.

In order to explain these two methods the following symbols will be used.

Let p_y = the yield stress in the material
p_f = the working or design stress based on the "factor of safety" method
n = the factor of safety
P_y = the load at which yield stress is reached
A = cross-sectional area of strut
p_1 = the working stress based on the load factor method
N = the load factor

By definition $\quad n = \dfrac{p_y}{p_f} \quad$ and $\quad N = \dfrac{P_y}{p_1 A}$

For equal values of n and N, the values of the stress p_f and p_1 will be equal only if $P_y = p_y \times A$. This can be the case only if the whole section reaches yield stress simultaneously. The combination of direct load and bending resulting from the eccentricity gives a non-uniform distribution of stress, hence P_y will be less than $p_y \times A$ and, for equality of n and N, the value p_1 must be less than p_f. Since a member is, in cases where the design is based on the elastic theory, assumed to have failed when the stress at any point on the section reaches yield stress, the design must be based on P_y and not p_y, consequently in struts the load-factor method is always used.

It has already been mentioned that the object of the strut formula is to give design stresses for those values of the l/k ratio where neither the Euler nor the direct compression methods apply. The commoner formulae are as follows.

The straight-line formula for pin ended steel struts, which gives the working stress, p, in N/mm² as

$$p = 125\left(1 - \frac{1}{300} \cdot \frac{l}{k}\right)$$

The figure 125 here was the value generally taken as the working stress in mild steel in direct compression. It is now general practice to use a higher value than this which should be substituted for 125 N/mm² in the formula. When applying this formula to struts with other than pinned ends an effective strut length (e.g. 0·7 × actual length if both ends are restrained) is used in the l/k term.

The Rankine–Gordon formula, which gives the load, P_y, to produce yield stress as

$$P_y = \frac{p_y A}{1 + a\left(\frac{l}{k}\right)^2}$$

where a is a dimensionless constant and $= \frac{p_y}{\pi^2 E}$ in the general case.

This formula is obtained from the assumption that—

$$\frac{1}{P_y} = \frac{1}{p_y A} + \frac{l^2}{\pi^2 EI}$$

It is obviously correct in the case of the short column when $l = 0$ and in the case of the long column the first term becomes small compared with the second giving a load equal to the Euler load for a pin-ended column.

For a mild steel column with pinned ends the value generally taken for a is 1/7,500.

The Perry–Robertson formula gives the design stress, p, as

$$Np = \frac{p_y + (\eta + 1)p_e}{2} - \left[\left\{\frac{p_y + (\eta + 1)p_e}{2}\right\}^2 - p_y p_e\right]^{\frac{1}{2}}$$

where N = load factor

p_y = yield stress

p_e = Euler stress $= \frac{\pi^2 E \,.\, k^2}{l^2}$

η = a constant for a given material and a given l/k ratio the magnitude of η being 0·003 l/k

This formula is based on the initial deflected form of the strut, being a cosine curve with a maximum departure from the straight of c_0 at mid-height where $c_0 = \frac{\eta k^2}{a}$.

Johnson's Parabolic Formula gives the yield load, P_y, from

$$P_y = A\left\{p_y - r\left(\frac{l}{k}\right)^2\right\}$$

If $\frac{P_y}{A}$ is plotted against l/k the resulting curve is a parabola. The value of r is chosen so as to make this parabola touch the Euler curve. For mild steel r = 0·006 67 N/mm², for values of l/k

less than 150 in the case of pin-ended struts, and 0·004 27 N/mm² for values less than 190 in the case of fixed-ended struts.

It would be outside the scope of this book to criticize and discuss the various formulae. It is proposed only to deal with the various examples on them which have occurred in examination and tutorial questions.

Example 12.3. Discuss the difference between designing on the "load factor" and "factor of safety" methods. The Perry formula for the intensity of end loading, p, which will cause the fibre stress in a pin-ended strut to reach yield point is

$$p = \frac{p_y + (\eta + 1)p_e}{2} - \left[\left\{\frac{p_y + (\eta + 1)p_e}{2}\right\}^2 - p_y p_e\right]^{\frac{1}{2}}$$

where p_y = yield stress

p_e = average stress at Euler load

η = a coefficient = $0{\cdot}003\, l/k$.

Taking $p_y = 225$ N/mm², $E = 200$ kN/mm² and $l/k = 120$, determine the permissible stress using: (*a*) a load factor of 2, and (*b*) a factor of safety of 2.

Solution. For $l/k = 120$, $p_e = \dfrac{\pi^2 \times 200}{14\,400} = 137{\cdot}5$ N/mm².

With a load factor of 2 and working stress = p_n

$$2 \times p_n = \frac{225 + 1{\cdot}36 \times 137{\cdot}5}{2} - \sqrt{\left(\frac{225 + 1{\cdot}36 \times 137{\cdot}5}{2}\right)^2 - 225 \times 137{\cdot}5}$$

$$= 206 - 108 = 98$$

Hence $\qquad p_n = 49$ N/mm²

With a factor of safety of 2, the value of $\dfrac{225}{2}$ must be substituted for p_y in the expression for p given in the question. This means that the maximum stress in the strut is then 113 N/mm². The average stress in the strut is then

$$p = \frac{113 + 1{\cdot}36 \times 137{\cdot}5}{2} - \sqrt{\left(\frac{113 + 1{\cdot}36 \times 137{\cdot}5}{2}\right)^2 - 113 \times 137{\cdot}5}$$

$$= 150 - 83{\cdot}8$$

$$= 66{\cdot}2 \text{ N/mm}^2$$

Example 12.4. A column shaft of length 6 m is built up from two 250 × 200 R.S.J.'s as shown in Fig. 12.6. If the column is pinned at both ends and subjected to an axial load equal to one

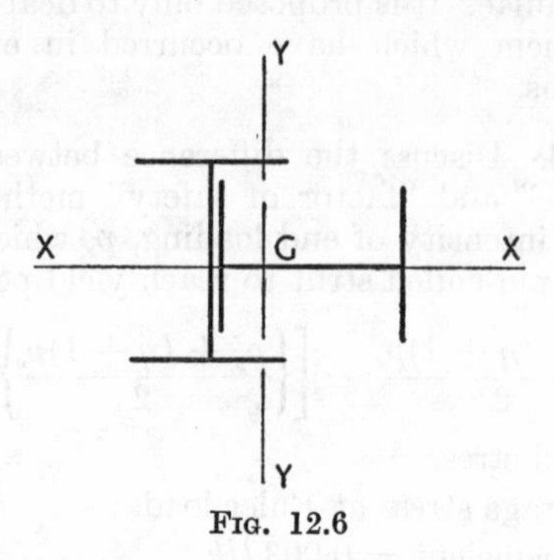

FIG. 12.6

quarter of the Euler critical load for the column, determine the load factor on the corresponding Perry–Robertson ultimate load.

If the load is displaced 12 mm along the Y-Y axis, determine now the factor of safety.

Take $E = 200$ kN/mm, $f_y = 225$ N/mm². For each R.S.J. $A = 1{\cdot}33 \times 10^4$ mm², $I_{\max} = 1{\cdot}43 \times 10^8$ mm⁴ units, $I_{\min} = 2{\cdot}97 \times 10^7$ mm⁴ units, thickness of web = 15 mm.

Solution. To get G take moments about the centre line of the joints shown upright on plan. If $\bar{x}$ is the distance of G from this centre line then

$$\bar{x} = \frac{1{\cdot}33 \times 132{\cdot}5}{1{\cdot}33 \times 2} = 66{\cdot}3 \text{ mm}$$

$$I_{XX} = (14{\cdot}3 + 2{\cdot}97)10^7 = 17{\cdot}3 \times 10^7 \text{ mm}^4$$

$$k_{XX} = \sqrt{\frac{17{\cdot}3 \times 10^7}{2 \times 1{\cdot}33 \times 10^4}} = 80{\cdot}7 \text{ mm}$$

I_{YY} is greater than I_{XX} and therefore not critical.

$$\text{Euler Load} = \frac{\pi^2 EI}{l^2} = \frac{\pi^2 \times 200 \times 17{\cdot}3 \times 10^7}{6 \times 6 \times 10^6} = 9450 \text{ kN}$$

$$\text{The Euler stress} = \frac{9450 \times 10^3}{2 \times 1{\cdot}33 \times 10^4} = 356 \text{ N/mm}^2$$

Actual load on column = 2363 kN

The Perry–Robertson constant $= 0{\cdot}003\,\dfrac{l}{k}$

$$= \frac{0{\cdot}003 \times 6 \times 10^3}{80{\cdot}7} = 0{\cdot}223$$

If P_y = load at which the fibre stress reaches yield point then

$$\frac{P_y}{2{\cdot}66 \times 10^4} = \frac{225 + 1{\cdot}223 \times 356}{2} - \sqrt{\left(\frac{225 + 1{\cdot}223 \times 356}{2}\right)^2 - 356 \times 225}$$

$$= 330 - 157 = 173 \text{ N/mm}^2$$

Hence $P_y = 173 \times 2{\cdot}66 \times 10^4 = 4600$ kN

Load factor $= \dfrac{4600}{2363} = 2{\cdot}03$

The Perry–Robertson formula is based on an initial departure from the straight of c_0 at the centre of the strut where c_0 is given by $\eta\,\dfrac{k^2}{a}$. Substituting the appropriate values in this case gives

$$c_0 = \frac{0{\cdot}223 \times 80{\cdot}7 \times 80{\cdot}7}{125} = 11{\cdot}6 \text{ mm}$$

When the load is displaced 12 mm along the Y–Y axis the resultant eccentricity is $12{\cdot}0 + 11{\cdot}6 = 23{\cdot}6$ mm.

The strut effect increases the central deflexion in the ratio $\dfrac{P_e}{P_e - P}$, i.e. the final deflexion $= \dfrac{4}{3} \times 23{\cdot}6 = 31{\cdot}4$ mm.

Note. A more correct value for the initial eccentricity is $c_o + 1{\cdot}2e$, where e is the added eccentricity (i.e. 0·5 in. in the example in question). The factor of 1·2 however varies with the ratio $\dfrac{P}{P_e}$ and the student could therefore be excused if it were not included.

The maximum fibre stress $= \dfrac{2363 \times 10^3}{2{\cdot}66 \times 10^4}\left(1 + \dfrac{31{\cdot}4 \times 125}{81{\cdot}7 \times 81{\cdot}7}\right)$

$$= 141{\cdot}4 \text{ N/mm}^2$$

Hence the factor of safety is $\dfrac{225}{141{\cdot}4} = 1{\cdot}59$

Example 12.5. A stanchion 5 m long consists of two channels 380 mm × 100 mm placed back-to-back connected by 400 × 20 plates across the 100 mm legs. The clear distance between the backs of the channels is 200 mm. Calculate the safe axial load on the member for (*a*) pinned ends and (*b*) fixed ends, using the formula for pin-ended struts

$$f = 140\left\{1 - 0{\cdot}0054\left(\frac{l}{k}\right)\right\}$$

where f = safe compressive stress in N/mm².

$\frac{l}{k}$ = slenderness ratio

Compare the values thus obtained with the buckling loads given by Euler's theory. ($E = 200$ kN/mm²).

The properties of one channel are—

Principal moments of inertia $= 1{\cdot}45 \times 10^8$ and $5{\cdot}56 \times 10^6$ mm⁴ units

Distance of centroid from back $= 24{\cdot}5$ mm

Area of cross-section $= 6{\cdot}9 \times 10^3$ mm²

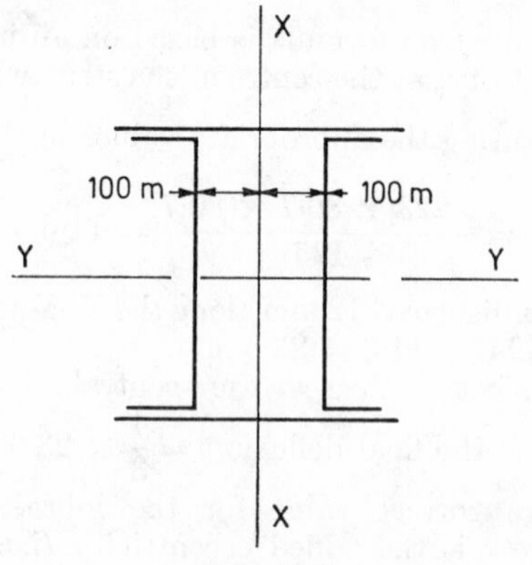

FIG. 12.7

Solution. The stanchion is shown diagrammatically in Fig. 12.7. The second moments of area about the axes are as follows—

Axis Y–Y

$$2 \times 145 \times 10^6 = 290 \times 10^6$$
$$2 \times 400 \times 20 \times 200^2 = 640 \times 10^6$$
$$\underline{\underline{930 \times 10^6}}$$

Axis X–X

$$2 \times 20 \times (400)^3 \times \tfrac{1}{12} = 214 \times 10^6$$
$$2 \times 6{\cdot}9 \times 10^3 \times 124{\cdot}5^2 = 214 \times 10^6$$
$$2 \times 5{\cdot}56 \times 10^6 = 11 \times 10^3$$
$$\underline{\underline{439 \times 10^6}}$$

$$\text{Area} = 2 \times 6{\cdot}9 \times 10^3 = 13{\cdot}8 \times 10^3$$
$$2 \times 400 \times 20 = 16{\cdot}0 \times 10^3$$
$$\underline{\underline{29{\cdot}8 \times 10^3}}$$

$$\therefore \quad k_{\min} = \sqrt{\frac{43{\cdot}9 \times 10^7}{29{\cdot}8 \times 10^3}} = 121 \text{ mm}$$

For pinned ends $l/k = 41{\cdot}3$
For fixed ends $l/k = 20{\cdot}7$

$\therefore$ Permissible stresses are—

Pinned ends, 108·5 N/mm²
Fixed ends, 126 N/mm²

The loads are—

Pinned ends, $108{\cdot}5 \times 29{\cdot}8 = 3250$ kN
Fixed ends, $126 \times 29{\cdot}8 = 3750$ kN

The Euler load for pinned ends

$$= \frac{\pi^2 EI}{l^2} = \frac{\pi^2 \times 200 \times 439 \times 10^6}{5 \times 5 \times 10^6} = 34\,900 \text{ kN}$$

$$\text{Euler load for fixed ends} = \frac{4\pi^2 EI}{l^2} = 139\,600 \text{ kN}$$

Worked Examples on Laterally Loaded Struts

There are two types of problem connected with laterally-loaded struts. They are—(i) the determination of the bending moment at any section due to the combination of lateral and direct loading, and (ii) the determination of a section with a suitable load factor to resist the combined loadings.

The first type of problem is generally dealt with geometrically by use of Polar Diagrams, sometimes referred to as Howard diagrams after their originator.

The second type of problem involves an awkward trigonometrical equation if a rigorous analysis is applied, consequently an approximation due to Perry is generally used. This approximation is based on the assumption that the final deflected form of the strut is a cosine curve. This then gives the intensity of end loading, p, required to produce yield stress, p_y, in the strut from the formula

$$p_y = p_b \left(\frac{p_e}{p_e - p}\right) + p$$

where p_e = Euler stress
p_b = stress due to bending caused by lateral loads alone
p = direct stress.

Example 12.6. A pin-ended strut of length l carries an axial load equal to $\frac{1}{4}$ of the Euler critical load and a uniformly distributed lateral load of w/unit length. Show that if moments of $\left(\frac{\sqrt{2}-1}{\sqrt{2}+1}\right)\frac{4wl^2}{\pi^2}$ are applied at each end the central bending moment is equal in magnitude to that at the ends.

Solution. This problem, and others similar, are best solved geometrically. An angle is set off of magnitude μl where $\mu = \sqrt{\frac{P}{EI}}$. This angle represents the strut length. In the example in question $P = \frac{\pi^2 EI}{4l^2}$

$$\therefore \qquad \mu^2 = \frac{\pi^2 EI}{4l^2} \cdot \frac{l}{EI} = \frac{\pi^2}{4l^2}$$

hence
$$\mu = \frac{\pi}{2l}$$

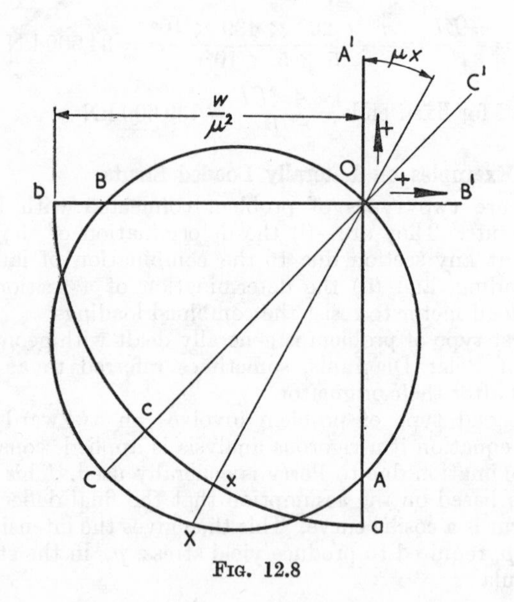

FIG. 12.8

The angle $\mu l = \frac{\pi}{2l} \cdot l = \frac{\pi}{2}$ and this is shown as A′OB′ in Fig. 12.8.

An arc of radius $\frac{w}{\mu^2}$, where w is the intensity of uniformly distributed lateral load, is struck with O as centre. Let ab be this arc. On Fig. 12.8 the positive direction of measuring moments is indicated and it should be noted that the arc is in the negative direction. This arc is used as a base line from which to draw the bending moment diagram. The moments at the ends are drawn from a and b. Let aA and bB be the end moments of magnitude M. They must be positive in sign in order to give a mid-height moment of equal magnitude as the lateral load produces a negative moment which is greater at mid-height than at the ends. The polar circle is then drawn through O, A and B; the distance between the arc ab and the polar circle gives the bending moment, the moment at a point distant x from the end A being given by xX.

Let OC be the diameter of the polar circle. Then, since the end moments are equal, A′OC′ represents the mid-height of the stanchion and Cc the central bending moment.

The problem is to show that if bB = cC then each is equal to

$$\left(\frac{\sqrt{2}-1}{\sqrt{2}+1}\right)\frac{4wl^2}{\pi^2}$$

Since OC is a diameter of circle OBCAO the angle CBO is a right angle and the angle BOC = 45°.

Hence $$\mathrm{OC} = \sqrt{2}\,\mathrm{OB}$$

But $$\mathrm{OC} = \frac{w}{\mu^2} + M$$

and $$\mathrm{OB} = \frac{w}{\mu^2} - M$$

Hence $$\frac{w}{\mu^2} + M = \sqrt{2}\left(\frac{w}{\mu^2} - M\right)$$

giving $$M = \left(\frac{\sqrt{2}-1}{\sqrt{2}+1}\right)\frac{w}{\mu^2}$$

Substituting for μ^2 gives $$M = \left(\frac{\sqrt{2}-1}{\sqrt{2}+1}\right)\frac{4wl^2}{\pi^2}$$

$$= 0{\cdot}0694wl^2$$

Example 12.7. A pin-ended strut 2 m long carries an axial load equal to one-quarter of the Euler critical load, a concentrated lateral load of 1100 N at mid-height and positive end couples of 200 and 400 N m at the top and bottom respectively. Determine the bending moment under the load.

Solution. This is an example of a polar diagram applied to a concentrated load. The strut is shown diagrammatically in Fig. 12.9.

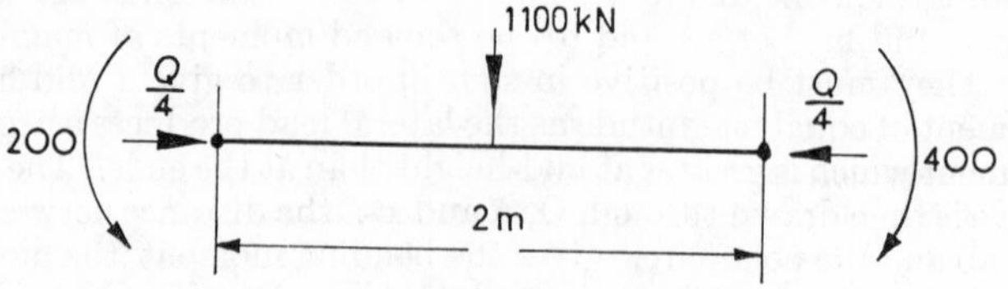

Fig. 12.9

$$\mu = \sqrt{\frac{P}{EI}} = \sqrt{\frac{\pi^2 EI}{4l^2} \cdot \frac{1}{EI}}$$

$$= \frac{\pi}{2l} = 0{\cdot}785 \text{ m}^{-1} \text{ units}$$

The polar angle μl is therefore $\frac{\pi}{2}$ and this is set out as $\widehat{AOB}$ in Fig. 12.10. AO is made 400 units and BO 200 units to some convenient scale. The concentrated load is at mid-height, hence the polar angle representing this load is at $\frac{\pi}{4}$ from either end. Thus $\widehat{AOC}$ is drawn at 45°. From B a line is drawn perpendicular to BO. From any point K on this perpendicular a line KK′ is drawn perpendicular to OC, the length KK′ being equal to W/μ, where W is the magnitude of the concentrated load. In the example, $W = 1100$ N and $\mu = 0{\cdot}785$, thus $W/\mu = 1390$ N m units. From A a line is drawn perpendicular to AO and from K′ a line drawn parallel to BK. Let these two lines intersect in J′. The line J′J is now drawn parallel to K′K to meet BK in J and CO produced in C′. The two polar circles for the left- and right-hand portions of the beam are AOC′ and BOC′ respectively, the shaded area giving the bending moment diagram for the strut. The bending moment under the load is OC′ and scaling off gives this as 270 N m approximately.

Example 12.8. A steel tube 50 mm in diameter and 1·42 mm thick is used as a pin-ended strut 3 m long to carry an axial load of 4·5 kN and a uniformly distributed lateral load of 80 N/m run. The final deflected form of the strut can be assumed a cosine curve and $E = 200$ kN/mm². Calculate the maximum fibre stress.

Solution. Since the final deflected form of the strut is a cosine curve, Perry's approximation applies.

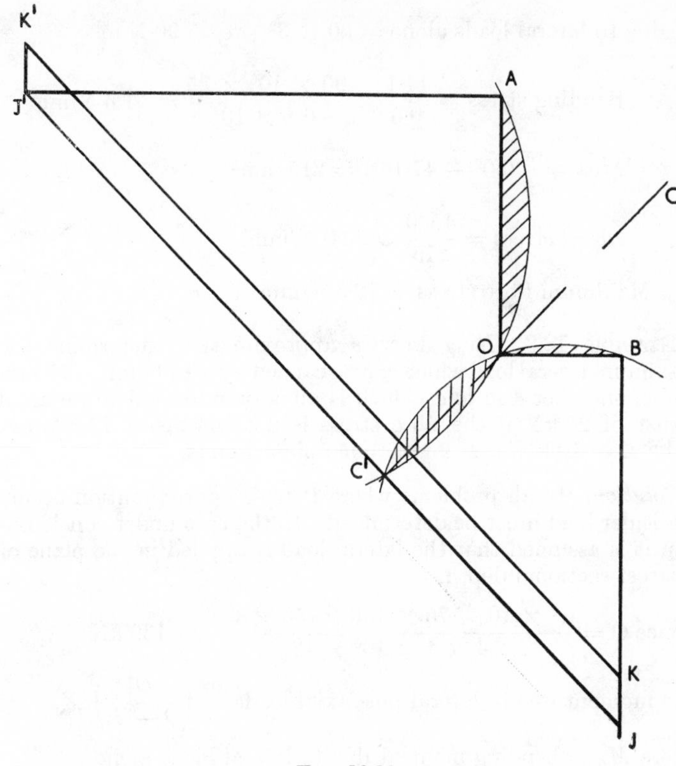

FIG. 12.10

With this approximation the stress due to bending caused by the lateral load plus the eccentricity of the axial load is given by

$$\left(\frac{p_e}{p_e - p}\right) p_b$$

where p_e = Euler stress
p = axial stress
p_b = bending stress due to lateral loads alone

In the example

$$I = \frac{\pi}{64}(50^4 - 47{\cdot}16^4) = 6{\cdot}4 \times 10^4 \text{ mm}^4$$

$$Q = \frac{\pi^2 EI}{l^2} = \frac{\pi^2 \times 200 \times 6{\cdot}4 \times 10^4}{3 \times 3 \times 10^6} = 14{\cdot}1 \text{ kN}$$

M due to lateral loads alone $= 80 \times 3^2 \times \frac{1}{8} = 90$ N m

$$\therefore \quad \text{Bending stress} = \frac{14{\cdot}1}{9{\cdot}6} \times \frac{90 \times 10^3 \times 25}{6{\cdot}4 \times 10^4} = 51{\cdot}6 \text{ N/mm}^2$$

$$\text{Area} = \frac{\pi}{4}(50^2 - 47{\cdot}16^2) = 215 \text{ mm}^2$$

$$\therefore \quad \text{Direct stress} = \frac{4500}{215} = 20{\cdot}9 \text{ N/mm}^2$$

$$\therefore \quad \text{Maximum fibre stress} = 72{\cdot}5 \text{ N/mm}^2$$

Example 12.9. Using Perry's approximation, determine the maximum lateral load which can be carried by a 150 mm × 75 mm timber member 4 m long which is already subjected to an axial thrust of 25 kN if the fibre stress is not to exceed 15 N/mm². Take $E = 10$ kN/mm² and assume pinned ends.

Solution. In all problems where Perry's approximation occurs the Euler load must be determined. In the case under consideration it is assumed that the lateral load is applied in the plane of greatest sectional depth.

$$\text{Hence } Q = \frac{\pi^2 \times 10 \times 75 \times 150 \times 150 \times 150}{4 \times 4 \times 10^6 \times 12} = 136 \text{ kN}$$

$$\text{The moment due to lateral plus axial loads} = \left(\frac{Q}{Q - P}\right) M_0$$

where M_0 = bending moment due to lateral loads alone.

$$\frac{Q}{Q - P} = \frac{136}{111} = 1{\cdot}224$$

$$\text{The stress due to axial load} = \frac{25 \times 10^3}{75 \times 150}$$
$$= 2{\cdot}22 \text{ N/mm}^2$$

$\therefore$ Stress available to resist bending $= 12{\cdot}78$ N/mm²

$\therefore$ Bending moment taken by section

$$= \frac{12{\cdot}78 \times 75 \times 150^2}{6 \times 10^3} = 1{\cdot}224 \times w \times 4^2 \times \frac{1}{8} \text{ (N/m units)}$$

where w = intensity of lateral loading.

$$\text{Equating gives } w = \frac{12{\cdot}78 \times 75 \times 150 \times 150 \times 8}{1{\cdot}224 \times 4^2 \times 6 \times 10^3}$$
$$= 1465 \text{ N/m run.}$$

EXAMPLES FOR PRACTICE ON STRUTS

1. Two similar vertical struts are built-in at their lower ends and at their upper ends support a horizontal girder AB which carries a load W attached to it as shown in Fig. 12.11. The girder may be considered to be rigid and the joints at A and B pin joints.

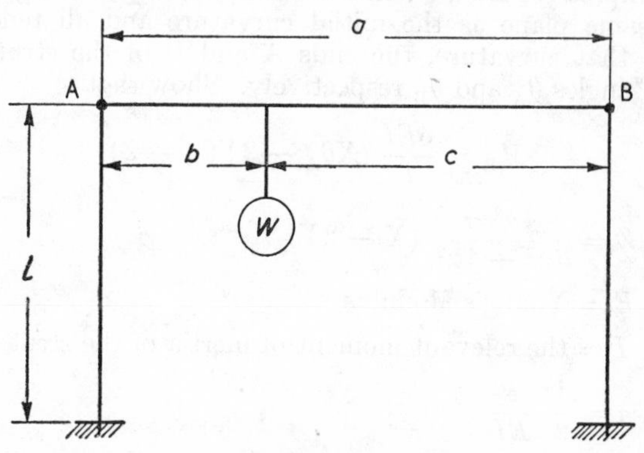

FIG. 12.11

Show that the deflexions in the vertical plane of the frame will tend to become large if W exceeds the value given by the solution of the equation

$$pla = c\sqrt{\frac{a}{b}}\tan pl\sqrt{\frac{b}{a}} + b\sqrt{\frac{a}{c}}\tan pl\sqrt{\frac{c}{a}}$$

where $p^2 = \frac{W}{EI}$, I being the relevant moment of inertia of the cross-section of the strut. (*Cambridge*)

2. A strut of length l is built-in at one end, its other end being supported in such a way that if any transverse displacement δ occurs there, a transverse restraining force $\mu\delta$ acts on the strut at that end.

Show that the buckling load, P, is given by the solution of the equation

$$\tan \alpha l = \alpha l - \frac{\alpha P}{\mu}$$

where $\alpha^2 = \frac{P}{EI}$ (*Cambridge*)

3. A strut AB of uniform cross-section and length l with hinged ends fixed in position has a small initial curvature defined by the equation

$$y = e \sin \frac{\pi x}{l}$$

x being the distance from the end A. Under the influence of an axial compressive load, P, and end couples, M_A and M_B, applied in the same plane as the initial curvature and all tending to increase that curvature, the ends A and B of the strut rotate through angles θ_A and θ_B respectively. Show that

$$M_B = \frac{6EI}{l}(X\theta_A - 2Y\theta_B - Z)$$

where $Z = \dfrac{4\alpha^2 e\pi}{l(\pi^2 - 4\alpha^2)}(X - 2Y)$

$E =$ Young's Modulus

$I =$ the relevant moment of inertia of the strut

$\alpha = \dfrac{l}{2} \cdot \dfrac{P}{EI}$

and X and Y are functions of α. (*Cambridge*)

4. The ends A and B of a strut of uniform cross-section and of length l are prevented from any rotation and the middle point C of the strut is maintained by a guide which prevents any lateral deflexion but offers no restraint to rotation. Prove that the axial load which will just produce buckling is given by $\dfrac{81EI}{l^2}$ nearly, where I is the least moment of inertia of a cross-section and E is Young's modulus.

The equation $\tan x = x$ is satisfied by $x = 4{\cdot}49$. (*Cambridge*)

5. A uniform strut of length l freely hinged at the ends is subjected to an axial force, P, and to a transverse force, Q, which is applied at the centre of the strut perpendicular to the axis and in the plane in which the buckling would occur. The relevant moment of inertia of the cross-section is I.

Find an expression for the deflexion at the centre of the strut and prove that the greatest bending moment has the value $\dfrac{Q}{2\alpha} \tan \dfrac{\alpha l}{2}$ where $\alpha^2 = \dfrac{P}{EI}$. (*Cambridge*)

6. A vertical stanchion of uniform cross-section is built-in at the base and carries a load, P, on a bracket attached to the stanchion as shown in Fig. 12.12. Bending is resisted by a force,

F, applied in the plane of bending, which varies directly as the deflexion, b, of the top. Find an expression for the bending

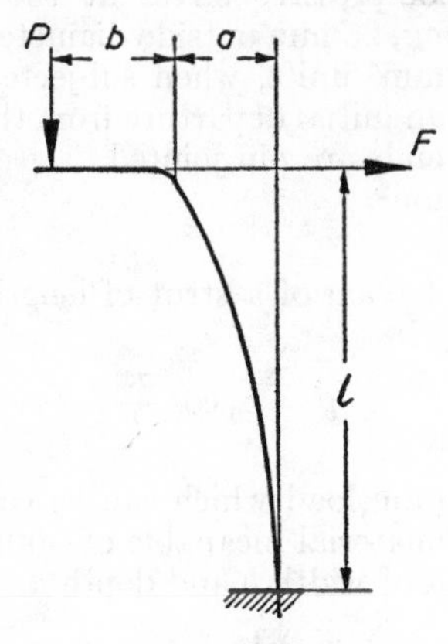

FIG. 12.12

moment at the base and show that if $\dfrac{F}{b} = \dfrac{2P}{l}$, this bending moment is

$$Pa\left\{\frac{2 \sin \alpha l - \alpha l}{2 \sin \alpha l - \alpha l \cos \alpha l}\right\}$$

where $\alpha^2 = \dfrac{P}{EI}$ and E and I have the usual significance.

(*Cambridge*)

7. A pin-ended column 6 m long is made of steel tube 200 mm outside diameter and 6 mm wall thickness. It carries an axial load of 120 kN. Calculate what lateral load, uniformly distributed, it can carry in addition if the maximum stress is limited to 110 N/mm². $E = 200$ kN/mm². Prove any formula you use.

(*London*)

Answer. 15·5 kN.

8. A column AB carrying an axial load consists of three rigid parts AC, CD and DB. The joints at C and D consist of springs which allow bending at the rate of k radians per N/m. The lengths of the sections are: AC $= l$, CD $= 2l$, DB $= 2l$. Assuming the column to be pin-jointed at A and B, find the value of the critical load which will cause buckling. (*Oxford*)

Answer. $0{\cdot}69/kl$.

9. Deduce Perry's formula for a long column which has an initial departure from the straight at the centre of c_0.

Find the total compressive stress at the centre of a steel tubular strut 2 m long, 25 mm outside diameter, area = 107 mm² and $I = 8{\cdot}3 \times 10^3$ mm⁴ units, when subjected to an axial load of 1·8 kN if there is an initial departure from the straight of 3 mm at the centre. The ends are pin-jointed.

Answer. 32·3 N/mm².

10. The initial deflexion of a strut of length l is given by the equation

$$y = c_0 \cos \frac{\pi x}{l}$$

Calculate the maximum load which can be carried by this strut if it is made from a material incapable of taking tension and has a rectangular section of width b and depth d.

Answer. $\left[\frac{\pi E b d^3}{12 l^2}\left(1 - \frac{6c_0}{d}\right)\right]$. *(St. Andrews)*

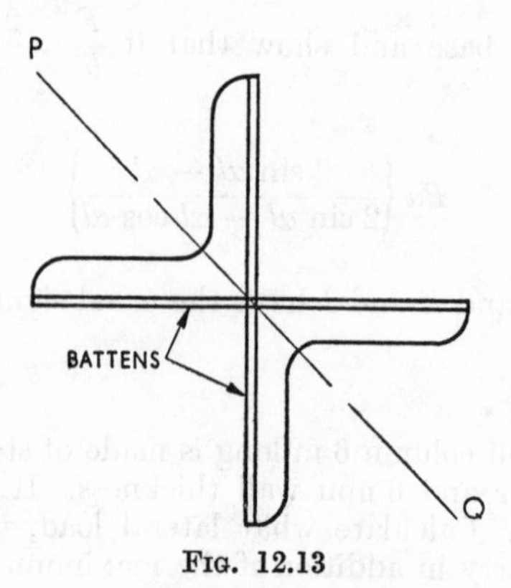

FIG. 12.13

11. Two 75 mm × 75 mm × 12 mm angles are battened together at intervals to give the strut section shown in Fig. 12.13, which has a gross area of $3{\cdot}55 \times 10^3$ mm² and a radius of gyration about PQ of 26 mm. The strut has pin-jointed ends and a length of 3 m. Calculate the working load from the following formula—

Rankine–Gordon with $a = \dfrac{1}{7500}$

Perry–Robertson with $\eta = 0{\cdot}003 l/k$

Johnson's Parabolic with $r = 6{\cdot}70 \times 10^{-3}$ N/mm²

Straight line — $p = 120\ \text{N/mm}^2 \left(1 - \dfrac{l}{300k}\right)$

In each case take $p_y = 230$ N/mm² and $N = 2{\cdot}0$.

Answer. Rankine, 147; Perry, 188; Parabolic, 250; Straight line, 262. (All loads in kN.)

12. Using the Rankine–Gordon formula for pinned-ended struts, determine the outside diameter required for a thin-walled mild steel column 10 m high to enable it to carry a load of 80 kN with a load factor of 2. Take the yield stress in the material as 230 N/mm², the wall thickness as 3 mm and $a = \dfrac{1}{7500}$.

Answer. 227 mm.

13. The Perry–Robertson formula states that the stress, p, in an axially loaded column, which causes the fibres to yield, is given from the expression

$$p = \frac{p_y + (\eta + 1)p_e}{2} - \sqrt{\left\{\frac{p_y + (\eta + 1)p_e}{2}\right\}^2 - p_y p_e}$$

Using this formula calculate the fibre stress when a steel tube 50 mm outside diameter, 1·5 mm thick and 3 m long is used as a strut to carry an axial load of 4·5 kN. Calculate also the maximum load which can be carried by such a tube, using (*a*) a load factor of 2, (*b*) a factor of safety of 3.

Take $p_y = 230$ N/mm², $\eta = 0{\cdot}003l/k$ and $E = 200$ kN/mm². (*St. Andrews*)

Answer. 32·4 N/mm²; 6·66 kN; 8·5 kN.

14. A thin steel tube of 50 mm diameter and 1·5 mm thick is used as a strut 3 m long with pinned ends to carry an axial load equal to one-ninth of the Euler critical load and a lateral load of 400 N/m run. If the final deflected form of the strut is a cosine curve, determine the maximum fibre stress. Take $E =$ 230 kN/mm². (*London*)

Answer. 198 N/mm².

15. A strut is subjected to a longitudinal thrust equal to 4/9 of its Euler critical load and has a length of 2 m. End moments of 200 N m are applied at each end so as to augment each other's effect at the centre of the strut where a point load is applied which reduces the moment there to zero. What is the value of the point load?

Answer. 485 N.

16. A tube 70 mm external diameter and 2·5 mm thick, 2 m long, carries a compressive load of 40 kN. The load is offset 5 mm from the centre line of the strut at one end and 6 mm at the other. Find the maximum stress in the tube taking E as 230 kN/mm^2.

Answer. 98 N/mm^2.

17. A pin-ended strut of length l carries an axial load equal to one-quarter of the Euler critical load and a lateral load of w/unit length. Couples of equal magnitude are applied at the ends and the result is that there is no bending moment at mid-height. Determine the magnitude of these end couples.

Answer. $0{\cdot}119wl^2$.

18. A strut of length l carries an axial load equal to one-ninth of the Euler critical load and a lateral load of intensity w/unit length. It is pinned at one end and a moment is applied to the other which produces a point of contraflexure at mid-height of the stanchion. Calculate the magnitude of this moment.

(*Glasgow*)

Answer. $0{\cdot}245wl^2$.

19. A pin-ended strut 3 m long carries an axial load equal to one-sixth of the Euler critical load, positive end moments of 200 and 100 N m at the upper and lower ends respectively, and lateral loads of 135 N/m run and 270 N/m run over the upper 1·2 m and lower 1·8 m respectively. Determine the maximum negative moment, and the bending moment and shearing force at the centre.

Answer. 135 N m; 100 N m; 81 N.

20. A strut 1·8 m long carries an axial load of one-ninth of the Euler critical load, a concentrated load of 180 N at 0·6 m from the left end and terminal couples of 30 and 60 N m at the left- and right-hand end respectively. Find the bending moment under the load and the position of the points of contraflexure.

Answer. − 320 N m; 0·38 m; 1·10 m.

Chapter 13

The Suspension Bridge

THE SUSPENSION BRIDGE consists essentially of a cable of light weight from which a number of hangers are suspended. A girder forming the bridge deck is attached to the hangers as shown diagrammatically in Fig. 13.1.

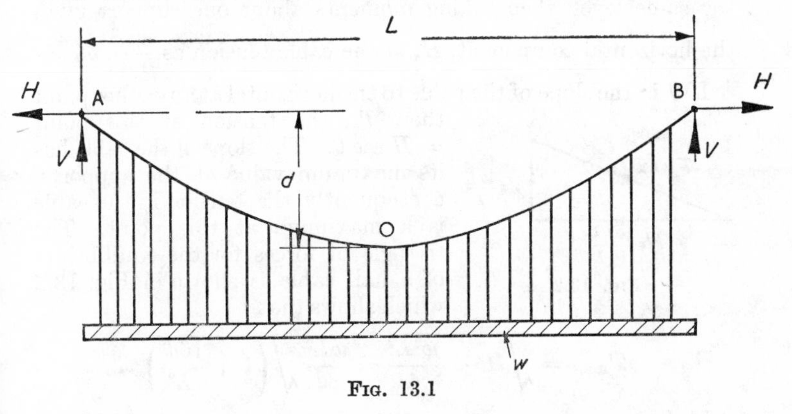

FIG. 13.1

The exact theory of the suspension bridge is complicated and beyond the scope of most first degree examinations. The simplified theory which is generally taught up to this stage is based on the following assumptions.

1. The cable does not change its shape under load and its weight is negligible.
2. The load on the stiffening girder produces an evenly distributed load in all the hangers.
3. The hangers produce a uniformly distributed load on the cable.

These assumptions, whilst giving a simple theory, are far from the truth. If a load on the stiffening girder is to produce a uniform load in the hangers without the cable changing shape, it implies that the girder is infinitely stiff. If this is so, then there is obviously no point in introducing a cable to carry the load for the infinitely stiff girder will carry it unaided.

Any true theory of the suspension bridge must take into account the deflexion of the stiffening girder under load. Most of the deflexion theories are difficult, but the student will find that due to Professor A. G. Pugsley, published in *The Structural Engineer* for March, 1953, quite easy to follow. It will give him a picture which is far nearer the truth than the "uniform load" theory which is dealt with in this chapter.

The suspension bridge has two main components, one is the cable and the other the stiffening girder.

The cable can take no bending and the force is a direct tension. Consider the cable shown in Fig. 13.1 which supports a load of w/unit length. The cable is horizontal at its lowest point, consequently the force at this point is horizontal. If the span of the cable is L, the dip at the lowest point d and the supports are at the same level, then taking moments about one support gives the horizontal component, H, of the cable tension as $\dfrac{wL^2}{8d}$.

If θ is the slope of the cable to the horizontal at any other point then T_θ, the tension at this point $= H \sec \theta$. The slope of the cable has its maximum value at the supports, consequently the tension in the cable is a maximum at this point. The triangle of forces for the equilibrium of a half cable is shown in Fig. 13.2 which shows that

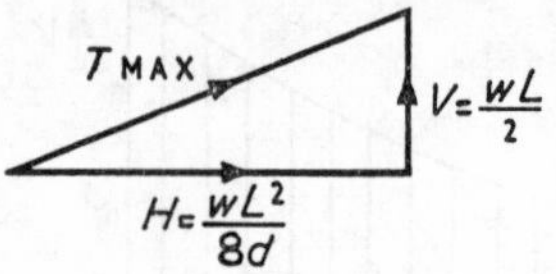

FIG. 13.2

$$T_{\max} = \sqrt{H^2 + \frac{w^2L^2}{4}} = \frac{wL^2}{8d}\sqrt{\left(1 + \frac{16d^2}{L^2}\right)}$$

If the two supports are at different levels, the first problem is to find the position of the lowest point on the cable relative to the supports. Consider the cable shown in Fig. 13.3 where the support B is at a height h above A. Let the lowest point be at a horizontal distance x from A.

Equating the values of the horizontal tensions, obtained by considering each portion separately, at the lowest point gives

$$\frac{(L - x)^2}{(h + d)} = \frac{x^2}{d}$$

a quadratic equation from which the value of x can be obtained.

Having obtained x, the value of $H = \dfrac{wx^2}{2d}$ can be found. Then by using H, the tension in the cable at any other point can be obtained.

The cable is attached to the tops of towers which are in turn anchored by cables to foundations. There are two main methods

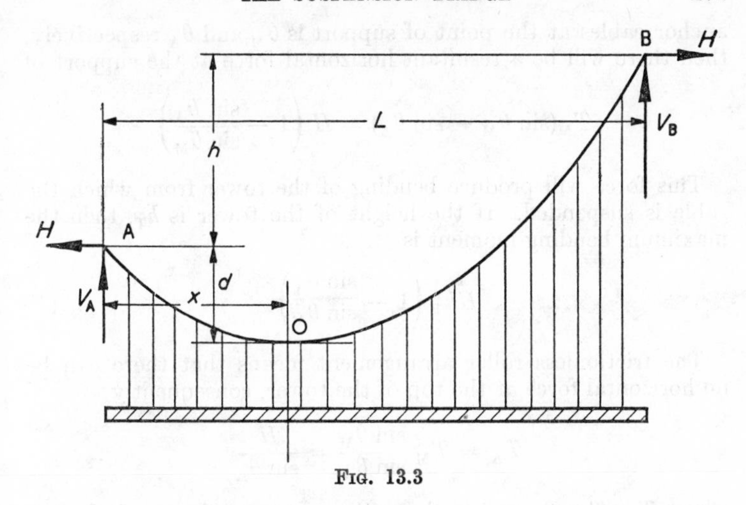

Fig. 13.3

by which the main and anchor cables are connected: (i) over a pulley arrangement as shown in Fig. 13.4, or (ii) by a frictionless roller as in Fig. 13.5.

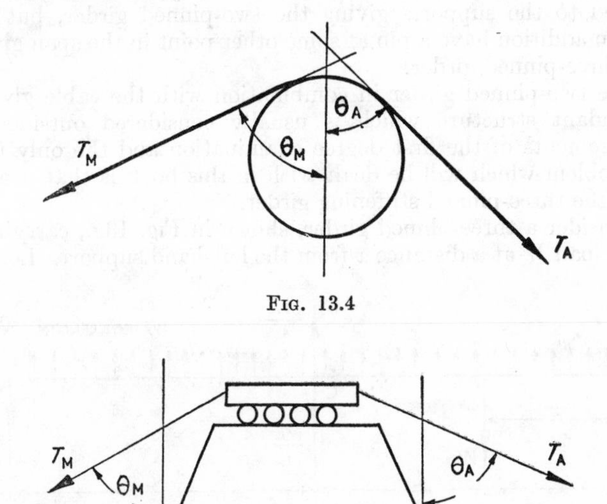

Fig. 13.4

Fig. 13.5

The pulley arrangement means that the tension in the anchor cable, T_A, must be equal to the maximum tension in the main cable, T_M. If the inclination to the horizontal of the main and anchor cables at the point of support is θ_M and θ_A respectively, then there will be a resultant horizontal force at the support of

$$T_M(\sin\theta_M - \sin\theta_A) = H\left(1 - \frac{\sin\theta_A}{\sin\theta_M}\right)$$

This force will produce bending of the tower from which the cable is suspended. If the height of the tower is h_T, then the maximum bending moment is

$$Hh_T\left(1 - \frac{\sin\theta_A}{\sin\theta_M}\right)$$

The frictionless roller arrangement means that there can be no horizontal force at the top of the tower, consequently

$$T_A = T_M\frac{\sin\theta_M}{\sin\theta_A} = \frac{H}{\sin\theta_A}$$

where T_A, T_M, θ_A and θ_M have the same meanings as before.

The use of the frictionless roller eliminates bending in the towers.

The stiffening girder can be of two main types. It is generally pinned to the supports giving the two-pinned girder, but can also in addition have a pin at some other point in the span giving the three-pinned girder.

The two-pinned girder in combination with the cable gives a redundant structure which is usually considered outside the requirements of the first degree examination and the only type of problem which will be dealt with in this book is that dealing with the three-pinned stiffening girder.

Consider a three-pinned girder, shown in Fig. 13.6, carrying a point load W at a distance x from the left-hand support. Let the

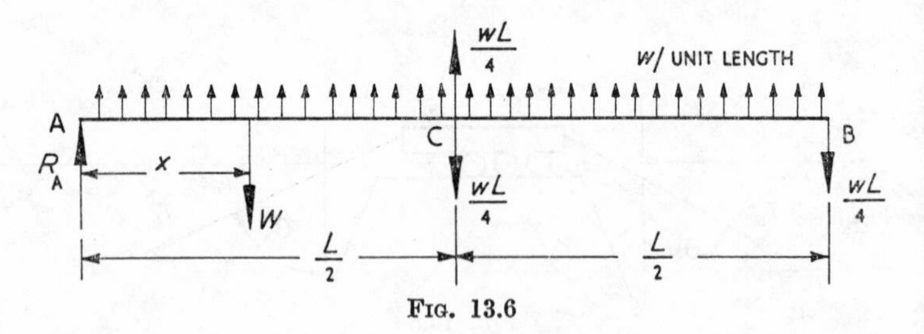

FIG. 13.6

span of the girder be L, the third pin at mid-span, and let w be the intensity of uniform load which the girder transmits to the cable. The problem is to determine w. When this is known then the bending moment and shearing force at any point on the girder can be calculated.

A useful method of dealing with the stiffening girder problem is to consider the equilibrium of each portion separately. Take first the right-hand portion CB. This is in equilibrium under the action of w acting upwards over its whole length together with downward reactions at B and C. These two reactions must each be equal to half the load on BC or expressed symbolically

$$R_B = R_C = \frac{wL}{4}$$

Consider now the left-hand portion AC. The forces acting on this are W downwards, w upwards and the reactions at A and C. But pin C is in equilibrium, therefore R_C in span AC is equal and opposite to R_C in span CB. In other words, in the span AC,

$$R_C = \frac{wL}{4}$$

acting upwards. Taking moments about A gives

$$W \times x = \frac{wL}{4} \cdot \frac{L}{2} + \frac{wL}{2} \cdot \frac{L}{4}$$

leading to
$$w = \frac{4Wx}{L^2}$$

Finally the reaction at A can be obtained by vertical resolution of the forces acting on AC giving

$$R_A = \tfrac{3}{4}wL - W = W\left(\frac{3x}{L} - 1\right)$$

Worked Examples on Suspension Bridges

Example 13.1. A suspension cable hangs between two points A and B separated horizontally by 100 m and with B 15 m above A. The lowest point in the cable is 3 m below A.

The cable supports a stiffening girder weighing 10 kN/m run which is hinged vertically below A, B and the lowest point of the cable. Calculate the maximum tension which occurs in the cable when a 200 kN wheel load crosses the girder from A to B.

(*St. Andrews*)

Solution. The question is shown diagrammatically in Fig. 13.7.

Let x be the distance of the lowest point of the cable from the left-hand support.

Consideration of the equilibrium of the left-hand portion of the cable gives

$$H = \frac{wx^2}{6}$$

whilst the right-hand portion gives

$$H = \frac{w(100 - x)^2}{36}$$

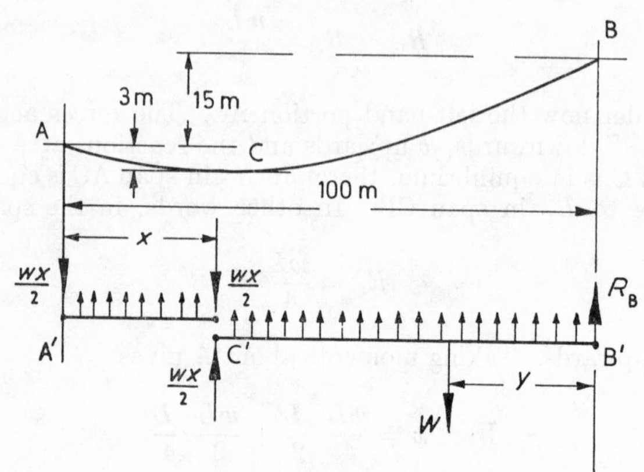

FIG. 13.7

Equating these values gives

$$\frac{x^2}{6} = \frac{(100 - x)^2}{36}$$

leading to $x = 28{\cdot}9$ m.

Now consider the two portions of the stiffening girder and let the load be on the portion C′B′ at a distance y from B′.

Taking moments about B′ gives

$$\frac{w(100 - x)^2}{2} + \frac{wx(100 - x)}{2} = 200 \times y$$

Here the variables are w and y; w will obviously have its maximum value when y is a maximum. This occurs when

$$y = (100 - x)$$

The equilibrium equation then becomes

$$w(100 - x) + wx = 2 \times 200$$

giving
$$w = \frac{400}{100} = 4 \text{ kN/m run}$$

The total weight transmitted to the cable is therefore $10{\cdot}0 + 4{\cdot}0 = 14{\cdot}0$ kN/m.

The maximum tension in the cable will occur at B. Consideration of the triangle of forces for this portion of the cable gives

$$T = \sqrt{H^2 + w^2(100 - x)^2}$$

where $H = \dfrac{wx^2}{6}$.

Substituting $x = 28{\cdot}9$ and $w = 14{\cdot}0$ gives $T_{\max} = 2{,}150$ kN.

Example 13.2. A suspension cable has a span L and a central dip d and is stiffened by a girder which is pinned to the abutments and hinged at a point $\dfrac{L}{4}$ from the left-hand abutment. Show that if $\dfrac{d}{L}$ is small the maximum tension in the cable when a unit load is placed anywhere on the girder is approximately $\dfrac{L}{4d}$.

Draw the bending moment diagram for the girder when the load is in the position causing this maximum tension. (*London*)

Solution. The cable and girder are shown diagrammatically in Fig. 13.8.

Consider the stiffening girder ACB. Let the load be at a distance x from A and on the portion AC.

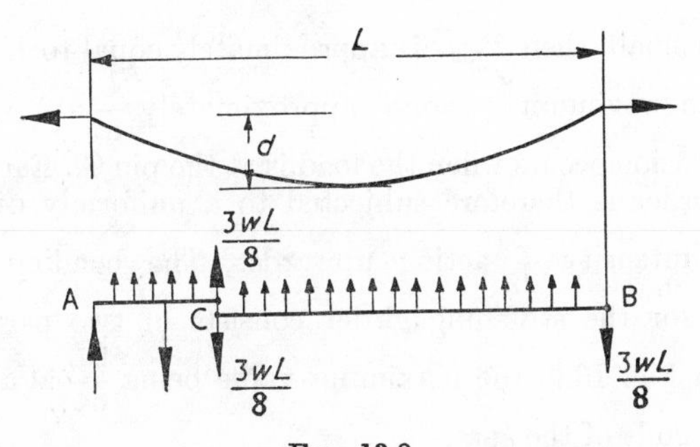

Fig. 13.8

The equilibrium of the girder CB gives the vertical reaction on the pin at C as $\dfrac{3wL}{8}$. This acts downwards on span CB and upwards on AC.

Taking moments about A for the girder AC gives

$$1 \times x = \frac{wL}{4} \cdot \frac{L}{8} + \frac{3wL}{8} \cdot \frac{L}{4} = \frac{wL^2}{8}$$

Hence $$w = \frac{8x}{L^2}$$

The maximum value of w therefore occurs when x has its maximum value. This is when

$$x = \frac{L}{4}$$

Hence $$w_{\text{max}} = \frac{8}{L^2} \cdot \frac{L}{4} = \frac{2}{L}$$

The horizontal component of the cable tension has been shown to be $\frac{wL^2}{8d}$, and putting $w = \frac{2}{L}$ gives

$$H = \frac{L}{4d}$$

The maximum tension in the cable is obtained by considering the triangle of forces for half the cable from which

$$T^2_{\text{max}} = H^2 + \frac{w^2L^2}{4} = H^2\left(1 + \frac{16d^2}{L^2}\right)$$

If $\frac{d}{L}$ is small, then T_{max} is approximately equal to H; in other words the maximum tension is approximately $\frac{L}{4d}$.

This tension occurs when the load is at the pin C. Each portion of the girder is therefore subjected to a uniformly distributed load of intensity $\frac{2}{L}$ acting upwards. The bending moment diagram for the stiffening girder consists of two parabolas as shown in Fig. 13.9, the maximum value being $\frac{9L}{64}$ at a point $\frac{3L}{8}$ from the ends of the span.

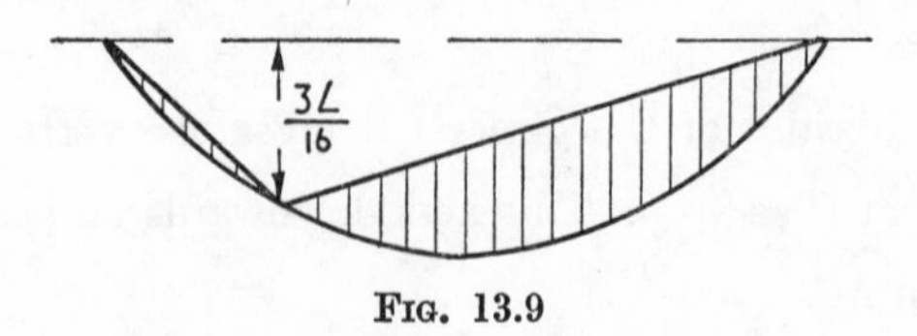

FIG. 13.9

Example 13.3. A suspension cable is suspended from two piers 140 m apart, one support being 3·5 m above the other. The cable carries a uniformly distributed horizontal load of 15 kN/m and has its lowest point 7 m below the lower support. The ends of the cable are attached to saddles on rollers at the top of the piers and the backstays which may be assumed straight are inclined at 60° to the vertical.

Determine the maximum tension in the cable, the tension in the backstays and the thrust on each pier. (*London*)

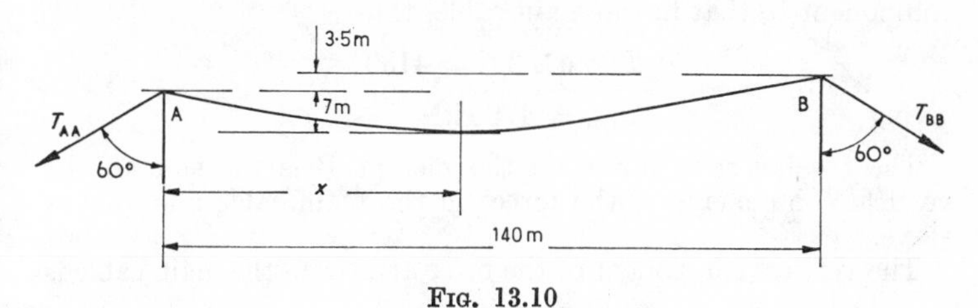

FIG. 13.10

Solution. The cable is shown diagrammatically in Fig. 13.10. Let x be the distance of the lowest point of the cable from the left-hand support. Consideration of the forces acting on the left-hand portion of the cable gives

$$H = \frac{wx^2}{14}$$

whilst consideration of the forces on the right-hand portion gives

$$H = \frac{w(140 - x)^2}{21}$$

These two values of H must be equal, hence

$$\frac{x^2}{14} = \frac{(140 - x)^2}{21}$$

or

$$x^2 = \frac{14}{21}(140 - x)^2$$

Taking the square root of both sides gives

$$x = \pm\, 0{\cdot}818(140 - x)$$

hence

$$x = 62{\cdot}1 \text{ m}$$

The maximum tension in the cable occurs at B. If its value is T_B, this is given by the following equation

$$T_B{}^2 = H^2 + w^2 \,.\, 77{\cdot}9^2$$

Substituting for w and x in the previous expression for H gives

$$H = 15 \cdot \frac{62{\cdot}1^2}{14} = 4130 \text{ kN}$$

thus $$T_B = \sqrt{4130^2 + 15^2 \times 77{\cdot}9^2} = 4270 \text{ kN}$$

The cable is supported on frictionless rollers, consequently there is no horizontal force at the top and the tension in the anchor chain at B, T_{BB}, is found by equating its horizontal component to that in the main cable, thus

$$T_{BB} \cos 30° = 4130$$

giving $$T_{BB} = 4780 \text{ kN}$$

The total vertical force on the pier at B is the sum of the vertical components of the forces in the main cable and anchor stays.

The vertical component of the force at B from the main cable is

$$15 \times 77{\cdot}9 = 1170 \text{ kN}$$

whilst that in the anchor stays is

$$T_{BB} \cos 60° = 2390 \text{ kN}$$

Thus the total thrust in the pier at B is

$$1170 + 2390 = 3560 \text{ kN}$$

Since both anchor chains are inclined at the same angle to the vertical and H is constant throughout the cable, the tension in the anchor chain at A, T_{AA}, is the same as the tension in the anchor chain at B, i.e. $T_{AA} = 4780$ kN.

The vertical component of the load in the main cable at A is 933 kN, therefore the total thrust in the pier at A is

$$933 + 2390 = 3323 \text{ kN}$$

Examples for Practice on the Suspension Bridge

1. The roadway for a suspension bridge of span $2l$ is stiffened by longitudinal girders of length l pin-jointed together at the centre of the span and hinged at their outer ends to the abutments.

The girders are supported by a large number of vertical tie rods attached to suspending chains, the lengths of the tie rods being such that each chain is in the form of a parabola with the axis vertical.

Show that the greatest hogging and sagging bending moments set up in the bridge by a concentrated load W advancing across the span are $\frac{Wl}{8}$ and $\frac{Wl}{3\sqrt{3}}$. (*Cambridge*)

2. A small suspension bridge is stiffened by a three-hinged girder to carry light traffic over a river.

Dimensions and Data

Span = 50 m
Central Dip of Cable = 5 m
Self weight of bridge carried by one cable = 10 kN/m
Live load per cable = 15 kN/m
Working stress in girder = 150 N/mm²
Allowable stress in cable = 300 N/mm²

Answers Required

(*a*) The sectional area of one cable.
(*b*) The required section modulus for the stiffening girder.

(*Durham*)

Answer. $5{\cdot}6 \times 10^3$ mm²; $4{\cdot}71 \times 10^6$ mm³ units.

3. A pipe line CD weighing 1500 kN is carried over a river by means of a suspension cable AB as shown (Fig. 13.11). The lowest point of the cable, X, is 2 m below A and 9 m below B.

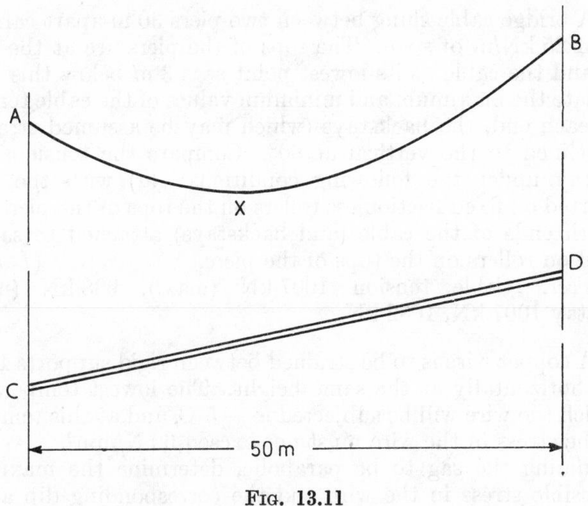

FIG. 13.11

If the pipe line may be considered to have no flexural rigidity, calculate the position of X and the horizontal component of tension in the cable. (*London*)

Answer. 16 m from A; $H = 1920$ kN.

4. An unstiffened suspension bridge spans a horizontal distance of 30 m between piers A and B, A being 1 m below the level of B and the lowest point of the suspension cables 4 m below A.

The cables and anchor chains are attached to saddles and are thus free to move horizontally on the piers and the anchor chains are inclined at 45° and may be assumed straight.

Each cable carries a load of 15 kN/m horizontally. Calculate: (*a*) the tension at the lowest point of the cables, (*b*) the tension at the higher pier B, and (*c*) the thrust on the pier B. (*London*)

Answer. 380 kN; 450 kN; 617 kN.

5. A suspension cable is supported at points A and B which are 160 m apart measured horizontally; B is 8 m above A and the maximum dip of the cable below A is 6 m. The cable is stiffened by a girder hinged at the three points vertically below A, B and the point of maximum dip.

Plot the bending moment diagram on the girder when a load of 10 kN is placed half way between A and the point of maximum dip.

Calculate the values of the maximum positive and negative bending moments. (*London*)

Answer. 73 kN m; 127 kN m.

6. A bridge cable slung between two piers 30 m apart carries a load of 25 kN/m of span. The tops of the piers are at the same level and the cable at its lowest point sags 3 m below this level. Calculate the maximum and minimum values of the cable tension.

At each end, the backstays (which may be assumed straight) are inclined to the vertical at 60°. Compare the tensions in a backstay under the following conditions: (*a*) with the cable supported on fixed frictionless rollers on the tops of the piers, and (*b*) with ends of the cable (and backstays) attached to saddles carried on rollers on the tops of the piers. (*London*)

Answer. Cable tension 1007 kN (max.), 938 kN (min.); Backstay 1007 kN, 1082 kN.

7. A copper wire is to be strained between rigid supports 100 m apart horizontally at the same height. The lowest temperature to which the wire will be subjected is -5°C, and at this temperature the stress in the wire must not exceed 60 N/mm^2.

Assuming the sag to be parabolic, determine the maximum permissible stress in the wire and the corresponding dip at the middle of the span when the wire is erected at a temperature of 20°C.

Weight of wire $= 8{\cdot}55 \times 10^{-5}$ N/mm^3.

Coefficient of linear expansion $= 1{\cdot}87 \times 10^{-5}$ per °C. (*London*)

Answer. 48·1 N/mm^2; 2·22 m.

8. One section of a suspension bridge has a span of 30 m with towers of 5 m and 8 m effective height above the road level, and the lowest point on the two cables is 2 m above road level.

Determine the slope of the cables at each tower and the force in each cable at those points if there is a uniform load of 15 kN/m run covering the bridge. (*London*)

Answer. 25° 35′, 434 kN; 28° 50′, 446 kN.

9. A suspension cable of span L and dip d is stiffened by a girder which is pinned to the supports and at mid-span.

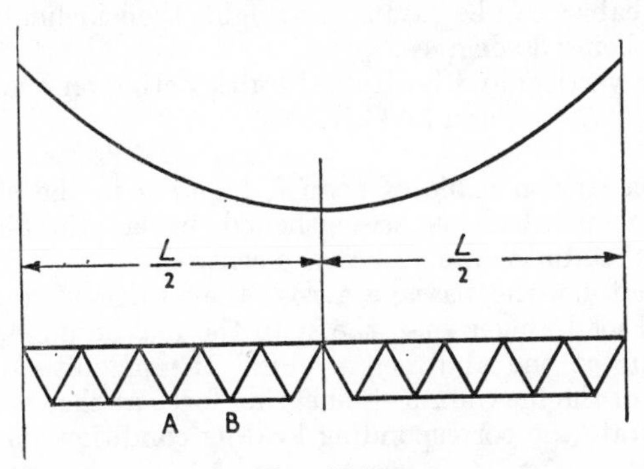

Fig. 13.12

The stiffening girders consist of five-panel inverted Warren girders formed from 60° triangles as shown in Fig. 13.12. (The suspension rods are omitted in the drawing.)

Draw the influence line for the load in AB. (*London*)

Answer. Force varies from $+ 0{\cdot}97W$ to $- 0{\cdot}695W$.

10. A suspension cable having a span of 30 m and a central dip of 6 m is stiffened by a girder which is pinned at its ends to the cable towers and which has a hinge at the centre of the span. It may be assumed that the cable retains its parabolic shape so that the tension in each of the numerous suspension rods is the same whatever the nature or position of the load. A point load of 100 kN moves from one end to the centre of the span.

Calculate: (*a*) the maximum horizontal component of the tension cable, and (*b*) the maximum bending moment in the stiffening girder and the point at which this occurs. (*London*)

Answer. 125 kN; 288 kN m.

11. A bridge of total weight 3600 kN is supported across a span of 100 m by two sets of suspension cables hanging in parabolic form with a central dip of 10 m and strengthened on each side by a stiffening girder hinged at the centre and ends.

Find for a live load of 30 kN/m longer than the span, the necessary section modulus for the stiffening girders and the required cross-sectional area of one set of cables.

Permissible stresses in girder and cables, 125 and 150 N/mm² respectively. (*London*)

Answer. $Z = 2{\cdot}66 \times 10^{7}$ mm³ units; $A = 3{\cdot}03 \times 10^{4}$ mm².

12. An unstiffened suspension cable carries a load of 25 kN/m of span. The span is 60 m and the central dip of the cable 5 m. At the piers the cable is supported on smooth fixed rollers. The anchoring cables can be assumed straight, their inclination to the horizontal being 30 degrees.

Find the vertical and horizontal forces acting on one pier.

Answer. 2370 kN and 300 kN.

13. A suspension cable of span l, hanging in the shape of a symmetrical parabola is strengthened by a stiffening girder pinned at the abutments and at the centre.

Show that for the passage across the bridge of a uniformly distributed load longer than the span the maximum $\pm$ shearing forces occur at the abutments. Find the magnitude of these forces and of the maximum $\pm$ shearing force at the centre of the span and state the corresponding loading conditions in all cases. (*London*)

Answer. $\pm\dfrac{wl}{6}$; $\pm\dfrac{wl}{8}$.

Chapter 14

Riveted and Welded Connexions

THE DESIGN of riveted and welded connexions is a problem which occurs very frequently in engineering construction. The main interest for the structural engineer is in the connexion of beams to stanchions, particularly where such connexion introduces eccentric load on the stanchion.

Consider the bracket shown in Fig. 14.1 in which a load W is transferred from the joist to the stanchion through the group of six rivets. This group is subjected to a direct load W together with a bending moment of We, where e is the distance of the line of action of W from the centre of rotation of the rivet group. This centre of rotation is assumed to coincide with the centre of area of the rivet group.

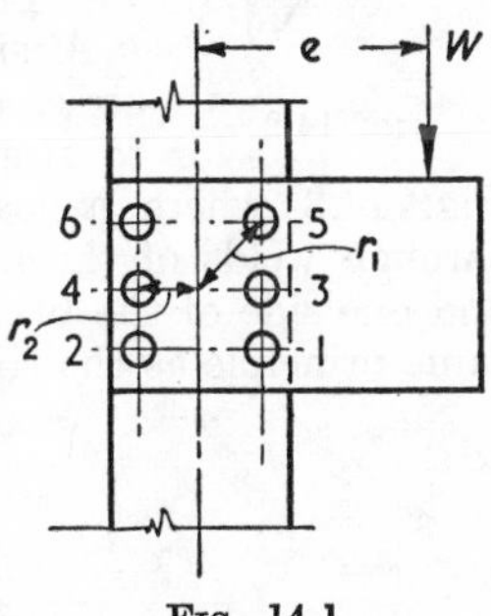

FIG. 14.1

In calculating the load taken by any rivet it is further assumed

1. That the direct load is equally shared by the rivets.
2. That the stress in each rivet due to the bending moment We is directly proportional to the distance of that rivet from the centre of rotation of the group.

Applying these assumptions to the rivet group in Fig. 14.1, and letting P_d and P_b denote the direct and bending load respectively in any rivet, then P_d is obviously equal to $\frac{W}{6}$.

The rivets 1, 2, 5 and 6 are at a distance r_1 from the centre of rotation of the group whilst 3 and 4 are at a distance r_2. Therefore the stress due to bending in rivet 1 is $\frac{r_1}{r_2}$ times the stress due to bending in rivet 3. But the total bending moment of the group about its centre of rotation must equal the applied bending moment We. If the load due to bending in rivets 1 and 3 is denoted by P_{b1} and P_{b3} respectively then

$$4P_{b1} \times r_1 + 2P_{b3} \times r_2 = We$$

But
$$P_{b1} = \frac{r_1}{r_2} P_{b3}$$

hence
$$\frac{P_{b1}}{r_1}(4r_1^2 + 2r_2^2) = We$$

From this relationship the load in the rivet due to bending can be obtained. This load acts perpendicularly to the line joining the rivet in question to the centre of the group. The actual load on any rivet is the resultant of the P_b and P_d forces for that rivet.

The most heavily loaded rivets in the example shown are 1 and 5, the load P_5 being obtained from the triangle of forces for those rivets (see Fig. 14.2).

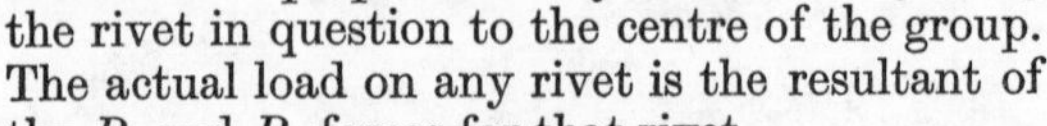

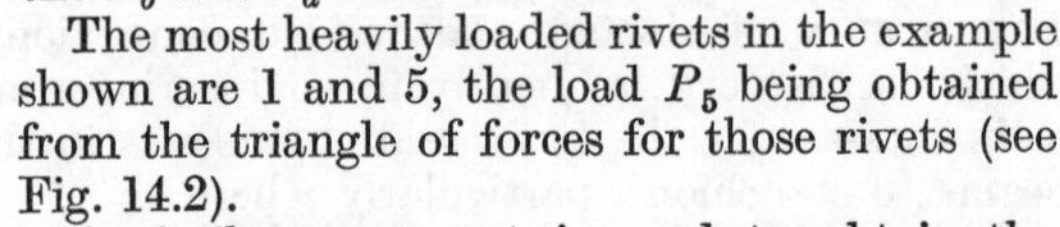

FIG. 14.2

A similar argument is used to obtain the stresses in the case of brackets which are welded to stanchions. Consider the bracket shown in Fig. 14.3 where a load W is transmitted to the stanchion through welds of throat thickness t running along top, bottom and one side of the bracket plate. This joint is designed on the same principle as the riveted one previously described.

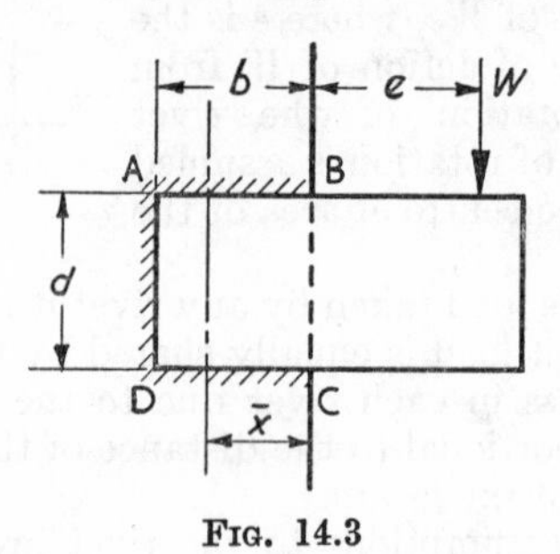

FIG. 14.3

The direct throat stress

$$p_d = \frac{W}{t(d + 2b)}$$

In order to determine the bending stress the position of the centre of area of the welds must be determined. If the distance of this centre is $\bar{x}$ from the front of the stanchion,

$$\bar{x} = \frac{bd + b^2}{2b + d}$$

The bending moment on the weld is then $W(e + \bar{x})$. The second moment of area of the weld about an axis through its centre of area and perpendicular to the plane must be determined. If this is I_{zz}, then its value is given by

$$I_{zz} = I_{yy} + I_{xx}$$

where $I_{xx} = 2 \times bt \times \frac{d^2}{4} + \frac{td^3}{12}$

and $\quad I_{yy} = \frac{2tb^3}{12} + 2bt\left(\bar{x} - \frac{b}{2}\right)^2 + td\left(b - \bar{x}\right)^2$

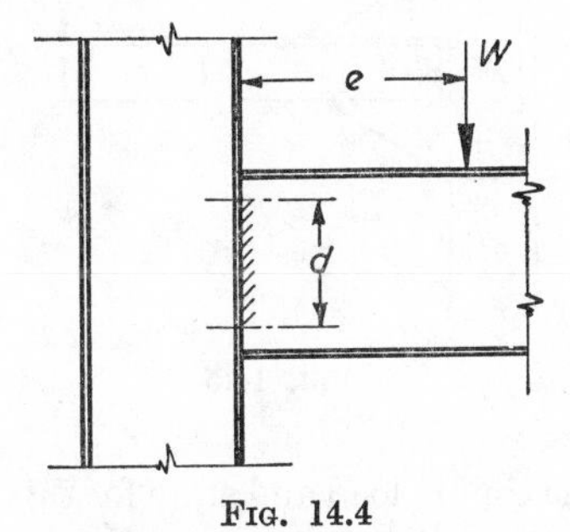

FIG. 14.4

The maximum stress due to bending occurs at B and C. If r = the distance of either of these points from the centre of rotation, then the throat stress due to bending, p_b, is given by

$$p_b = \frac{W(e + \bar{x}) \times r}{I_{zz}}$$

This stress acts perpendicular to the line joining B or C to the centre of rotation. The final stress at either of these points is the resultant of p_b and p_d which is obtained by drawing the triangle of forces for that point.

Another type of welded connexion is shown in Fig. 14.4. If this joint has to carry a load W acting at a distance e from the face of the stanchion, then the direct stress, p_d, is given by $\frac{W}{2dt}$. The stress due to bending, p_b, is given by

$$p_b = \frac{We \times 6}{2td^2}$$

The stresses p_d and p_b act at right angles to each other and their resultant is given by

$$p = \sqrt{p_d^{\,2} + p_b^{\,2}}$$

Worked Examples on Riveted and Welded Joints

Example 14.1. A horizontal member A consisting of two 100 × 100 × 12 mm angles is connected to a vertical member B by means of a gusset plate 9 mm thick fitted between the angles. If the load on the member A is 10 kN acting as shown, determine

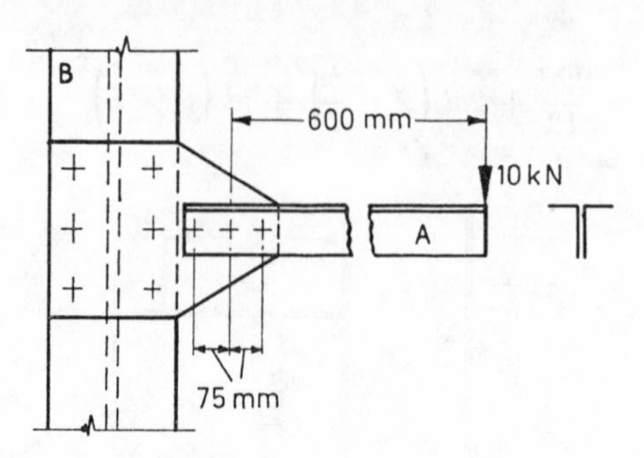

Fig. 14.5

the rivet under maximum load and state for this rivet the shearing and bearing stresses if 19 mm diameter rivets are used (see Fig. 14.5).

Solution. Shear load per rivet $= \frac{10}{3}$ kN

Let $P_b =$ bending load per rivet.

The moment of the applied load about the centre of rotation of the group = 6000 kN mm.

The bending resistance of the rivet group $= 2 \times P_b \times 75$.

$$\text{Thus } P_b = \frac{6000}{150} = 40 \text{ kN}.$$

This load is acting vertically downwards on rivet 1 and upwards on rivet 3, whilst the shear load of $\frac{10}{3}$ kN acts downwards on each rivet.

The maximum load is therefore 43·3 kN

$$\text{The shearing stress} = \frac{130}{3} \times \frac{1}{2} \times \frac{4 \times 10^3}{\pi \times 19^2} = 76{\cdot}5 \text{ N/mm}^2$$

$$\text{The bearing stress on the 9 mm plate} = \frac{130 \times 10^3}{3 \times 19 \times 9}$$

$$= 25{\cdot}3 \text{ N/mm}^2$$

Example 14.2. Calculate the diameter of rivet required in the group shown in Fig. 14.6 if the maximum shear stress is limited to 80 N/mm². The load of 60 kN is applied with an eccentricity of 250 mm. The dimensions in the figure are given in mm.

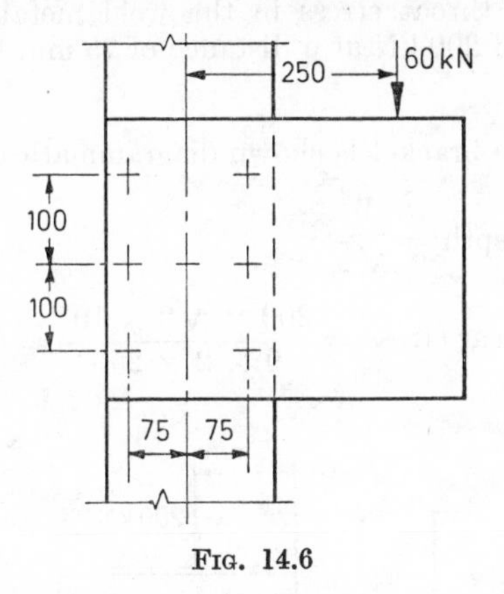

FIG. 14.6

(*St. Andrews*)

Solution. The direct load per rivet = 10 kN.
Let the maximum bending load per rivet be P_b.

Then $$\frac{2P_b}{125}(125^2 + 75^2 + 125^2) = 60 \times 250$$

giving $$P_b = 25{\cdot}4 \text{ kN}$$

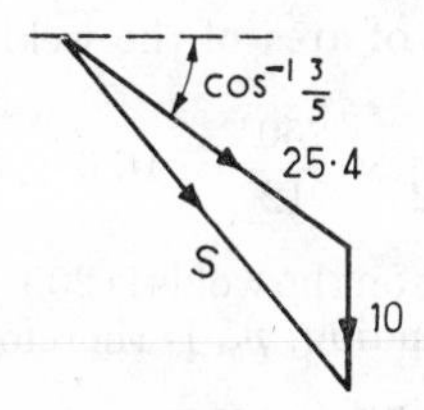

FIG. 14.7

The triangle of forces for the most heavily loaded rivet is shown in Fig. 14.7. The angle between the forces is $\cos^{-1} - \frac{3}{5}$, hence the maximum load, S, is obtained from the relationship

$$S^2 = 10^2 + 25{\cdot}4^2 + 2 \times 10 \times 25{\cdot}4 \times \tfrac{3}{5}$$

giving

$$S = 30{\cdot}1 \text{ kN}$$

The area of rivet required is therefore 376 mm² and a 22 mm diameter rivet should be used.

Example 14.3. An I section bracket is connected to a vertical member by two side fillet welds 9 mm wide by 250 mm deep. Determine the throat stress in the weld metal if the bracket carries a load of 200 kN at a distance of 75 mm from the face of the stanchion.

Solution. The bracket is shown diagrammatically in Fig. 14.8.

$$\text{The throat depth} = \frac{9}{\sqrt{2}}$$

$$\text{The direct shear stress} = \frac{200 \times \sqrt{2} \times 10^3}{9 \times 2 \times 250} = 62{\cdot}9 \text{ N/mm}^2.$$

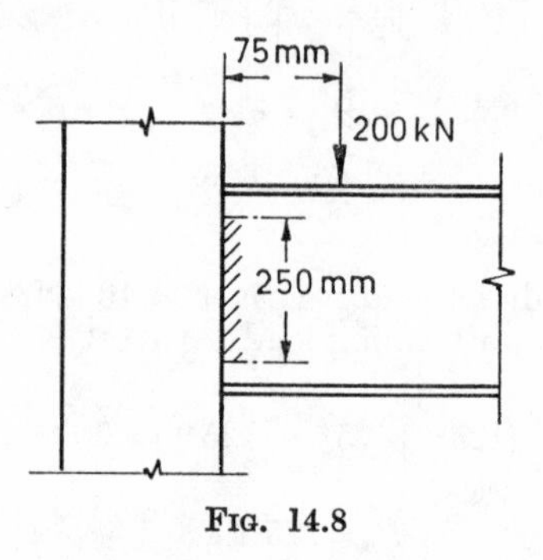

FIG. 14.8

The second moment of area of the welds, I_{xx}, is given by

$$I_{xx} = 9 \times \frac{2}{\sqrt{2}} \times \frac{250^3}{12} = 16{\cdot}6 \times 10^6 \text{ mm}^4 \text{ units}$$

The bending moment on the welds is $200 \times 75 = 15\,000$ kN mm
The stress due to bending, p_b, is therefore given by

$$p_b = \frac{200 \times 75 \times 125 \times 10^3}{16{\cdot}6 \times 10^6} = 113 \text{ N/mm}^2$$

The resultant stress $p_r = \sqrt{p_b^2 + p_s^2}$ and is in this case given by

$$\sqrt{113^2 + 62{\cdot}9^2} = 129 \text{ N/mm}^2$$

Example 14.4. The bracket shown in the Fig. 14.9 is welded to a stanchion by side fillet welds on three sides indicated by heavy lines. Calculate the maximum force per mm of weld metal when the bracket carries the load of 200 kN acting as shown. The dimensions are given in mm. *(Oxford)*

Solution. Let the centre of rotation be $\bar{x}$ from AD.

Then $$\bar{x} = \frac{2 \times 150 \times 75}{2 \times 150 + 1 \times 300} = 37{\cdot}5 \text{ mm}$$

$\therefore$ Eccentricity of load $= 150 + 125 - 37{\cdot}5 = 237{\cdot}5$ mm.

Considering unit throat thickness

$$I_{xx} = 2 \times 150 \times 150^2 + \frac{300^3}{12} = 9 \times 10^6 \text{ mm}^4 \text{ units}$$

$$I_{yy} = \frac{2 \times 150^3}{12} + 2 \times 150 \times 37{\cdot}5^2 + 300 \times 37{\cdot}5^2$$

$$= 1{\cdot}40 \times 10^6 \text{ mm}^4 \text{ units}$$

$$\therefore \quad I_{zz} = I_{xx} + I_{yy} = 10{\cdot}40 \times 10^6 \text{ mm}^4 \text{ units}$$

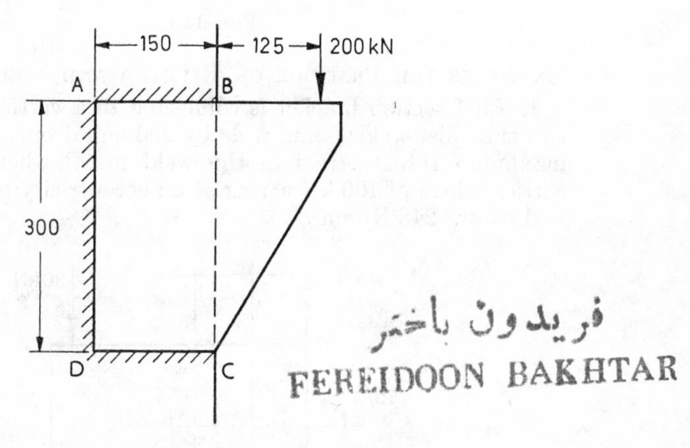

FIG. 14.9

Maximum bending stress occurs at B and C. The distance, r, of either of these points from the centre of rotation is given by

$$r = \sqrt{150^2 + 112{\cdot}5^2} = 187{\cdot}5 \text{ mm}$$

$$\therefore \quad P_b = \frac{200 \times 237{\cdot}5 \times 187{\cdot}5 \times 10^3}{10{\cdot}40 \times 10^6} = 858 \text{ N/mm run}$$

The direct load

$$P_d = \frac{200}{300 + 2 \times 150} = 333 \text{ N/mm run}$$

The vector diagram for the stress at C is shown in Fig. 14.10. The angle between P_b and P_d is $\cos^{-1} - \frac{3}{5}$. The resultant maximum load, P, is therefore given by

$$P = \sqrt{P_s^2 + P_b^2 - 2P_sP_b \cos\theta}$$
$$= \sqrt{333^2 + 858^2 + 2 \times 0{\cdot}6 \times 333 \times 858}$$
$$= 1065 \text{ N/mm run}$$

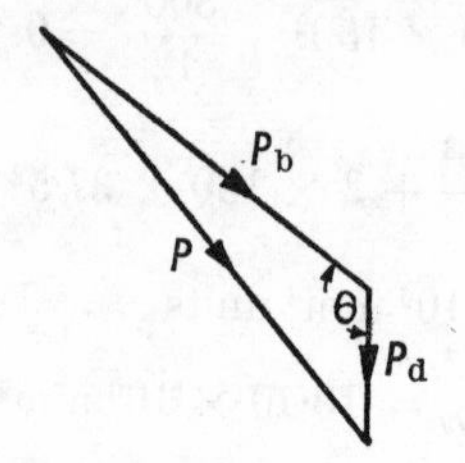

FIG. 14.10

EXAMPLES FOR PRACTICE ON RIVETED AND WELDED JOINTS

1. An I section bracket is connected to a vertical member by two side fillet welds 9 mm wide by 200 mm deep. Determine the maximum throat stress in the weld metal when the bracket carries a load of 100 kN acting at an eccentricity of 200 mm.

Answer. 248 N/mm².

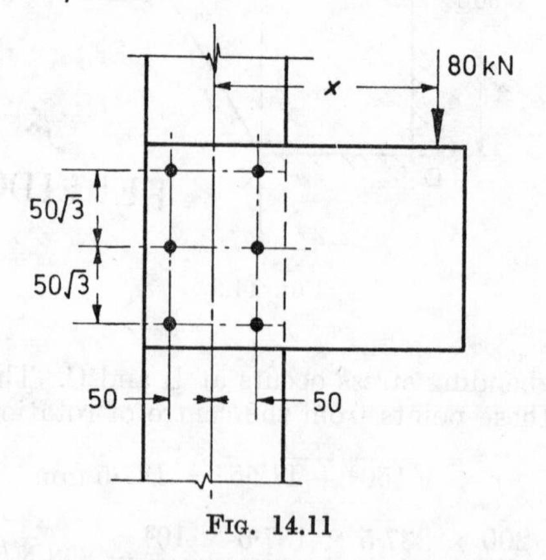

FIG. 14.11

2. Determine the maximum value of x for the group of 22 mm diameter rivets shown in Fig. 14.11 in order that the stress may not exceed 90 N/mm² when the load is 80 kN. Dimensions are in mm. *(Oxford)*

Answer. 146 mm.

3. Explain the basis of calculations usually made for determining the distribution of load amongst the individual rivets of a group.

A tie bar is attached to a gusset plate by four rivets arranged at the corners as shown in Fig. 14.12 the pitch of the rivets being

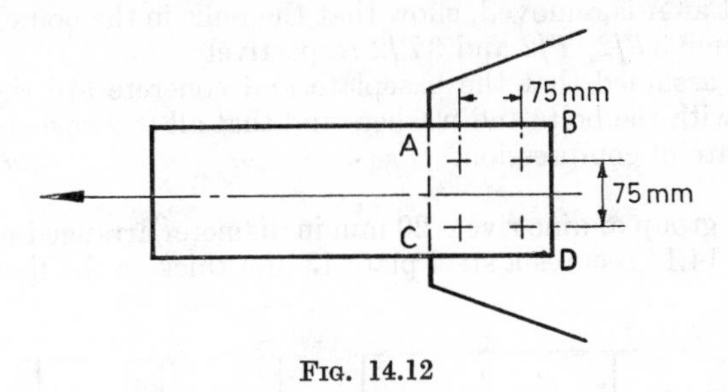

FIG. 14.12

75 mm. The pull is applied symmetrically. If the rivet at D is now removed, the pull being maintained at its former value, calculate by what percentage the load on each of the rivets is increased. *(Cambridge)*

Answer. A, 17·6 per cent; B, 21·2 per cent; C, 67·5 per cent.

4. State the assumptions usually made in estimating the load carried by each rivet in a riveted joint which is subject to both shear and bending in the plane of the joint.

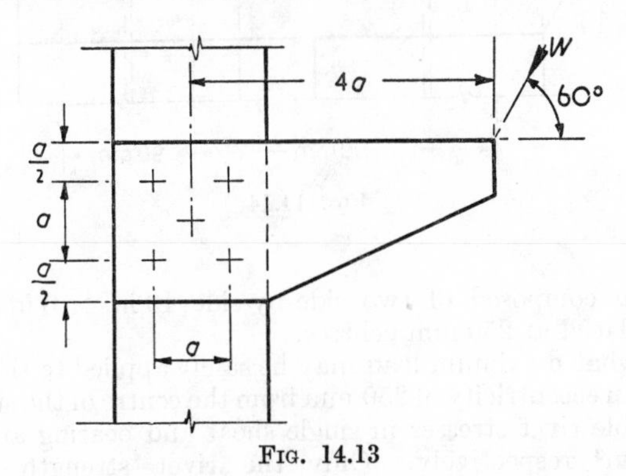

FIG. 14.13

In the arrangement shown in Fig. 14.13 the five rivets are symmetrically arranged at the corners and centre of a square of side a. Find, in terms of W, the load transmitted by the most heavily loaded rivet. *(Cambridge)*

Answer. $1{\cdot}24W$.

5. The baseplate of a machine rests on four elastic washers situated at the corners of a rectangle ABCD and is bolted through the washers to a concrete foundation. Initially each bolt has a tension T. The extension per unit tension of a bolt equals one quarter of the contraction per unit compression of a washer. If the nut at A is removed, show that the pulls in the bolts B, C and D become $3T/2$, $T/2$ and $3T/2$ respectively.

It is assumed that the baseplate and concrete are rigid compared with the bolts and washers and that all the washers remain in a state of compression. *(Cambridge)*

6. A group of nine rivets 20 mm in diameter arranged as shown in Fig. 14.14 secures a steel plate 13 mm thick to the flanges of a

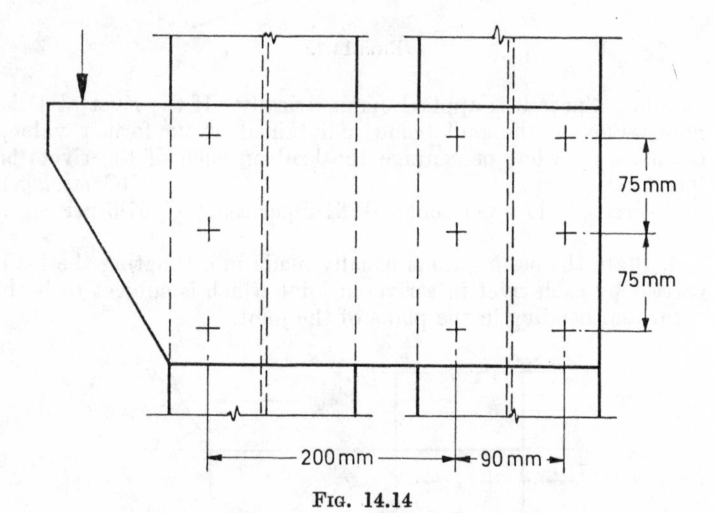

Fig. 14.14

stanchion composed of two side-by-side 10 in. × 6 in. × 40 lb. B.S.B. placed at 250 mm centres.

Find what maximum load may be safely applied to the bracket plate at an eccentricity of 250 mm from the centre of the stanchion. Permissible rivet stresses in single shear and bearing are 90 and 180 N/mm² respectively. Only the rivet strength need be investigated. *(London)*

Answer. 72·5 kN.

7. A riveted joint between two plates is made with seven rivets arranged in three horizontal rows at the corners and centre of a regular hexagon of side 100 mm. The joint is transmitting a vertical load of 100 kN acting at a distance of 300 mm from the centre of the hexagon.

Find the load carried by each rivet stating any assumptions you make. (*Oxford*)

Answer. 69·4; 58·5; 44·5; 14·3 (All in kN).

8. A tee section bracket is connected to a stanchion by welds along the top and sides of the tee as shown in Fig. 14.15, that at the top being 125 mm long and the sides 250 mm. Calculate the maximum load per mm run of weld metal when the bracket carries a load of 100 kN acting at a point distant 200 mm from the plane of the welds.

Answer. 0·704 kN/mm.

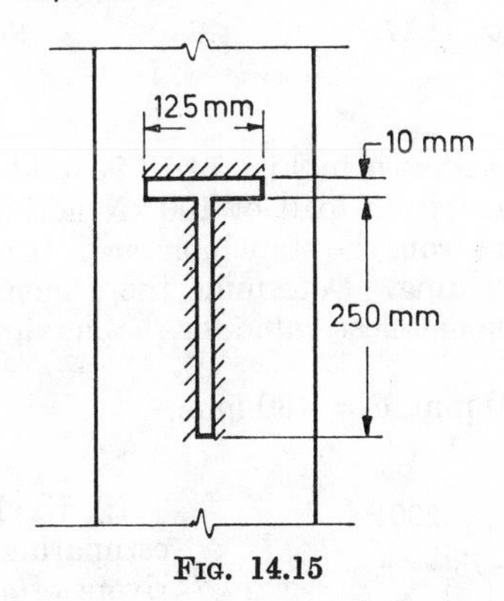

FIG. 14.15

9. Calculate the maximum eccentricity with which a 100 kN load can act in the case of the bracket shown in Fig. 14.16 if the load in the weld metal is not to exceed 1·2 kN/mm run.

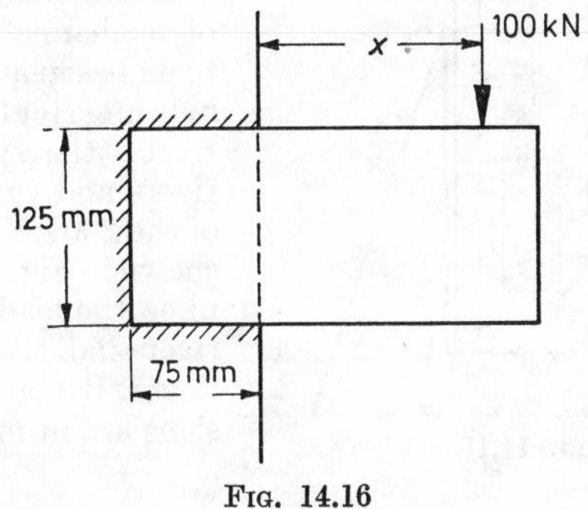

FIG. 14.16

Answer. 50 mm.

10. The angle bracket shown in Fig. 14.17 is connected to the stanchion by fillet welds of 10 mm throat. Calculate the maximum eccentricity at which a load of 200 kN can be placed if the stress in the weld metal is not to exeed 120 N/mm².

Answer. 118 mm.

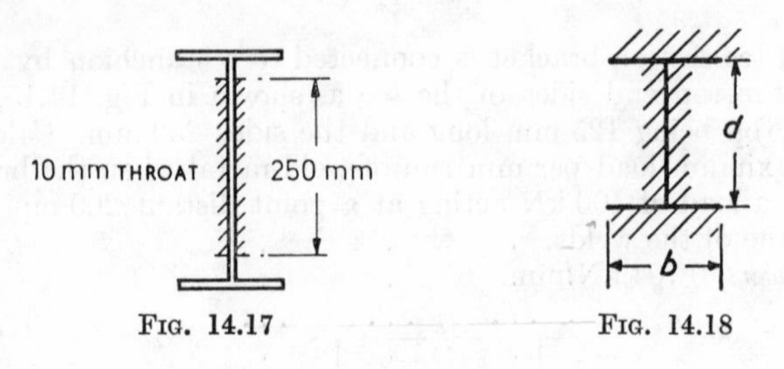

FIG. 14.17 FIG. 14.18

11. The bracket shown in Fig. 14.18 is welded to the side of a stanchion and carries a load of 100 kN acting vertically at a distance of 100 mm from the stanchion face. It is proposed to use 500 mm run of welding. Determine the dimensions b and d in order to obtain the smallest value for the maximum stress in the weld metal.

Answer. $b = 60$ mm, $d = 190$ mm.

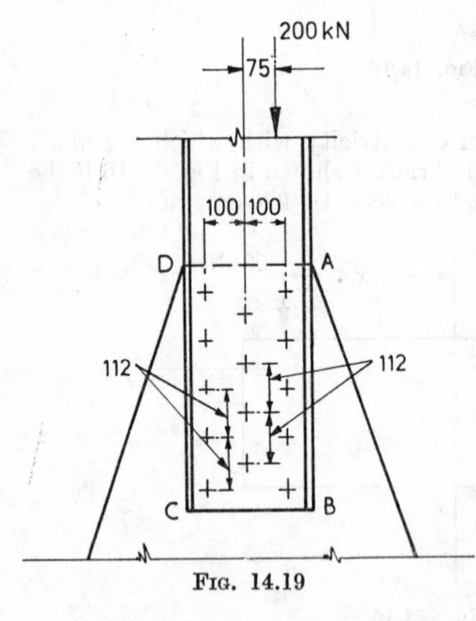

FIG. 14.19

12. Explain a method for estimating the loads on the rivets of a joint when the resultant force to be resisted does not pass through the centroid of the rivet centres.

A stanchion in the form of a channel is connected to its baseplate by fourteen symmetrically arranged rivets, the spacing of the rivets and the eccentricity of the 200 kN load being as shown in Fig. 14.19. Determine the loads on the four rivets indicated by the letters A, B, C and D. Dimensions are in mm.

(*Cambridge*)

Answer. 19·6 kN (A and B); 13·45 kN (C and D).

13. A group of eight rivets is arranged at the corners and mid-points of the sides of a square as shown in Fig. 14.20. A load W

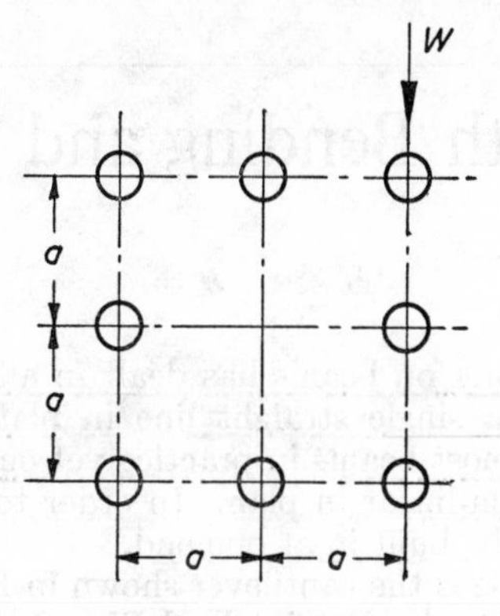

FIG. 14.20

acts along one of the sides of the square. Determine the ratio between the loads carried by those rivets which take the maximum and minimum loads.

Answer. 5·43.

Chapter 15

Beams with Bending and Torsion

THE PREVIOUS WORK on beams has dealt in all cases with those which consist of a single straight line in plan. Whilst this is generally true of most beams in practice yet occasionally a beam occurs which is non-linear in plan. In order to ensure stability, such beams must be built-in at one end.

The simplest case is the cantilever shown in Fig. 15.1, which is built-in at A and carries a point load W at B. The reactions at A are then

Shearing force $= W$
Bending moment $= Wa$
Twisting moment, or torque, $= Wb$

It is the presence of the torque which makes the problem of the curved beam differ from the straight one.

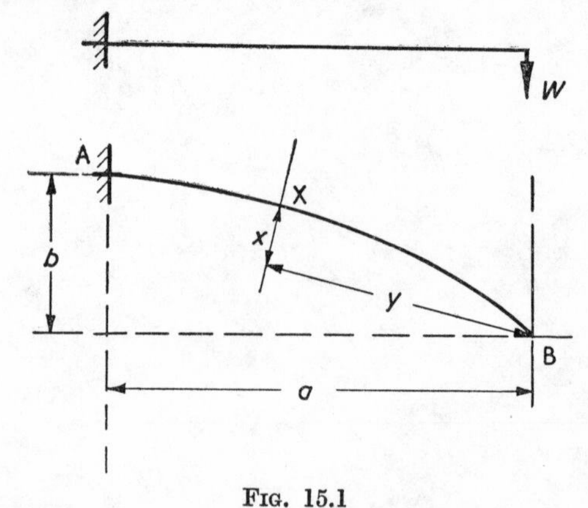

FIG. 15.1

At any point, X, on the cantilever shown in Fig. 15.1, the bending moment $M_X = Wy$ and the torque, $T_X = Wx$. The section at X must be designed to resist both the bending and twisting moments.

The curved cantilever is statically determinate but the introduction of a support at B makes the structure redundant.

The simplest support at B is a spherical seating which will permit rotation in any direction but prevent any vertical deflexion at B. This gives the member shown in Fig. 15.2, in which the reactions are a shearing force, a moment and a torque at A, and a shearing force at B. There are four reactions in all and, as static equilibrium equations give only three of these, the fourth

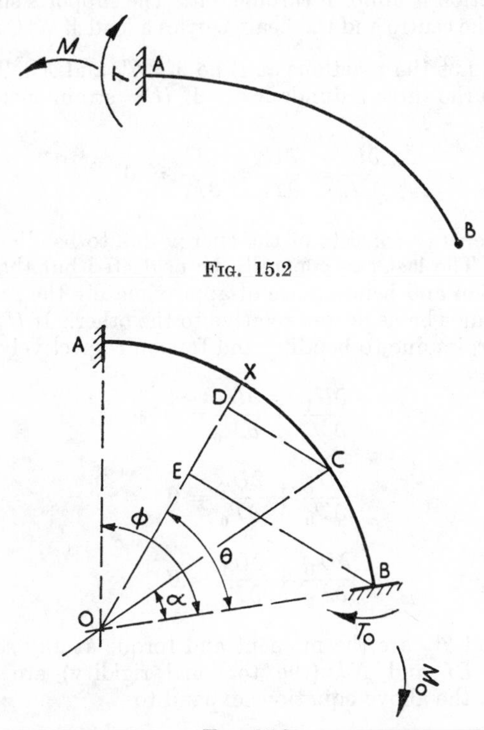

FIG. 15.2

FIG. 15.3

must be found by strain-energy, slope-deflexion or any other method used in redundant frame analysis.

The more usual type of support at B is as shown in Fig. 15.3, where the end B is completely built-in, thus providing not only restraint against deflexion but also against a change of slope in any direction. There are, therefore, three reactions at B, a shearing force, a bending moment, and a torque. These combined with similar reactions at A give six reactions in all, thus giving a three-redundancy problem.

A slope-deflexion method of solution to this problem was published by Gibson and Ritchie in 1913 whilst the strain-energy solution was given by Pippard and Barrow in 1926. The latter is, in the author's opinion, the neater solution.

As with all redundant structures, the properties of the section must be known before the analysis can be begun. In this problem both the torsional and the flexural rigidities are required.

Consider the bow girder shown in plan in Fig. 15.3. Assume that the section is uniform throughout. The supports subtend an angle ϕ at the centre and the beam carries a load W at C such that $\widehat{COB} = \alpha$. Let the reactions at B be M_0, T_0 and F_0. These can be taken as the three redundancies. If U = strain energy of the beam, then

$$\frac{\partial U}{\partial M_0} = \frac{\partial U}{\partial T_0} = \frac{\partial U}{\partial F_0} = 0$$

The strain energy consists of the energy due to bending, torsion and shear. The last can generally be neglected but the energies due to torsion and bending are of approximately the same order and one cannot be neglected relative to the other. If U_B and U_T are the energies due to bending and torsion respectively, then

$$\frac{\partial U_B}{\partial M_0} + \frac{\partial U_T}{\partial M_0} = 0$$

$$\frac{\partial U_B}{\partial T_0} + \frac{\partial U_T}{\partial T_0} = 0$$

$$\frac{\partial U_B}{\partial F_0} + \frac{\partial U_T}{\partial F_0} = 0$$

If M_X and T_X are the moment and torque at any section X, then since EI and NJ (the torsional rigidity) are constant throughout, the above equations expand to

$$\frac{1}{EI}\int M_X \frac{\partial M_X}{\partial M_0}\,ds + \frac{1}{NJ}\int T_X \frac{\partial T_X}{\partial M_0}\,ds = 0 \quad . \quad . \quad (1a)$$

$$\frac{1}{EI}\int M_X \frac{\partial M_X}{\partial T_0}\,ds + \frac{1}{NJ}\int T_X \frac{\partial T_X}{\partial T_0}\,ds = 0 \quad . \quad . \quad (1b)$$

$$\frac{1}{EI}\int M_X \frac{\partial M_X}{\partial F_0}\,ds + \frac{1}{NJ}\int T_X \frac{\partial T_X}{\partial F_0}\,ds = 0 \quad . \quad . \quad (1c)$$

The convention with regard to torque is that, looking along the positive direction of S, a positive torque is one which tends to turn the section clockwise.

At section X, the angle $\widehat{XOB}$ being θ,

$$M_X = M_0 \cos\theta + T_0 \sin\theta - F_0 \times BE + W \times CD$$
$$T_X = -M_0 \sin\theta + T_0 \cos\theta - F_0 \times XE + W \times XD$$

If θ is less than α, the terms with W are omitted.

$$\frac{\partial M_X}{\partial M_0} = \cos\theta \qquad \frac{\partial T_X}{\partial M_0} = -\sin\theta$$

$$\frac{\partial M_X}{\partial T_0} = \sin\theta \qquad \frac{\partial T_X}{\partial T_0} = \cos\theta$$

$$\frac{\partial M_X}{\partial F_0} = -BE = -R\sin\theta, \quad \frac{\partial T_X}{\partial F_0} = -XE = R(\cos\theta - 1)$$

$$CD = R\sin(\theta - \alpha) \qquad XD = R\{1 - \cos(\theta - \alpha)\}$$

The set of equations (1) now become

$$\frac{1}{EI}\int\{M_0\cos\theta + T_0\sin\theta - F_0R\sin\theta + WR\sin(\theta-\alpha)\}\cos\theta \, . \, d\theta$$
$$-\frac{1}{NJ}\int[-M_0\sin\theta + T_0\cos\theta - F_0R(1-\cos\theta) + WR\{1-\cos(\theta-\alpha)\}]\sin\theta \, . \, d\theta = 0 \quad . \quad . \quad (2a)$$

$$\frac{1}{EI}\int\{M_0\cos\theta + T_0\sin\theta - F_0R\sin\theta + WR\sin(\theta-\alpha)\}\sin\theta \, . \, d\theta$$
$$+\frac{1}{NJ}\int[-M_0\sin\theta + T_0\cos\theta - F_0R(1-\cos\theta) + WR\{1-\cos(\theta-\alpha)\}]\cos\theta \, . \, d\theta = 0 \quad . \quad . \quad (2b)$$

$$-\frac{1}{EI}\int\{M_0\cos\theta + T_0\sin\theta - F_0R\sin\theta + WR\sin(\theta-\alpha)\}\sin\theta \, . \, d\theta$$
$$+\frac{1}{NJ}\int[-M_0\sin\theta + T_0\cos\theta - F_0R(1-\cos\theta) + WR\{1-\cos(\theta-\alpha)\}](1-\cos\theta) \, . \, d\theta = 0 \quad . \quad . \quad (2c)$$

The limits for integration are 0 to ϕ for all terms in M_0, T_0 and F_0, and α to ϕ for those in W.

Integrating and putting in the limits gives the following three simultaneous equations.

$$
\begin{aligned}
M_0 &\left[\frac{1}{EI}(2\phi + \sin 2\phi) + \frac{1}{NJ}(2\phi - \sin 2\phi)\right] \\
&+ T_0\left[\left(\frac{1}{EI} - \frac{1}{NJ}\right)(1 - \cos 2\phi)\right] \\
&+ F_0R\left[\frac{1}{EI}(\cos 2\phi - 1) - \frac{1}{NJ}(4\cos\phi - \cos 2\phi - 3)\right] \\
&- WR\left[\frac{1}{EI}\{\cos(2\phi - \alpha) + 2(\phi - \alpha)\sin\alpha - \cos\alpha\}\right. \\
&\qquad - \frac{1}{NJ}\{4\cos\phi - 3\cos\alpha + 2(\phi - \alpha)\sin\alpha \\
&\qquad\qquad \left. - \cos(2\phi - \alpha)\}\right] = 0 \quad . \quad . \quad (3a)
\end{aligned}
$$

$$
\begin{aligned}
M_0 &\left(\frac{1}{EI} - \frac{1}{NJ}\right)(1 - \cos 2\phi) \\
&+ T_0\left[\frac{1}{EI}(2\phi - \sin 2\phi) + \frac{1}{NJ}(2\phi + \sin 2\phi)\right] \\
&+ F_0R\left[\frac{1}{EI}(\sin 2\phi - 2\phi) - \frac{1}{NJ}(4\sin\phi - \sin 2\phi - 2\phi)\right] \\
&+ WR\left[\frac{1}{EI}\{2(\phi - \alpha)\cos\alpha + 2\cos\phi\sin(\alpha - \phi)\}\right. \\
&\qquad + \frac{1}{NJ}\{(4\sin\phi - 3\sin\alpha - 2(\phi - \alpha)\cos\alpha \\
&\qquad\qquad \left. - \sin(2\phi - \alpha)\}\right] = 0 \quad . \quad . \quad (3b)
\end{aligned}
$$

$$
\begin{aligned}
M_0(1 - \cos\phi) &- T_0\sin\phi + F_0R(\phi - \sin\phi) \\
&- WR\{\phi - \alpha - \sin(\phi - \alpha)\} = 0 \quad . \quad . \quad (3c)
\end{aligned}
$$

These equations can be solved if the values of EI, NJ, ϕ and α are given. This information must be included in the data for any particular problem.

If the girder carries a uniformly distributed load of intensity w per unit length, the form of the equations naturally differs from those for a point load. The coefficients of the terms containing M_0, T_0 and F_0 in equations 1, 2 and 3 remain unchanged but the terms in W come out and are replaced by terms in w.

Consider the girder shown in plan in Fig. 15.4 which carries a uniformly distributed load of w per unit length. X is in the same

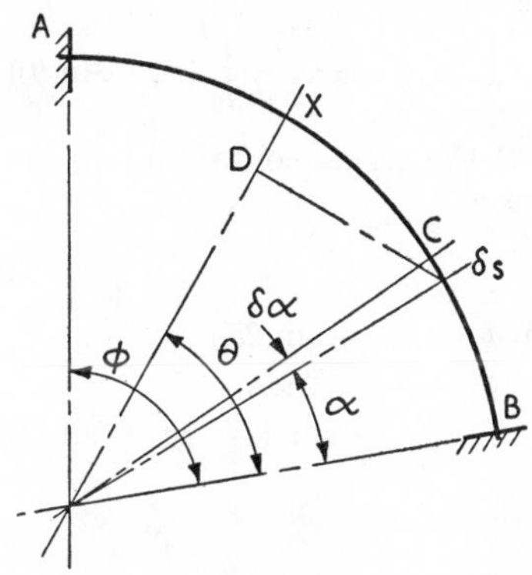

FIG. 15.4

position as before and the problem is to find the moment and torque at X due to the load covering the portion XB.

Let δm_{X} and δt_{X} be the moment and torque at X due to the element of load $w\delta s$ at C. Then

$$\delta m_{\mathrm{X}} = \mathrm{CD} \times w\delta s = wR^2 \sin(\theta - \alpha)\delta\alpha$$

$$\delta t_{\mathrm{X}} = + \mathrm{XD} \times w\delta s = + wR^2\{1 - \cos(\theta - \alpha)\}\delta\alpha$$

Thus the moment and torque at X due to the load on the portion XB are

$$m_x = wR^2\int_0^\theta \sin(\theta - \alpha)\mathrm{d}\alpha = wR^2(1 - \cos\theta)$$

$$t_x = + wR^2\int_0^\theta \{1 - \cos(\theta - \alpha)\}\mathrm{d}\alpha = + wR^2(\theta - \sin\theta)$$

The terms in W in equations (2) are, therefore, replaced by the following—

In equation $2a$,

$$wR^2\left[\frac{1}{EI}\int_0^\phi (1 - \cos\theta)\cos\theta\,\mathrm{d}\theta - \frac{1}{NJ}\int_0^\phi (\theta - \sin\theta)\sin\theta\,\mathrm{d}\theta\right]$$

In equation 2*b*,

$$wR^2\left[\frac{1}{EI}\int_0^{\phi}(1-\cos\theta)\sin\theta\,d\theta+\frac{1}{NJ}\int_0^{\phi}(\theta-\sin\theta)\cos\theta\,d\theta\right]$$

In equation 2*c*

$$-wR^2\left[\frac{1}{EI}\int_0^{\phi}(1-\cos\theta)\sin\theta\,d\theta - \frac{1}{NJ}\int_0^{\phi}(\theta-\sin\theta)(1-\cos\theta)d\theta\right]$$

In equations (3) the terms containing W are replaced by the following terms in w.

In equation 3*a*,

$$wR^2\left[\frac{1}{EI}(4\sin\phi-2\phi-\sin 2\phi) - \frac{1}{NJ}(4\sin\phi-2\phi+\sin 2\phi-4\phi\cos\phi)\right]$$

In equation 3*b*,

$$wR^2\left[\frac{1}{EI}(3+\cos 2\phi-4\cos\phi) + \frac{1}{NJ}(\cos 2\phi+4\cos\phi-5+4\,\phi\sin\phi)\right]$$

There is no need to use equation 3*c* in the case of a uniformly distributed load because F_0 must by symmetry be equal to half the load on the beam, hence there are only two unknowns.

The equations given in this fundamental analysis are somewhat complicated due to the fact that they have been drawn up for the general case. In most examples the data given produces simpler equations. In all cases of symmetrical loading the reactions at each end must be the same and, since the vertical reactive forces must balance the vertical load, F_0 can be obtained straight away without solving simultaneous equations.

In the case of the point load when

$$\alpha=\frac{\phi}{2},\quad F_0=\frac{W}{2}$$

For a distributed load $F_0=\dfrac{wR\phi}{2}$

For symmetrical loading the torque and shear are both zero at the centre. It is therefore easier to work from the centre where there is only one unknown, M.

Worked Examples on Torsion and Bending

Example 15.1. The beam shown in plan in Fig. 15.5 carries a uniformly distributed load of w per unit length. It is built-in

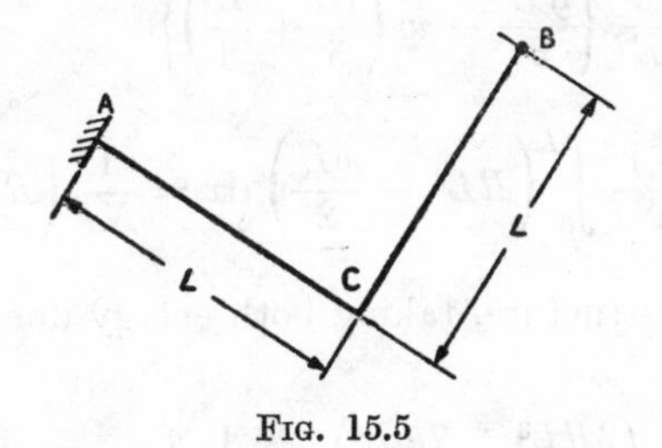

FIG. 15.5

at A and supported on a spherical seating at B. If NJ for the beam $= 0{\cdot}8EI$, determine the reactions at the supports.

Solution. The redundancy is the reaction at B. Let this be R. Then for the whole beam

$$\frac{\partial U}{\partial R} = 0,$$

where U is the strain energy. (Only that due to bending and torsion will be considered.) For BC, considering a section distant x from B,

$$M_x = -Rx + \frac{wx^2}{2}$$

$$\frac{\partial M_x}{\partial R} = -x$$

If $\left[\frac{\partial U}{\partial R}\right]^{M}_{BC}$ denotes the partial differential of the strain energy due to bending for the portion BC, then

$$\left[\frac{\partial U}{\partial R}\right]^{M}_{BC} = \frac{1}{EI}\int_0^L \left(Rx^2 - \frac{wx^3}{2}\right) dx = \frac{1}{EI}\left(\frac{RL^3}{3} - \frac{wL^4}{8}\right)$$

For CA, considering a section distant x from C,

$$M_x = -Rx + wLx + \frac{wx^2}{2} \qquad \frac{\partial M_x}{\partial R} = -x$$

$$T_x = +RL - \frac{wL^2}{2} \qquad \frac{\partial T_x}{\partial R} = +L$$

$$\left[\frac{\partial U}{\partial R}\right]_{\mathrm{CA}}^{\mathrm{M}} = \frac{1}{EI}\int_0^L \left\{Rx^2 - w\left(Lx^2 + \frac{x^3}{2}\right)\right\} \mathrm{d}x$$

$$= \frac{1}{EI}\left\{\frac{RL^3}{3} - w\left(\frac{L^4}{3} + \frac{L^4}{8}\right)\right\}$$

$$\left[\frac{\partial U}{\partial R}\right]_{\mathrm{CA}}^{\mathrm{T}} = \frac{1}{NJ}\int_0^L \left(RL^2 - \frac{wL^3}{2}\right) \mathrm{d}x = \frac{1}{NJ}\left(RL^3 - \frac{wL^4}{2}\right)$$

For the whole structure, taking both energy due to torsion and bending

$$\frac{\partial U}{\partial R} = \frac{1}{EI}\left(\frac{2RL^3}{3} - \frac{7wL^4}{12}\right) + \frac{1}{NJ}\left(RL^3 - \frac{wL^4}{2}\right) = 0$$

Putting $\dfrac{NJ}{EI} = 0{\cdot}8$ gives

$$\frac{1{\cdot}6R}{3} - \frac{5{\cdot}6wL}{12} + R - \frac{wL}{2} = 0$$

giving $R = 0{\cdot}63wL$.

Having found the reaction at B, the reactions at A are given by

$$F_{\mathrm{A}} = 2wL - R_{\mathrm{B}} = 1{\cdot}37wL$$

$$M_{\mathrm{A}} = -R_{\mathrm{B}}L + \frac{3}{2}wL^2 = +0{\cdot}87wL^2$$

$$T_{\mathrm{A}} = -R_{\mathrm{B}}L + \frac{wL^2}{2} = -0{\cdot}13wL^2$$

Example 15.2. The girder shown in plan in Fig. 15.6 carries a uniformly distributed load of w per unit length on the straight

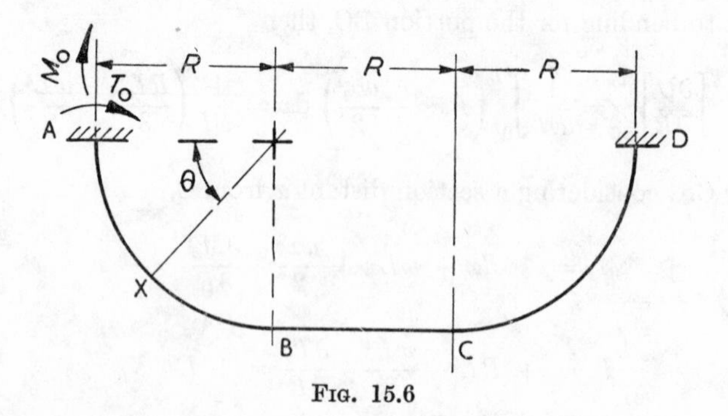

FIG. 15.6

portion BC. If the flexural rigidity is α times the torsional rigidity, determine the fixing moments at A.

Solution. The moments at A will be a moment, M_0, acting at right angles to the face of the support, and a moment, T_0, acting along the supporting face. Since the girder is symmetrical the moments at D will be the same.

The vertical reactions, R_A and R_D, must each be equal to half the total load on the beam, i.e. $\frac{wR}{2}$.

The two fixing moments, M_0, must together equal the moment of the external load about the face, i.e. $2M_0 = wR^2$, hence $M_0 = \frac{wR^2}{2}$. Thus the only redundancy is T_0. This can be obtained by application of the second theorem of Castigliano, giving

$$\frac{\partial U}{\partial T_0} = \frac{1}{NJ}\int T_X \frac{\partial T_X}{\partial T_0}\, dx + \frac{1}{EI}\int M_X \frac{\partial M_X}{\partial T_0}\, dx$$

For AB the moment, M_X, and the torque, T_X, at any point X subtending an angle θ with AD are given by

$$M_X = \frac{wR^2}{2}\cos\theta + T_O\sin\theta - \frac{wR^2}{2}\sin\theta$$

$$T_X = T_O\cos\theta - \frac{wR^2}{2}\sin\theta + \frac{wR^2}{2}(1 - \cos\theta)$$

$$\left[\frac{\partial U}{\partial T_O}\right]_{AB} = \frac{R}{EI}\int_0^{\pi/2}\left(\frac{wR^2\sin 2\theta}{4} + T_O\sin^2\theta - \frac{wR^2}{2}\sin^2\theta\right)d\theta$$

$$+ \frac{R}{NJ}\int_0^{\pi/2}\left\{T_O\cos^2\theta - \frac{wR^2}{4}\sin 2\theta\right.$$

$$\left. + \frac{wR^2}{2}(\cos\theta - \cos^2\theta)\right\}d\theta$$

$$= \frac{R}{EI}\left[\frac{-wR^2\cos 2\theta}{8} + T_O\left(\frac{\theta}{2} - \frac{\sin 2\theta}{4}\right)\right.$$

$$\left. - \frac{wR^2}{2}\left(\frac{\theta}{2} - \frac{\sin 2\theta}{4}\right)\right]_0^{\pi/2}$$

$$+ \frac{R}{NJ}\left[T_O\left(\frac{\sin 2\theta}{4} + \frac{\theta}{2}\right) + \frac{wR^2\cos 2\theta}{8}\right.$$

$$\left. + \frac{wR^2}{2}\left(\sin\theta - \frac{\sin 2\theta}{4} - \frac{\theta}{2}\right)\right]_0^{\pi/2}$$

$$= \frac{R}{EI}\left\{\frac{wR^2}{4}\left(1 - \frac{\pi}{2}\right) + \frac{T_O\pi}{4}\right\}$$

$$+ \frac{R}{NJ}\left\{\frac{T_O\pi}{4} - \frac{wR^2}{4}\left(1 - \frac{\pi}{2}\right)\right\}$$

A similar expression will be obtained for the portion of the beam CD.

For BC the moment, M_X, and the torque, T_X, at a point distant x from B are given by

$$M_X = T_0 - \frac{wR}{2}(R + x) + \frac{wx^2}{2}$$

$$T_X = \frac{-wR^2}{2} + \frac{wR^2}{2} = 0$$

$$\left[\frac{\partial U}{\partial T_O}\right]_{BC} = \frac{1}{EI}\int_0^R \left\{T_0 - \frac{wR}{2}(R + x) + \frac{wx^2}{2}\right\} dx$$

$$= \frac{1}{EI}\left[T_O x - \frac{w}{2}\left(R^2 x + \frac{Rx^2}{2} + \frac{x^3}{3}\right)\right]_0^R$$

$$= \frac{R}{EI}\left\{T_0 - \frac{11wR^2}{6}\right\}$$

For the whole structure, putting $EI = \alpha NJ$,

$$\left[\frac{\partial U}{\partial T_0}\right] = \frac{wR^2}{2}\left(1 - \frac{\pi}{2}\right) + \frac{T_O\pi}{2} + \frac{\alpha T_O\pi}{2}$$

$$- \frac{\alpha wR^2}{2}\left(1 - \frac{\pi}{2}\right) + T_0 - \frac{11wR^2}{6} = 0$$

$$T_0\left\{1 + \frac{\pi}{2}(1 + \alpha)\right\} = \frac{wR^2}{2}\left\{\frac{11}{6} - 1 + \frac{\pi}{2} + \alpha\left(1 - \frac{\pi}{2}\right)\right\}$$

Hence
$$T_0 = \frac{wR^2}{2}\left\{\frac{5 + 3\pi + 6\alpha\left(1 - \frac{\pi}{2}\right)}{6 + 3\pi + 3\alpha\pi}\right\}$$

EXAMPLES FOR PRACTICE ON TORSION AND BENDING

1. A beam, ABCD, is bent to form in plan three sides of a rectangle having AB = ½BC = CD = L. It is built-in at A and D and carries a uniformly distributed load of w per unit length.

Determine the deflexion at the mid-point of BC if the torsional rigidity of the beam is one quarter of its flexural rigidity.

(Oxford)

Answer. $\dfrac{0{\cdot}636wL^4}{EI}$.

2. Determine the twisting moment at the supports and mid-span when a beam shown in plan in Fig. 15.7 carries a central

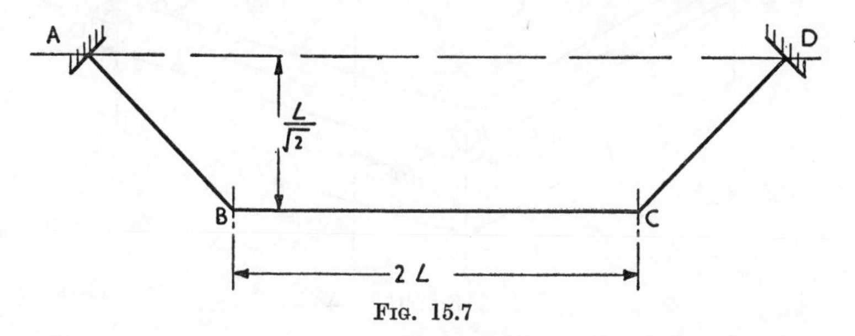

Fig. 15.7

point load of W. Take the section constant throughout with a flexural rigidity α times the torsional rigidity. *(St. Andrews)*

Answer. At support, $M = \dfrac{WL}{4}\dfrac{\sqrt{2}-1}{3+\alpha}$; at mid-span, $M = 0$.

3. Two thin-walled circular tubes, AB, CD, (Fig. 15.8) of length l, diameter d, are fully fixed horizontally at A and C. They lie in the same horizontal plane distant h apart. At B and D they are rigidly connected to a beam BD.

A load, W, is applied to BD at E, where BE $= n$BD. Show that the fraction of the load W carried by the tube AB is

$$\frac{1 - n + k/2}{1 + k}$$

where $k = \dfrac{8}{3}\dfrac{Cl^2}{Eh^2}$

C and E being respectively the modulus of rigidity and Young's Modulus for the material. The beam BD may be treated as rigid. Neglect deflexions due to shear.

Show that the ratio of the shear stress due to torsion in the tube AB to the greatest longitudinal normal stress due to bending is

$$\frac{1}{2}\cdot\frac{h}{l}\cdot\frac{(1-2n)k}{1-n+k/2}$$

(Cambridge)

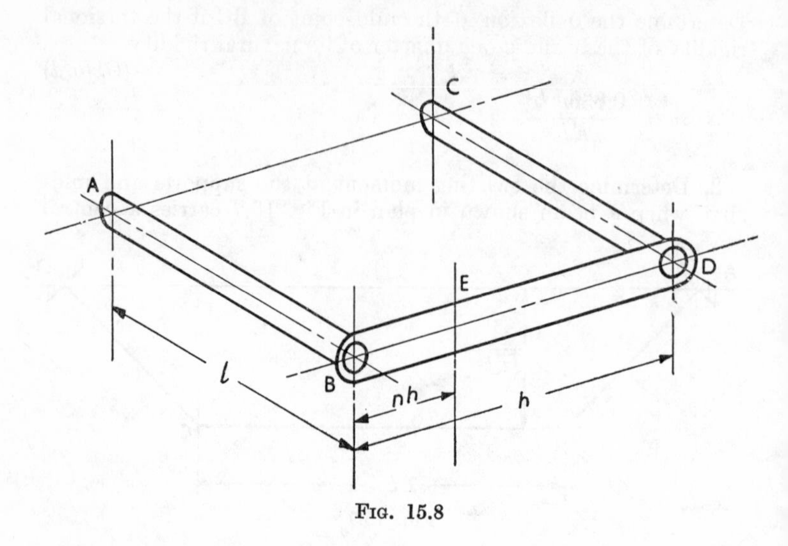

FIG. 15.8

4. A rod of circular section radius r projects, in the form of a semicircle of radius R, from a vertical wall into which it is built-in horizontally, as shown in plan and elevation in Fig. 15.9. The rod is subjected to a vertical force at the section C

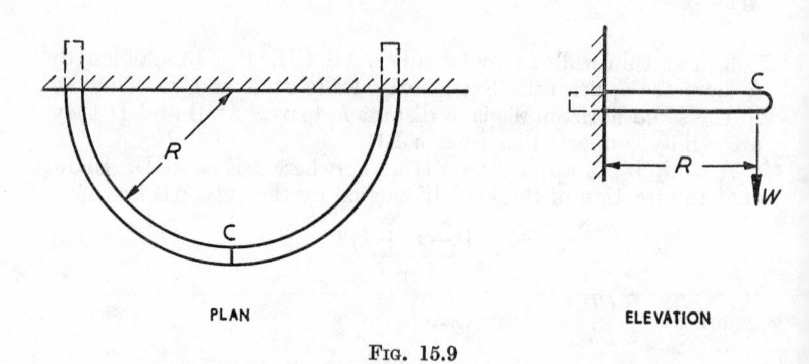

FIG. 15.9

farthest from the wall. Show that at C the sagging bending moment in the rod is $\dfrac{WR}{\pi}$ and find an expression for the deflexion at this section. *(Cambridge)*

Answer. $\Delta = \dfrac{0{\cdot}296 WR^3}{Er^4}$.

5. An L-shaped bracket consists of two 1·5 m lengths of 25 mm diameter steel rod rigidly connected to one another at right angles. The ends are built-in to a wall so that both portions lie in a horizontal plane. A force of 450 N is applied vertically to the free end. Calculate the deflexion under the load if $N = \frac{2}{5} E =$ 80 kN/mm².

Answer. 100 mm.

6. Two girders, AC, BCD, having a flexural rigidity EI and a torsional rigidity CJ are rigidly connected at C to form a T of which the arms CB, CA and CD have equal horizontal length l. The ends B, A, D are clamped so that the T lies in a horizontal plane, and a vertical load, W, is imposed at C.

Show that BC and CD (besides being bent) are subjected to twisting couples of magnitude T, where

$$\left(\frac{1}{3} + \frac{EI}{2JC}\right) T = \frac{1}{36} Wl \qquad \textit{(Oxford)}$$

7. Fig. 15.10 shows in horizontal plan (diagrammatically) a bracket formed of 1-in. diameter solid steel rod, bent to form

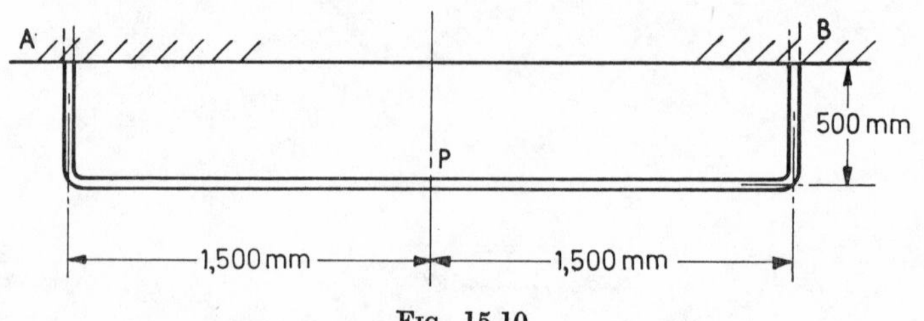

FIG. 15.10

three sides of a rectangle and clamped into a rigid wall at A and B. A vertical load of W kN is imposed at P.

What bending and twisting moments will be induced at the sections A and B? *(Oxford)*

Answer. T = 264 W kN mm; M = 250 W kN mm.

8. One end of a round steel rod is welded to one end of a similar rod so that the angle between them is 60°. The other ends are rigidly fixed into a wall so that the rods lie in a horizontal plane, the free length of each rod being 2 m. When a force of 100 N is carried by the welded ends, find the value of the torque in each rod and sketch the bending moment diagram for one of them. *(Oxford)*

Answer. M = 85·8 N m; T = 17·5 N m.

9. Calculate the fixing moments at the ends of a quadrantal circular arc bow girder carrying a uniformly distributed load if the torsional rigidity is one-half of the flexural rigidity.

Answer. $M = 0{\cdot}232wR^2$; $T = 0{\cdot}017wR^2$.

Chapter 16

Plastic Design

IN THE CHAPTER dealing with struts reference is made to the design of members using the load-factor method. This, up to the present time, has only been generally employed in the design of struts but researches carried out over the past forty years have shown that it is possible to use the load-factor method in the

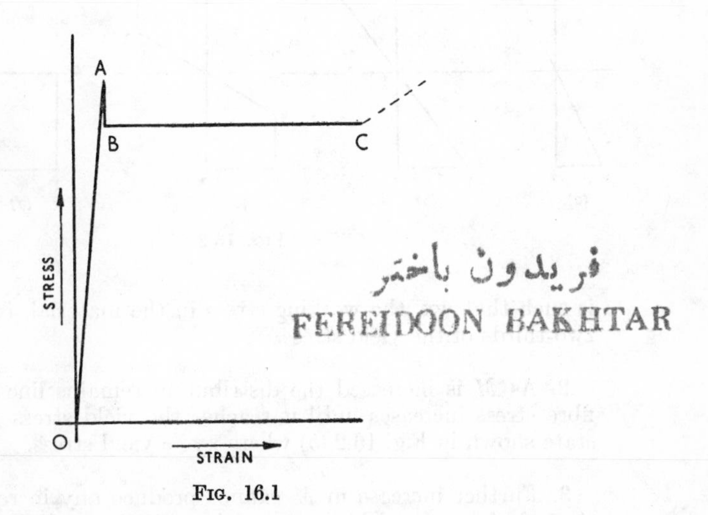

Fig. 16.1

design of other members. This application is generally known as the plastic theory of member design and it is now being frequently used, especially in the case of rigid-jointed frames.

The fundamental principle of the method is to determine the collapse load either of individual members or of a rigid-jointed frame. The working load is multiplied by the load factor and this factorized load made equal to the collapse load of the frame.

The basis of the theory is as follows. The stress-strain diagram for annealed mild steel is as shown in Fig. 16.1, A being the upper yield point and B the lower. The line BC represents the plastic strain which is very much greater than the elastic strain (which is given by the horizontal distance OA). In Fig. 16.1 the inclination of OA to the vertical is exaggerated for the sake of clarity. From C onwards strain hardening takes place. The plastic theory ignores the upper yield point and assumes that yield occurs at a stress equivalent to B. It also neglects the part of the diagram beyond C and assumes that the behaviour in compression is similar to that in tension.

If an increasing bending moment is applied to a section of a beam the stages through which the stress distribution passes are shown in Figs. 16.2 and enumerated as follows—

1. Fig 16.2 (*a*) shows the elastic distribution occurring at low values of M. Normally the moment, M, carried by the section

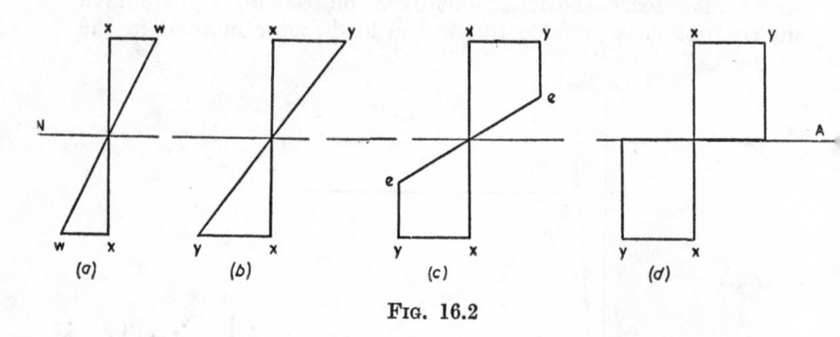

FIG. 16.2

is such that wx, the working stress in the material, is generally two-thirds of the yield stress.

2. As M is increased the distribution remains linear but the fibre stress increases until it reaches the yield stress giving the state shown in Fig. 16.2 (*b*) where xy = yield stress.

3. Further increase in M cannot produce any increased fibre stress but causes yield to spread into the inner fibres giving at some stage a stress distribution as shown in Fig. 16.2 (*c*) in which the depth of the beam, ye, has reached yield but the portion ee is still elastic.

4. The limit is reached when the point "e" has dropped to the neutral axis as shown in Fig. 16.2 (*d*). The whole section is now at yield stress and is carrying the maximum moment to which it can be subjected without strain hardening taking place. The moment which produces this state of stress is called the collapse moment.

Application to Simple Beams

Example 16.1. Determine the central point load which will cause collapse for a rectangular beam whose span is L, made from a material of yield stress, y. Determine also the length of the beam over which yielding will take place when the collapse load is applied.

Solution. Let the beam section be as shown in Fig. 16.3 The stress distribution is as shown in Fig. 16.2 (d).

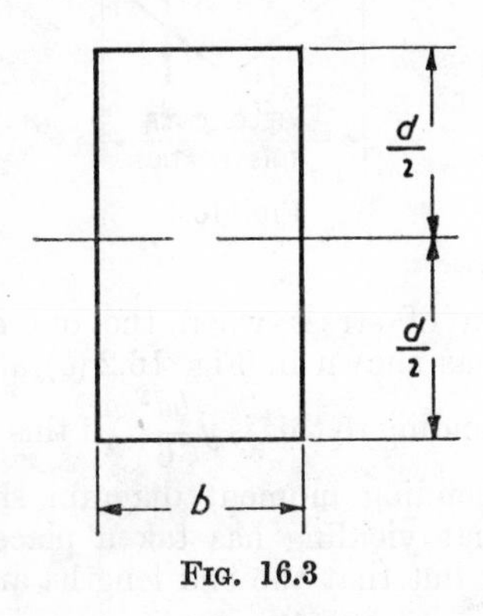

FIG. 16.3

The total tension, which equals the total compression, is $y\dfrac{bd}{2}$ and the lever arm (the distance between the centres of compressive and tensile forces) is $d/2$. If M_c = collapse moment, then

$$M_c = y\frac{bd}{2} \times \frac{d}{2} = y\frac{bd^2}{4}$$

But $\dfrac{bd^2}{4}$ = first moment of area of the rectangular section about an axis through its centroid. If this is symbolized by S then

$$M_c = yS$$

In order to have the analogy with the formula $M = fZ$ for elastic bending the term S is known as the plastic modulus.

The central collapse load

$$C = \frac{4M_c}{L} = \frac{ybd^2}{L}$$

and the bending moment diagram under the collapse load is as shown in Fig. 16.4.

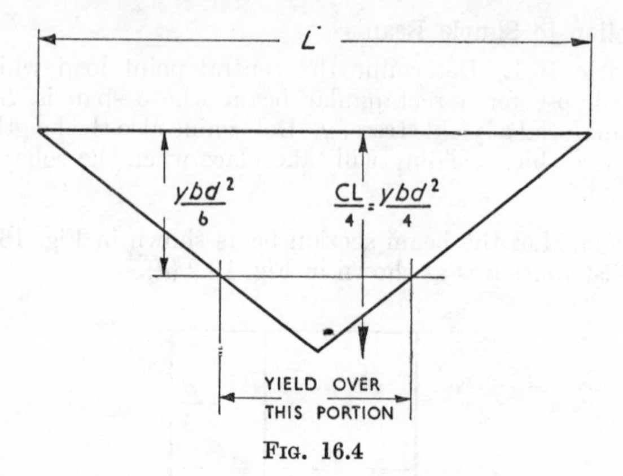

FIG. 16.4

The distribution of stress when the outer fibres have just begun to yield is as shown in Fig. 16.2 (b) and the moment of resistance corresponding to this is $y\,\frac{bd^2}{6}$. If this moment is marked on the collapse bending moment diagram shown in Fig. 16.4, then it is seen that yielding has taken place over the middle third of the beam but that the end lengths are still elastic over the whole section.

The factor S/Z is known as the shape factor and is equal to 1·5 in the case of a rectangular section. For most joist sections, however, the value is 1·15.

Example 16.2. The working stress in mild steel is 150 N/mm². If the yield stress is 230 N/mm², calculate the load factor in the case of a joist having a shape factor of 1·15.

Solution

$$\text{Collapse load} = 230S$$

$$\text{Working load} = 150\,Z$$

$$\therefore \quad \text{Load factor} = \frac{\text{collapse load}}{\text{working load}} = \frac{230\,S}{150\,Z}$$

$$= 1{\cdot}535 \times 1{\cdot}15$$

$$= 1{\cdot}76$$

The collapse in the case of a simply supported beam is due to the formation of a plastic hinge at the point of maximum bending moment. This hinge, together with the pins at each end support, gives three pins in the span which produce instability. This method of failure, i.e. production of three hinges, will be the same for all beams. It is applied in the following examples on built-in beams.

Application to Built-in Beams

Example 13.6. Calculate the collapse load and the load factor in the beam shown in Fig. 16.5 which is built-in at one end and carries a central point load.

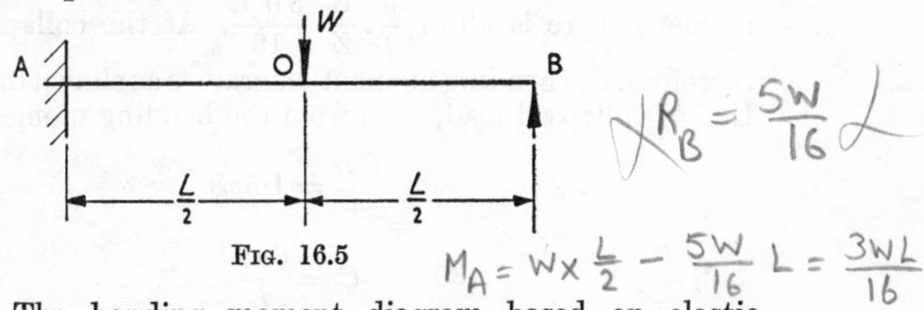

Fig. 16.5

Solution. The bending moment diagram based on elastic behaviour with a central load W is as shown in Fig. 16.6 (*a*), the maximum moment being $\frac{3}{16}WL$ at the fixed end. If f = working stress, then

$$W = \frac{16fZ}{3L}$$

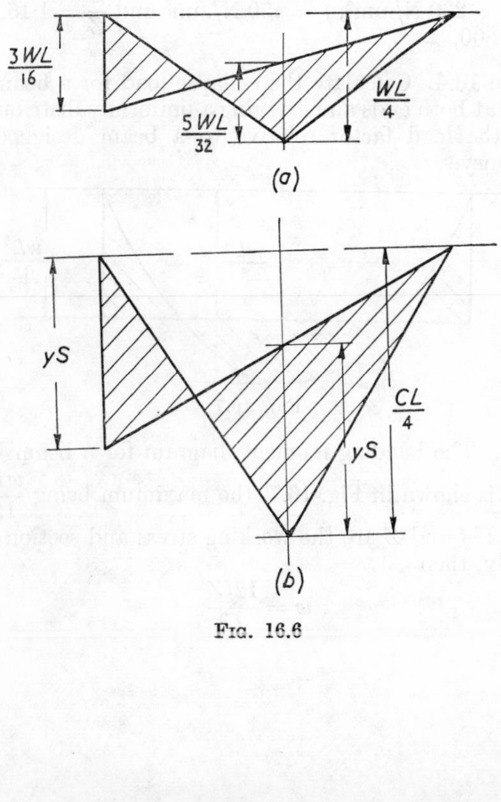

(*a*)

(*b*)

Fig. 16.6

If the load is increased, then a plastic hinge is first formed at A. This hinge forms when

$$M = yS = \frac{y}{f} \cdot \frac{S}{Z} \cdot \frac{3WL}{16}$$

Collapse, however, does not take place since there are only two hinges present and the beam is capable of carrying further load. It does so until a third hinge forms under the load when the moment there is yS or $\frac{y}{f} \cdot \frac{S}{Z} \cdot \frac{3WL}{16}$. At the collapse condition, therefore, the bending moment diagram is as shown in Fig. 16.6 (*b*). If C = collapsed load, then from the bending moment diagram

$$\frac{CL}{4} = 1{\cdot}5yS$$

$$\therefore \qquad C = \frac{6yS}{L}$$

The working load $\qquad W = \dfrac{16FZ}{3L}$

hence the load factor $= \dfrac{C}{W} = \dfrac{6yS}{L} \cdot \dfrac{3L}{16fZ} = \dfrac{9}{8} \cdot \dfrac{y}{f} \cdot \dfrac{S}{Z}$

With $y = 230$ N/mm², $f = 150$ N/mm² and $\frac{S}{Z} = 1{\cdot}15$, the load factor $= 2{\cdot}00$.

Example 16.4. Calculate the collapse load for a beam of span L built-in at both ends and carrying a uniformly distributed load. What is the load factor relative to a beam designed by the elastic theory.

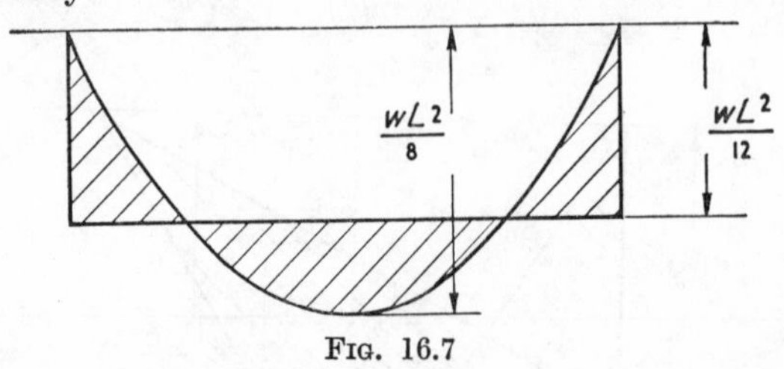

FIG. 16.7

Solution. The bending moment diagram for a beam behaving elastically is shown in Fig. 16.7, the maximum being $\frac{wL^2}{12}$ at the supports. If f and Z are the working stress and section modulus respectively, then

$$w = \frac{12fZ}{L^2}$$

Collapse will occur by the formation of three plastic hinges, one at each support and the other at mid-span. The collapse load bending moment diagram is shown in Fig. 16.8, M_H denoting the hinge moment and c the intensity of loading to cause collapse.

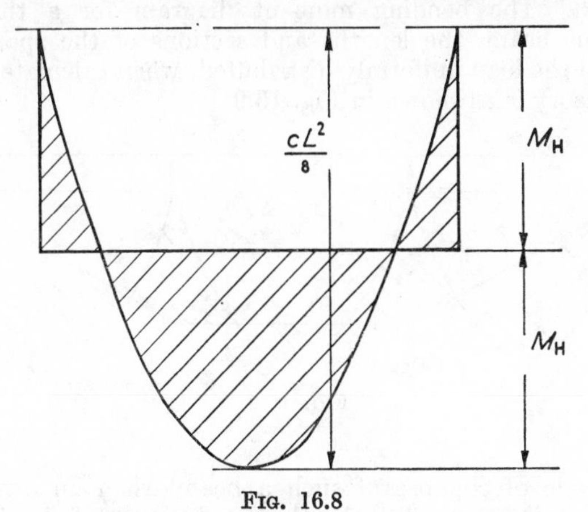

FIG. 16.8

Then $$2M_H = \frac{cL^2}{8}$$

But $$M_H = yS$$

hence $$c = \frac{16yS}{L^2}$$

The load factor $$= \frac{c}{w} = \frac{4}{3} \cdot \frac{yS}{fZ}$$

With the stress values given in Example 16.3 the load factoı in the above case is 2·36.

Application to Continuous Beams

Failure in the case of a continuous beam will take place by the same method as the built-in beam, namely, the development of hinges at the supports and mid-span for any *one* span of the continuous beam. The word one is italicized because failure in one span must mean that the beam as a whole has failed. It will obviously be most economical if the design is such that hinges are developed in all mid-span sections but this is seldom possible.

Example 16.5. A continuous beam consisting of three spans of 6 m carries a uniformly distributed load of 30 kN/m run.

Taking a load factor of 1·75, a yield stress of 230 N/mm² and a shape factor of 1·15, calculate a suitable joist section using collapse-load design.

Solution. The bending moment diagram for a three-span continuous beam, the lengths and sections of the spans being equal and the load uniformly distributed, when calculated on the elastic theory is as shown in Fig. 16.9.

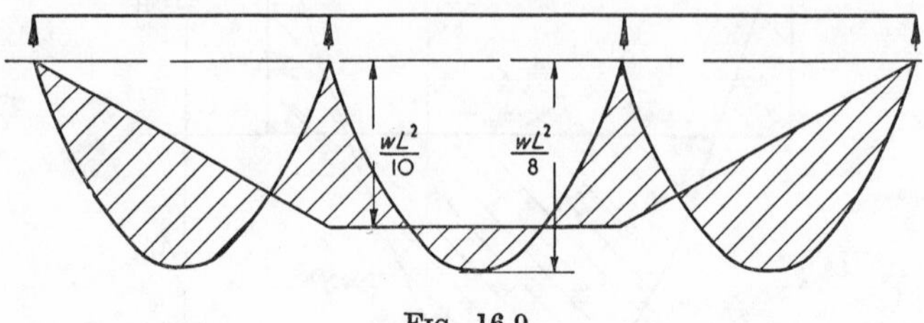

FIG. 16.9

The mode of collapse of such a beam when an overload is applied to this beam will be by the development of plastic hinges at the intermediate supports and at mid-span in the outer sections. The two outer spans will then have failed and therefore the beam has failed although the centre span could take more load before it collapsed.

The bending moment diagram at collapse is therefore as shown in Fig. 16.10. If c is the intensity of load per unit length at

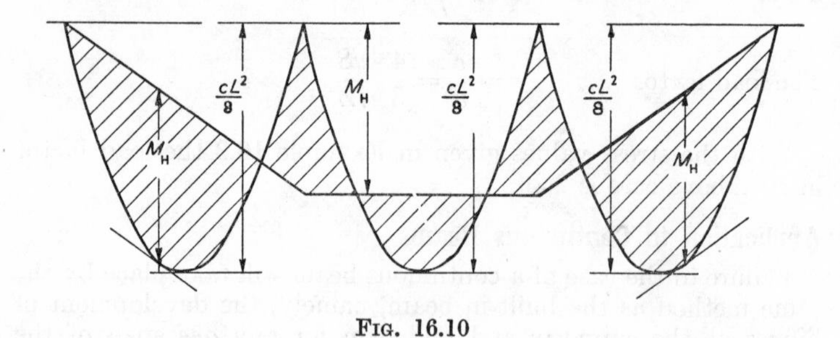

FIG. 16.10

collapse, then it can be shown from geometry (the student can check this by drawing the diagrams to scale) that the moment at the hinges,

$$M_H = 0{\cdot}685 \times \frac{cl^2}{8}$$

In the numerical example given

$$c = 1{\cdot}75 \times 30{\cdot}0 = 52{\cdot}5 \text{ kN/m run}$$

and $$l = 6 \text{ m}$$

hence $$M_H = 0{\cdot}685 \times 52{\cdot}5 \times 6^2 \times \frac{1}{8} = 162 \text{ kN m}$$

But $$M_H = yS$$

hence $$S = \frac{162 \times 10^6}{230} = 7{\cdot}05 \times 10^5 \text{ mm}^3 \text{ units}$$

The shape factor $= \dfrac{S}{Z} = 1{\cdot}15$

hence $$Z = \frac{7{\cdot}05 \times 10^5}{1{\cdot}15} = 6{\cdot}12 \times 10^5 \text{ mm}^3 \text{ units}$$

The standard joist giving the nearest approach to this is a 13 in. × 5 in., the section modulus of which is 43·62 in.³ or $7{\cdot}12 \times 10^5$ mm³ units.

This design is somewhat uneconomical since the plastic hinge has not been allowed to develop at mid-span in the centre span. If it is desired to produce a structure in which such a failure occurs, then a uniform section cannot be used throughout. A

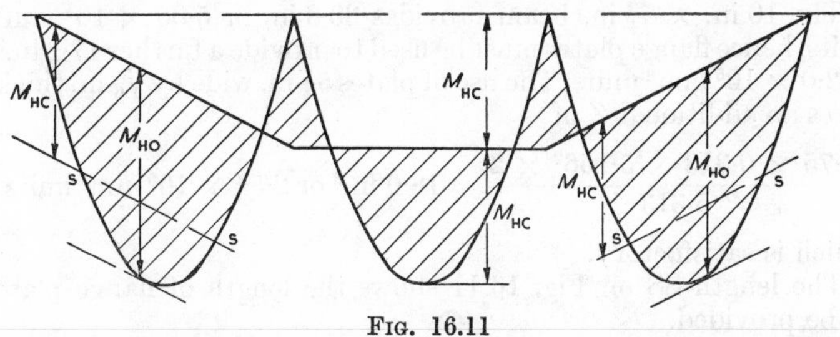

Fig. 16.11

lighter section is used for the centre span but the mid-span sections of the outer spans are strengthened to enable them to take a collapse moment greater than that on the centre span. The bending moment diagram is then as shown in Fig. 16.11.

The hinge moment, M_{HC}, for the lighter section is given by

$$M_{HC} = \frac{cl^2}{16}$$

where $c = 52{\cdot}5$ kN/m run

and $l = 6$ m

Hence $$M_{HC} = 52{\cdot}5 \times 6^2 \times \frac{1}{16} = 118 \text{ kN m}$$

$$S = \frac{M_{HC}}{y} = \frac{118 \times 10^6}{230} = 5{\cdot}13 \times 10^5 \text{ mm}^3$$

$$Z = \frac{5{\cdot}13 \times 10^5}{1{\cdot}15} = 4{\cdot}46 \times 10^5 \text{ mm}^3 \text{ units}$$

for which a 10 in. × $5\frac{3}{4}$ in. × 29 lb. universal beam with a section modulus of 30·8 in.^{3} or 5·00 × 10^5 mm^3 units is suitable.

The outer spans, however, must be strengthened so that their collapse moment is M_{HO}, where

$$M_{HO} = \frac{cl^2}{8} - 0{\cdot}43 M_{HC}$$

Substitution of the relevant figures gives

$$M_{HO} = 52{\cdot}5 \times 6^2 \times \frac{1}{8} - 0{\cdot}43 \times 118 = 186 \text{ kN m}$$

The section modulus required for the outer sections is therefore

$$\frac{186 \times 10^6}{230 \times 1{\cdot}15} = 7{\cdot}0 \times 10^5 \text{ mm}^3 \text{ units}$$

The 10 in. × $5\frac{3}{4}$ in. beam provides 30·8 in. or 5·00 × 10^5 mm^3 units, hence flange plates must be used to provide a further 17·2 in.^{3} or 2·0 × 10^5 mm^3 units. The use of plates $5\frac{3}{4}$ in. wide by $\frac{5}{16}$ in. thick gives an additional Z of

$$\frac{5{\cdot}75 \times 0{\cdot}313 \times 5{\cdot}156^2 \times 2}{5.313} = 18{\cdot}0 \text{ in.}^3 \text{ or } 2{\cdot}54 \times 10^5 \text{ mm}^3 \text{ units}$$

which is satisfactory.

The length SS on Fig. 16.11 shows the length of flange plate to be provided.

Application to Portals

The collapse of a single bay portal frame will occur when plastic hinges have been developed at sufficient points to produce instability. It has already been shown that a three-pinned portal is a statically determinate structure. The introduction of one further pin will be necessary before collapse occurs, hence in a single bay portal four plastic hinges must be formed at collapse.

In analysing portals by the elastic method the horizontal thrust at one of the base pins was taken as the redundancy and the value for H calculated which gave the minimum strain energy to the structure. In the plastic analysis this horizontal thrust is again taken as the unknown but its value is obtained not from strain energy calculations but from a consideration of the value required to produce a fourth plastic hinge in the structure.

Example 16.6. A portal with pinned bases consists of stanchions 6 m high and a beam of 10 m span. One of the stanchions carries a point load of 50 kN at mid-height and the beam carries a uniformly distributed load of 30 kN/m run. Using a load factor of 1·75, calculate the hinge moment at collapse.

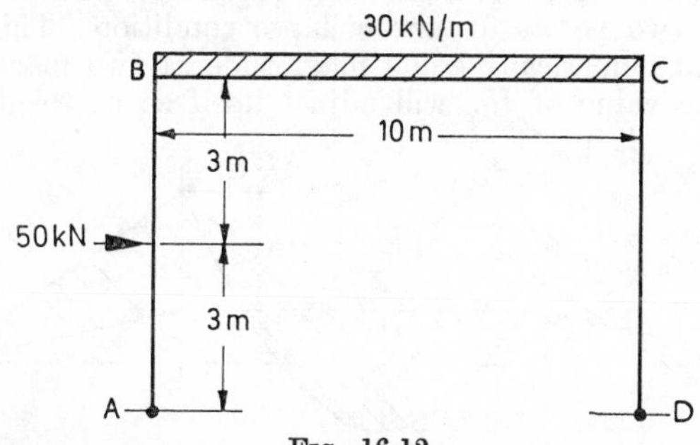

FIG. 16.12

Solution. The frame is shown diagrammatically in Fig. 16.12. If the pin at A is removed and a roller bearing placed there, the frame becomes statically determinate. The horizontal reaction at D is 50 kN and the bending moment diagram for the structure is as shown in Fig. 16.13. The crucial moments are

$$M_C = 50 \times 6 = 300 \text{ kN m}$$
$$M_B = 50 \times 3 = 150 \text{ kN m}$$
$$M_E = 152{\cdot}5, \text{ E being } 4{\cdot}5 \text{ m from B.}$$

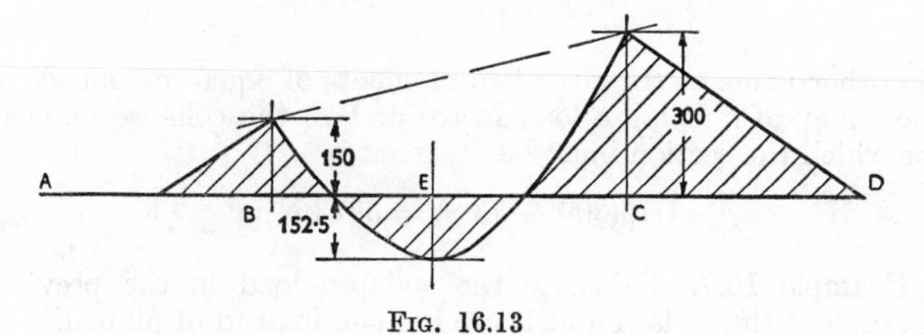

FIG. 16.13

If now a horizontal reaction is applied at A, then the bending moment diagram due to this load alone will be as shown in Fig. 16.14, the crucial values being

$$M_B = M_C = M_E = 6\,H_A \text{ kN m}$$

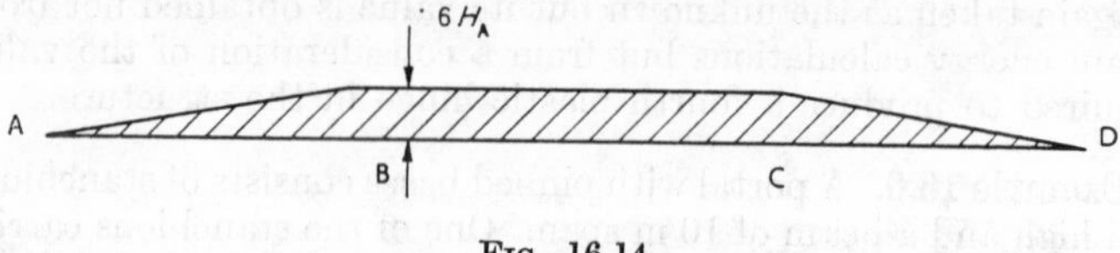

FIG. 16.14

The combined diagram for the applied loads together with the redundant force, H_A, is shown in Fig. 16.15. Hinges must be formed at two points for the collapse condition. This implies moments of numerically equal magnitude at two places on the beam. The value of H_A will adjust itself so as to give equal

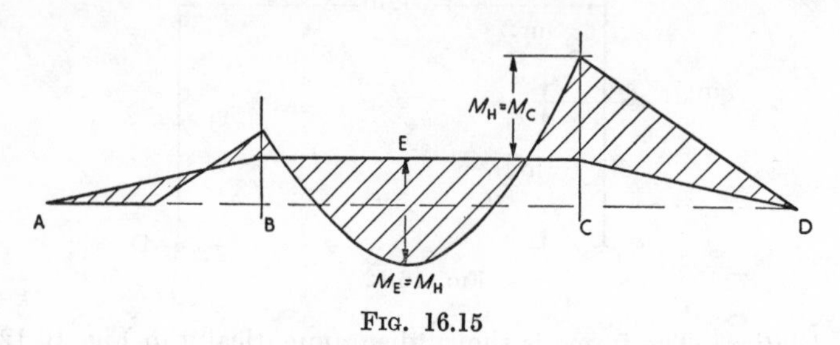

FIG. 16.15

moments at C and E which will be the collapse moments (neglecting the load factor)

$$M_C = 300 - 6\,H_A$$

$$M_E = 152{\cdot}5 + 6\,H_A$$

For these to be equal

$$H_A = \frac{147{\cdot}5}{12} = 12{\cdot}3 \text{ kN}$$

No other value of H_A gives two moments of equal magnitude on the beam BC. Using a load factor of 1·75 the collapse moment for which the section must be designed is

$$1{\cdot}75(300 - 73{\cdot}8) = 397 \text{ kN m}$$

Example 16.7. Calculate the collapse load in the previous example if the ends A and D are built-in instead of pinned.

Solution. The fact that the ends at A and D are fixed means that there are now two redundancies at each support, a moment and a thrust. If the support A is made a roller bearing and D a pin, then Fig. 16.13 still applies. If a moment, M_A, and a thrust, H_A, are applied at A and end D is fixed, the bending moment diagram for these two effects combined is as shown in Fig. 16.16, the critical values being

$$\text{At A and D, } M_A$$

$$\text{At B, } 6H_A - M_A$$

$$\text{At E, } 6H_A - \frac{M_A}{10}$$

$$\text{At C, } 6H_A + M_A$$

Fig. 16.16 must now be superimposed on Fig. 16.13 to give the diagram shown in Fig. 16.17. Four hinges must form when collapse of the portal as a whole occurs and they will generally be at A, E, C and D although a hinge may form at B instead of E. Alternatively the beam alone may fail by the development of

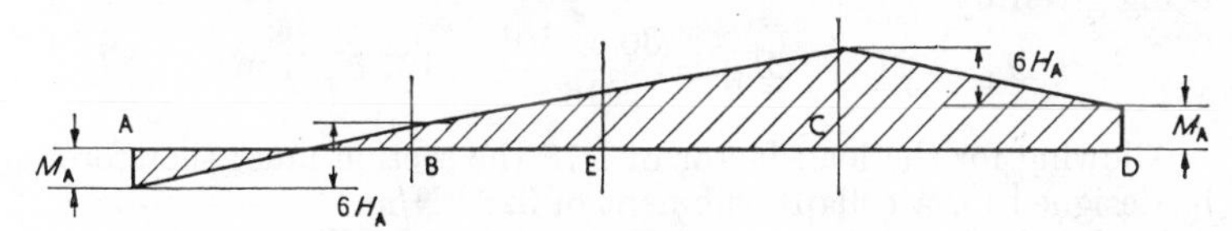

FIG. 16.16

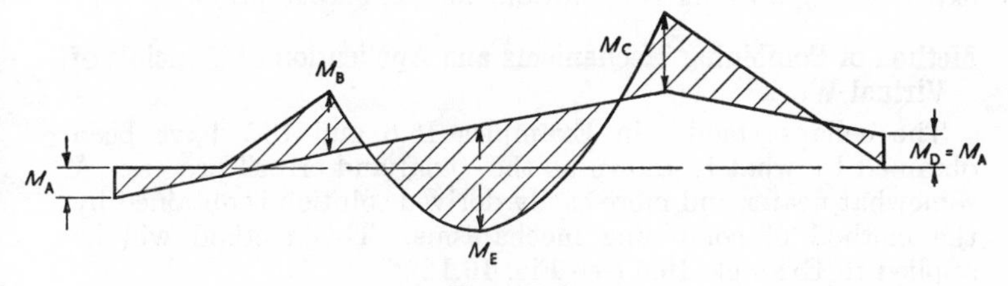

FIG. 16.17

hinges at B, E and C, and this implies the collapse of the portal although only three hinges have formed.

Assuming in the first case that in this example hinges develop at A, E, C and D, then the conditions given for the determination of the unknowns M_A and H_A are

$$M_A = 300 - (6H_A + M_A) = 152{\cdot}5 + \left(6H_A - \frac{M_A}{10}\right)$$

The solution of these equations gives $M_A = 146$ kN m and $H_A = 1{\cdot}5$ kN.

A check must now be made that M_B is smaller than M_E. Its value is

$$150 + M_A - 6H_A = 287 \text{ kN m}$$

This is greater than M_E, consequently the equations must now be built up and solved on the assumption that the hinges form at

A, B, C and D. The conditions now given for the determination of the unknowns are

$$M_A = 150 + (M_A - 6H_A) = 300 - (6H_A + M_A)$$

the solution of which is $M_A = 75$ kN m and $H_A = 25$ kN. The value of M_E under these conditions is

$$152{\cdot}5 + \left(6 \times 25 - \frac{75}{10}\right) = 295 \text{ kN m}$$

which is now greater than M_B. Consequently the second assumption is not valid and the only possibility is that failure occurs by the formation of hinges at B, E and C. Failure by three arithmetically equal moments at B, E and C corresponds to the built-in beam dealt with in Example 16.4, the values of the hinge moments being given by

$$M_B = M_C = M_E = \frac{30 \times 10^2}{16} = 187{\cdot}5 \text{ kN m}$$

Allowing for the load factor of 1·75 the section must therefore be designed for a collapse moment of 328 kN/m.

In the above example the various modes of collapse have been explained by working out solutions for each assumption.

Method of Combining Mechanisms and Application of Principle of Virtual Work

The collapse modes in Examples 16.6 and 16.7 have been obtained by what is known as the "trial and error" process. A somewhat neater and more easily derived solution is obtained by the method of combining mechanisms. This method will be applied to Example 16.6 (see Fig. 16.12).

Failure of this frame can take place by either: (1) sidesway mechanism, (2) beam mechanism, or (3) combined mechanism. These three collapse modes are indicated in Figs. 16.18 (*a*), (*b*) and (*c*), it being assumed in the first instance that the span hinge in the beam occurs in the exact centre.

If the sidesway mechanism shown in Fig. 16.18 (*a*) produces a rotation θ at the plastic hinges, and if M_P is the moment at each of these plastic hinges, then, equating the external to the internal work gives—

$$2M_P \times \theta = 50 \times 3 \times \theta \text{ leading to } M_P = 75 \text{ kN m.}$$

Similarly, the beam mechanism shown in Fig. 16.18 (*b*) gives the following equation—

$$4M_P\theta = \tfrac{1}{2} \times 300 \times 5\theta = 750\theta$$

(The factor of $\frac{1}{2}$ is inserted because the load is uniformly distributed.) This leads to a value of $M_P = 187{\cdot}5$ kN m.

When the sway and beam mechanisms are combined, as shown in Fig. 16.18 (c), it is noticed that no rotation takes place at the joint B, consequently no internal work is done there and when adding the internal and external work for the two mechanisms (a)

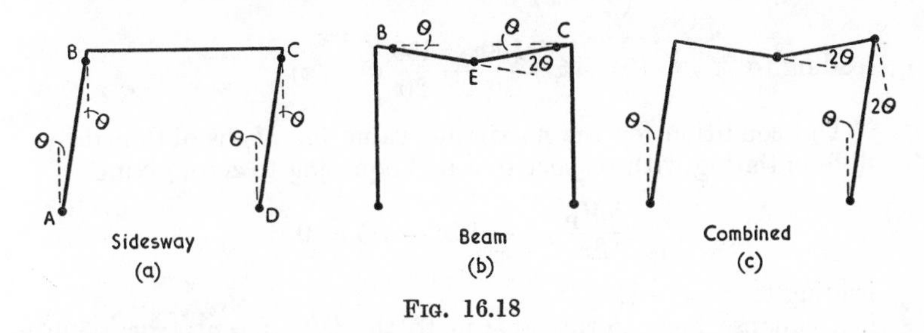

Fig. 16.18

and (b) to give (c) allowance must be made for this. The equation therefore for the combined mechanisms is

$$2M_P \times \theta + 4M_P \times \theta - 2M_P \times \theta = 150\theta + 750\theta$$

leading to $M_P = 225$ kN m.

Since this is the largest of the collapse moments it means that the frame will collapse by the formation of hinges at C and E, these together with the base hinges at A and B giving the four necessary for failure.

The design moment $= 1{\cdot}75 \times 225 = 393$ kN m. This figure is slightly less than the value of 397 kN m given on page 322, using the trial and error process. The reason is that the hinge in Figs. 16.17 (b) and (c) has been assumed to occur at the exact centre of the span, whereas in the previous working it was a short distance from the centre. The error involved in assuming the hinge at mid-span is small (of the order of 1 per cent) and for practical purposes could be neglected. The work involved in

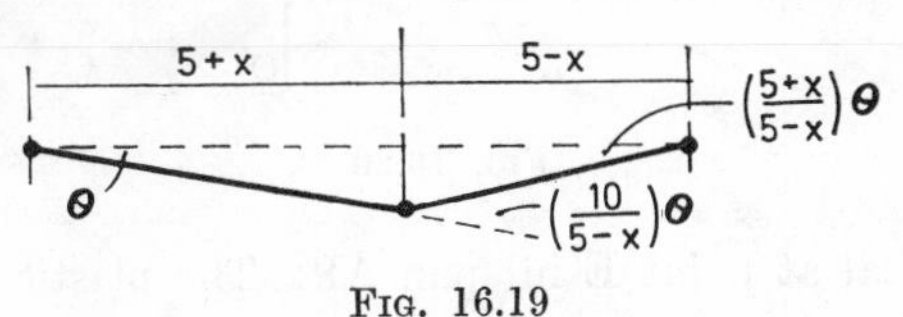

Fig. 16.19

getting an exact solution by combining mechanisms is, however, not difficult. Let the plastic hinge occur at a distance x metres to the right of the centre. Then the beam collapse mechanism is as shown in Fig. 16.19, and equating internal and external work gives

$$M_P\theta\left(1 + \frac{5 + x}{5 - x} + \frac{10}{5 - x}\right) = \tfrac{1}{2} \times 300 \times (5 + x)\,\theta$$

Equating internal and external work for the combined mechanism, Fig. 16.18 (*c*), gives

$$2M_P\theta + M_P\theta\left(\frac{20}{5-x}\right) - 2M_P\theta = 150\,(5+x)\,\theta + 150\theta$$

leading to $M_P = (5^2 - x^2)\dfrac{150}{20} + \dfrac{150}{20}(5-x)$

The condition for the maximum value for M_P is obtained by differentiating with respect to x and equating to zero, giving

$$\frac{\partial M_P}{\partial x} = -15x - 7{\cdot}5 = 0$$

leading to $x = -\frac{1}{2}$,
i.e., collapse hinge forms at $\frac{1}{2}$ m to the left of centre or 4·5 m from B, the same result as on page 321 using the "trial and error" method.

Substituting $x = -0{\cdot}5$ in the equation for M_P gives $M_P =$ 227 kN m and the design moment $= 1{\cdot}75 \times 227 = 397$ kN m, the same result as the "trial and error" method.

Example 16.8. A mild steel frame ABCD (see Fig. 16.20) is supported on rollers at A and C, and is rigidly fixed at D. It carries a collapse load of $\sqrt{2}\,\lambda\,W$ tons acting at an angle of 45°

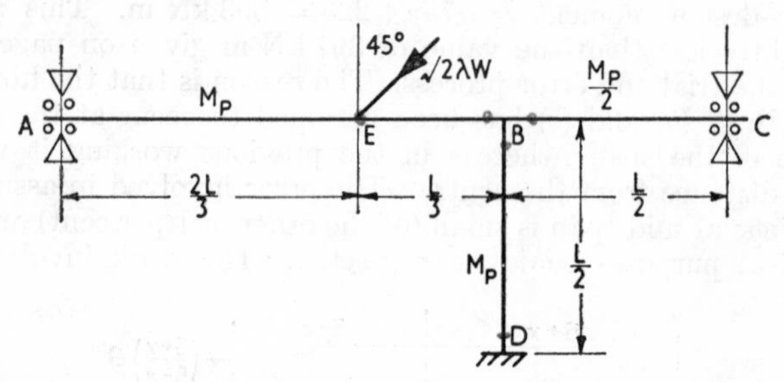

FIG. 16.20

to the vertical at point E in span AB. The plastic moment of resistance of members AB and BD $= M_P$ and that of BC $= \dfrac{M_P}{2}$. Calculate the load factor λ of the frame and draw the collapse bending moment diagram. (*Glasgow*)

Solution. The number of independent mechanisms must first be found from a knowledge of the number of possible hinges and the number of redundancies.

The number of possible hinges is 5, one each at D and E, and three at joint B. (Alternatively, it can be stated that if the bending moments at these 5 points are known, then the bending moment diagram for the whole frame can be drawn.)

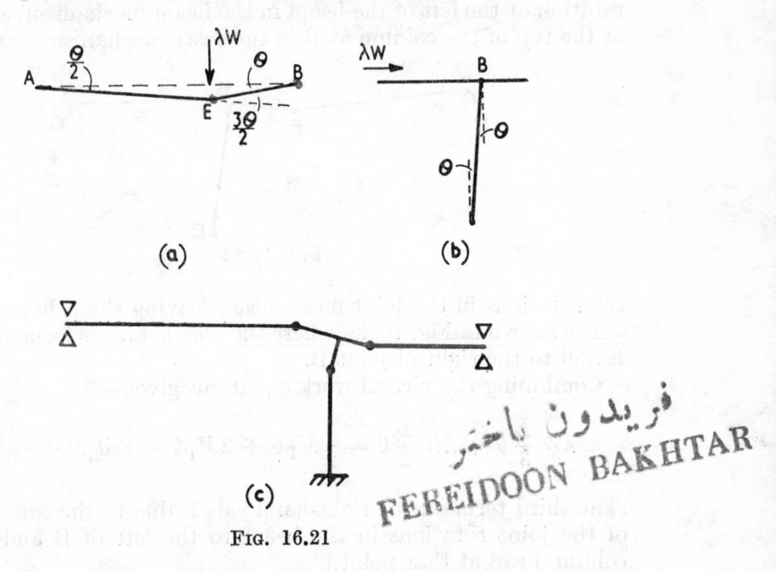

FIG. 16.21

The number of redundancies is 2, the vertical reactions at A and C.

The number of independent mechanisms is $5 - 2 = 3$. These are shown in Fig. 16.21 (*a*), (*b*) and (*c*). The first is a beam mechanism, the second a sway mechanism, and the third a joint mechanism at B.

Applying the principle of virtual work to the beam mechanism (Fig. 16.21 (*a*)) gives

$$\lambda W \,.\, \frac{L}{3} \,.\, \theta = M_P \theta \,(3/2 + 1)$$

giving

$$M_P = \frac{2\lambda W L}{15}$$

Applying it to the sway mechanism, Fig. 16.21 (*b*), gives

$$\lambda W \frac{L}{2}\, \theta = M_P \theta \,(1 + 1)$$

giving

$$M_P = \frac{\lambda W L}{4}$$

Since there is no displacement of loads during the rotation mechanism of Fig. 16.21 (*c*) no virtual work equation can be derived for this mechanism.

If, however, the three mechanisms are combined then the rotation at the left of the beam in the beam mechanism, and that at the top of the column at B in the sway mechanism, cancel out

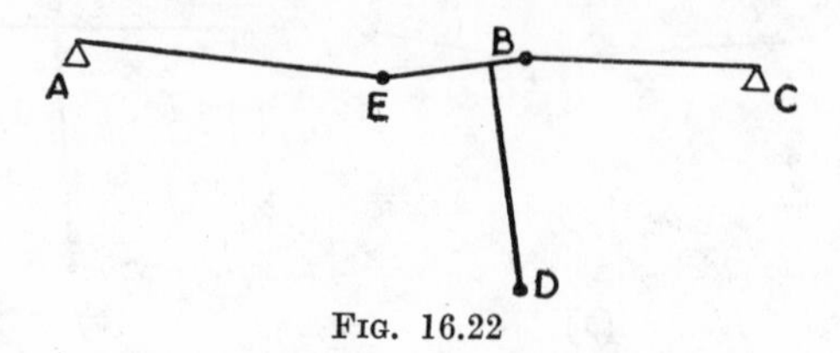

FIG. 16.22

the rotations in the joint mechanism, leaving the collapse mechanism shown in Fig. 16.22, where the plastic hinges form at D and E and to the right of joint B.

Combining the virtual work equations gives—

$$\lambda W \frac{L}{3} \theta + \lambda W \frac{L}{2} \theta = \frac{5}{2} M_P \theta + 2M_P \theta - 2M_P \theta + \frac{M_P}{2} \theta$$

(The third term on the right-hand side is due to the cancellation of the joint rotations in the beam to the left of B and at the column head at that point.)

or
$$\frac{5}{6} \lambda WL = 3M_P$$

leading to
$$M_P = \frac{5\lambda WL}{18}$$

The load factors for the three mechanisms are $7{\cdot}5 \dfrac{M_P}{WL}$, $4{\cdot}0 \dfrac{M_P}{WL}$ and $3{\cdot}6 \dfrac{M_P}{WL}$. Since the last is the smallest, this is the critical value, and the answer is $\lambda = \dfrac{3{\cdot}6\, M_P}{WL}$.

The ordinates at the critical points on the bending moment diagram are—

$$E\colon\ M_P = \frac{5\lambda WL}{18}$$

$$\text{D:}\ \ -M_P = \frac{5\lambda WL}{18}$$

$$\text{Right of B:}\ \frac{M_P}{2} = \frac{5\lambda WL}{36}$$

Column head at B $\quad \dfrac{5\lambda WL}{18} - \lambda W \cdot \dfrac{L}{2} = \dfrac{2}{9}\lambda WL$

Left of B: $\dfrac{2}{9}\lambda WL - \dfrac{5}{36}\lambda WL = \dfrac{\lambda WL}{12}$

The complete diagram is given in Fig. 16·23.

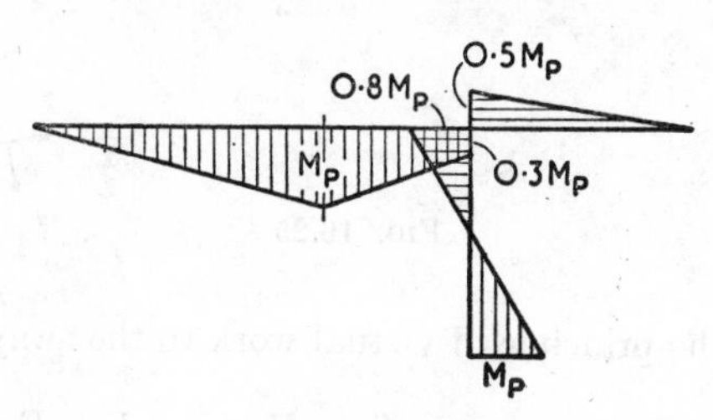

Fig. 16.23

Example 16.9. The frame shown in Fig. 16.24 is hinged at the points A, B and C and carries the loads shown. The plastic moments of resistance of the beams and columns are indicated. Using the method of combined mechanisms and taking a load factor of 1·75 calculate the plastic moduli of the sections if the yield stress in the material is 230 N/mm².

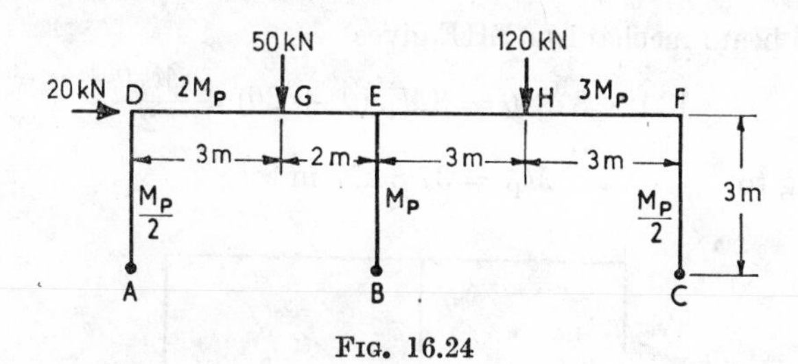

Fig. 16.24

Solution. The number of joints at which hinges can form is 7 (three at E and one each at D, G, H and F). The number of redundancies is 3, hence the number of independent mechanisms is 4. These are—

1. Sway mechanism, Fig. 16.25 (*a*).
2. Beam mechanism DGE, Fig. 16.25 (*b*).
3. Beam mechanism EHF, Fig. 16.25 (*c*).
4. Joint mechanism E, Fig. 16.25 (*d*).

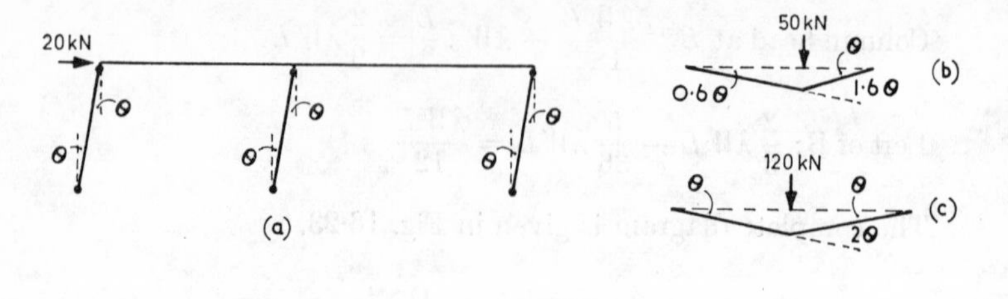

Fig. 16.25

Applying the principle of virtual work to the sway mechanism gives

$$20 \times 3 \times \theta = \theta \times M_P\,(\tfrac{1}{2} + 1 + \tfrac{1}{2})$$

leading to $\quad M_P = 30{\cdot}0 \text{ kN m}$

The beam mechanism DGE, gives

$$50 \times 3 \times \theta = 2M_P\,(\theta + 1{\cdot}6\theta) + \frac{0{\cdot}6M_P\theta}{2}$$

leading to $\quad M_P = \dfrac{150}{5{\cdot}5} = 27{\cdot}3 \text{ kN m}$

The beam mechanism EHF gives

$$120 \times 3 \times \theta = 3M_P\,(\theta + 2\theta) + \frac{M_P\theta}{2}$$

eading to $\quad M_P = 37{\cdot}8 \text{ kN m}$

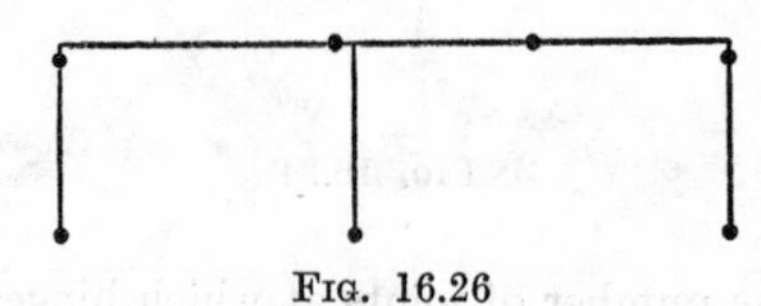

Fig. 16.26

The combination of the sway mechanism with beam mechanism EHF and joint rotation at E leads to the collapse mechanism shown in Fig. 16.26, and adding the virtual work equations gives

$$60\theta + 360\theta = 2M_P\theta + 9{\cdot}5M_P\theta - 3M_P\theta - M_P\theta + 2M_P\theta$$

giving $\quad M_P = \dfrac{420}{9{\cdot}5} = 44{\cdot}3 \text{ kN m}$

This combined mechanism gives the greatest value for M_P and is therefore the critical value.

$$\text{Plastic modulus } M_P \text{ required} = \frac{44{\cdot}3 \times 10^6 \times 1{\cdot}75}{230}$$

$$= 3{\cdot}36 \times 10^5 \text{ mm}^3 \text{ units.}$$

Hence, for AD plastic modulus $= 1{\cdot}68 \times 10^5$ mm³ units
BE ,, ,, $= 3{\cdot}36 \times 10^5$ mm³ units
CF ,, ,, $= 1{\cdot}68 \times 10^5$ mm³ units
DE ,, ,, $= 6{\cdot}72 \times 10^5$ mm³ units
EF ,, ,, $= 10{\cdot}08 \times 10^5$ mm³ units

The members of a portal, however, carry axial loads in addition to bending moments. Such loads are generally small and cause little alteration in the size of section required. It does, however, introduce the problem of combined bending and direct load and it is proposed to deal with this before concluding this chapter.

Combined Bending and Direct Load

Four possible stress distributions at failure are shown in Fig. 16.27. If the section fails under direct load only, then the distribution of stress is as in Fig. 16.27 (*a*). If the section is subjected to pure bending, then at failure the distribution is as in Fig. 16.27 (*d*). Two possible distributions for bending combined

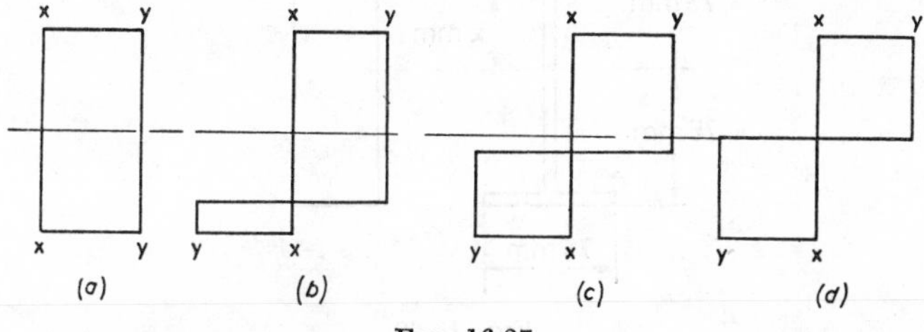

FIG. 16.27

with direct thrust are shown in Figs. 16.27 (*b*) and 16.27 (*c*), the former showing a small bending moment and the latter a larger one. If the section is symmetrical, then the amount of the section resisting each of the forces can be obtained as sketched in Fig. 16.28. The distance from the centre line to the point of zero stress is measured and an equal distance marked out on the other side of the centre line. This central area, being then in pure compression, can be considered as resisting the direct load whilst the outer areas, having in each of them an equal and opposite force, can be taken as resisting the bending.

If the central area $= a_d$ and each of the outer areas $= a_b$, the distances between the centroids of the outer areas being h, then the axial load at failure $= ya_d$ and the bending moment ya_bh.

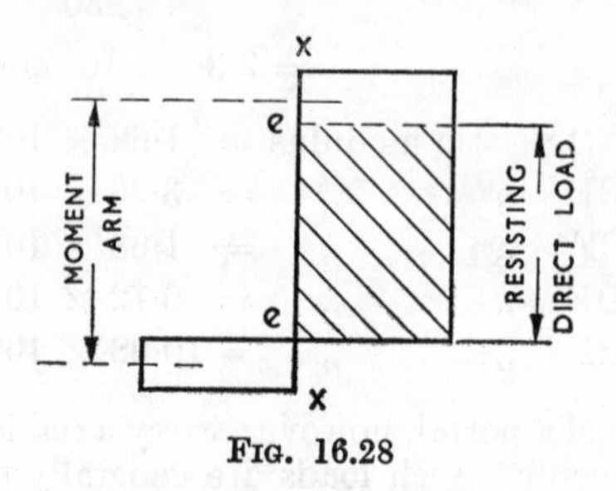

Fig. 16.28

Example 16.10. A 150 × 75 mm joist has a web 7 mm thick and a flange 7 mm thick. Calculate the maximum bending moment which in addition to a direct load of 200 kN will cause collapse in the section. Take the yield stress as 230 N/mm².

Solution. The section with the distribution of stress at failure is shown in Fig. 16.29. The area resisting the direct load is

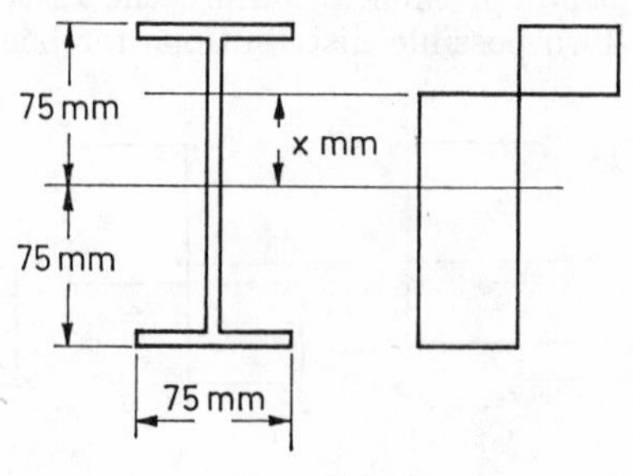

Fig. 16.29

$14x$ mm². This is stressed to 230 N/mm², hence $14x \times 230 = 200 \times 10^3$, thus $x = 62$ mm.

The area resisting bending is the flange together with the outer $(136 - 124) = 12$ mm of the web. It is easier, when considering the bending lever arm, to take flange and web separately giving the total resistance to bending in kN m as

$$230(75 \times 7 \times 143 + 6 \times 7 \times 130) \text{ N mm} = 18{\cdot}8 \text{ kN m}$$

Thus the bending moment which in addition to an axial load of 200 kN would cause collapse of the section is 18·8 kN m.

Examples for Practice on Plastic Design

1. A steel beam is 4 m long and of section shown (Fig. 16.30). It is rigidly built-in at both ends so that the clear span is 3·5m and loaded with a uniform load of w kN/m run. If the elastic limit stress is 230 N/mm², determine the following—

(*a*) The "shape factor."

(*b*) The maximum moment of resistance: (i) under elastic conditions, (ii) under plastic conditions.

(*c*) The load w_1 which can be supported when the fibre stress of 230 N/mm² is just reached at the outside of the section.

(*d*) The load w_2 which will induce total collapse of the beam.

Answer. (*a*) 2·00; (*b*) 245 kN m, 490 kN m; (*c*) 224 kN m; (*d*) 597 kN m.

2. A 150 × 75 mm joist whose section modulus is 1·155 × 10⁵ mm³ is used as a built-in beam to carry a load varying uniformly in intensity from zero at one end to a maximum at the other. If the joist has a shape factor of 1·15 and the yield stress in the material is 230 N/mm², determine the maximum load on a span of 3 m.

Answer. 159 kN.

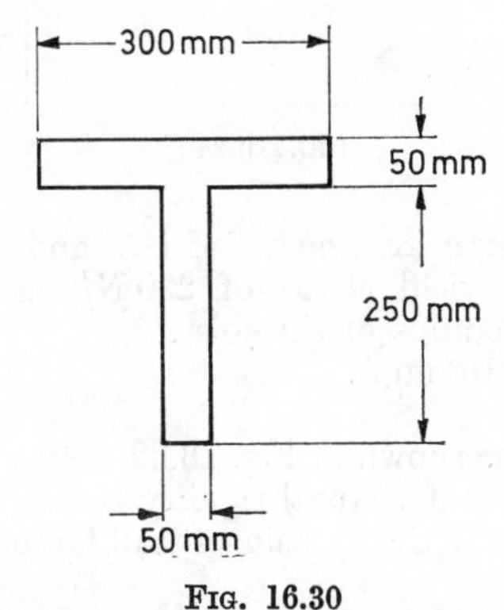

Fig. 16.30

3. A tie beam 150 mm wide and 50 mm thick carries a load placed with 25 mm eccentricity about the centre line. If it is made from a material having a yield stress of 250 N/mm², determine the load at which deformation of the whole section will take place. If the working stress in the material is 150 N/mm² calculate the load factor against deformation.

Answer. 1355 kN; 2·41.

4. A rectangular beam is built-in at one support and freely supported at the other. It has a span of 10 m and carries loads of 100 kN at each of the third points. It is made from a material whose yield stress is 300 N/mm². If the depth is to be three times the width, what section would be required if the beam has a load factor of 2·00?

Answer. $b = 90$ mm; $d = 270$ mm.

5. Outline briefly the usual procedure in designing a mild steel structure by the collapse load method.

A two-hinged portal carries loads as shown in Fig. 16.31 and is to have constant section throughout. It is to be designed on the

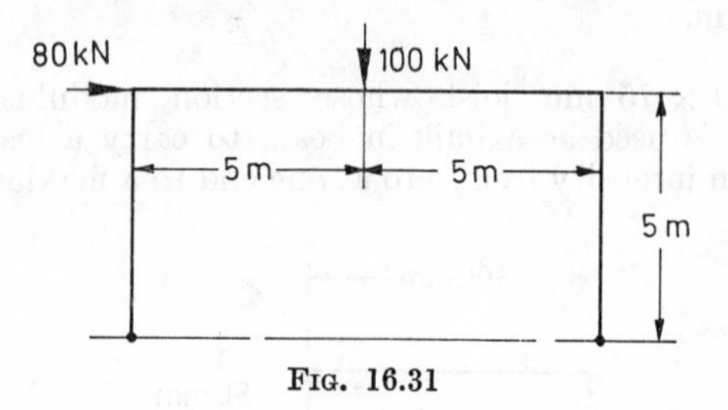

FIG. 16.31

plastic theory, using a load factor of 1·75 and a shape factor of 1·15. Adopting a yield stress of 230 N/mm², determine the required section modulus of the joist. (*Aberdeen*)

Answer. $1{\cdot}49 \times 10^6$ mm³.

6. The steel frame shown in Fig. 16.32 carries the loads shown and is to have constant flexural rigidity throughout. It is to be designed on the plastic theory using a load factor of 1·75. Taking

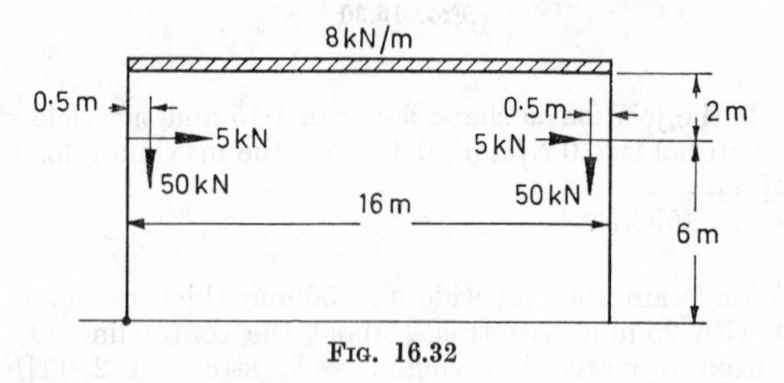

FIG. 16.32

a shape factor of 1·15 and a yield stress in the steel of 230 N/mm², calculate the required section modulus for the joist. (*St. Andrews*)

Answer. $9{\cdot}48 \times 10^5$ mm³.

7. Portal Frame.

Subject of Problem. A pitched portal frame, ABCDE, (Fig. 16.33), framing a workshop. A cable is to be tensioned between A and E to supply a horizontal reaction at each of these points.

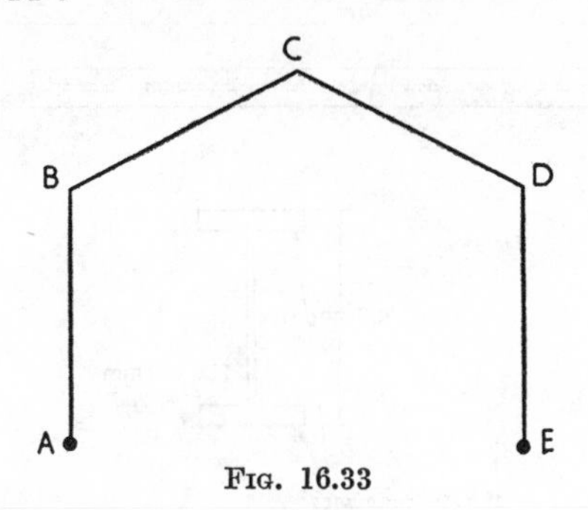

FIG. 16.33

Dimensions and Data

Span AE = 20 m
Vertical height AB = 8 m
Sloping length CD = 11 m
$I_{BC} = I_{DC} = 4$ units
$I_{AB} = I_{ED} = 1$ unit
Loading $\begin{cases} \text{2 kN/m from A to B} \\ \text{6 kN/horizontal m from B to D} \end{cases}$

Answers required

(*a*) The tension, T, in cable AE which will produce in the frame minimum values of maximum B.M.
(*b*) B.M. diagram with numerical values entered.
(*c*) S.F. diagram with numerical values entered.

(*Durham*)

Answer. 1·7 kN; $M_B = 85{\cdot}2$ kN m; $M_C = M_D = 141{\cdot}7$ kN m.

8. The beam shown in Fig. 16.34 is designed to carry a concentrated load W with a load factor of 2 at any point within the middle 1 m of the span. Find the maximum permissible value of W and find approximately the best position, x, for the bolt group. Make the usual assumptions for a fully ductile material with no drop of stress at the yield. The mean shear stress in the bolts may be assumed to behave in a similar manner.

Yield stress in tension is 240 N/mm², mean yield stress in shear is 120 N/mm.²

Briefly criticize the load-factor method of design.

(*Cambridge*)

Answer. $W = 103$ kN; $x = 236$ mm.

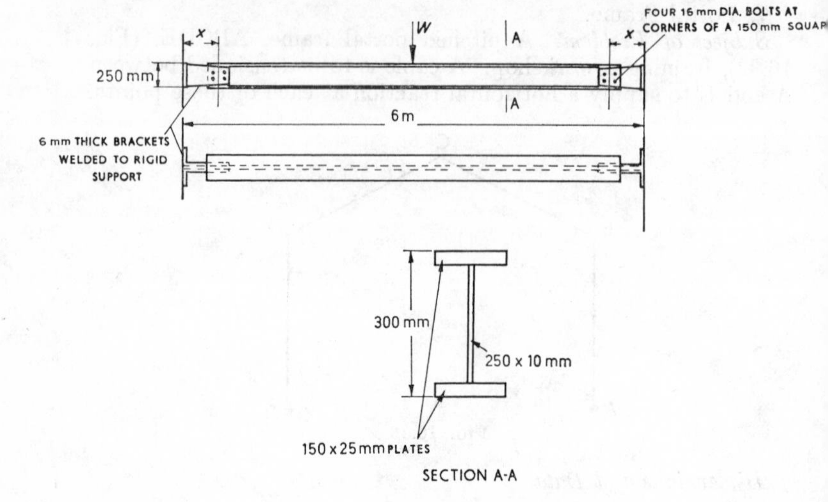

FIG. 16.34

9. A rigidly jointed rectangular portal frame span 3·75 m, height 2·5 m, is constructed throughout of 250 × 200 mm R.S.J., with a section modulus of $9{\cdot}47 \times 10^5$ mm³. The stanchion feet are to be considered fixed and the foundations able to sustain the full plastic moment of the section. The frame carries a central downward load W tons and an external horizontal load H tons at beam level.

Taking the yield stress of the material as 235 N/mm² and the shape factor of the section as 1·15, construct a curve showing anticipated values of W against H for plastic collapse.

Discuss, as far as time permits, the relative merits of the elastic and plastic theories in the design of structures. (*London*)

Answer. $H = 21 + \frac{3}{4}W$.

10. State the rule used in plastic design for determining the number of elementary mechanisms in a structure, and explain how it is derived. Use the method of combining mechanisms to determine the collapse load factor for the structure loaded as in Fig. 16.35. (*London*)

Answer. $\dfrac{3M_P}{L}$.

11. The portal frame ABCD in Fig. 16·36 is pinned to the supports A and D. It supports a dead + superimposed load of 25 kN/m uniformly distributed along BC, and a wind loading of 3 kN/m uniformly distributed along AB. The portal frame is to

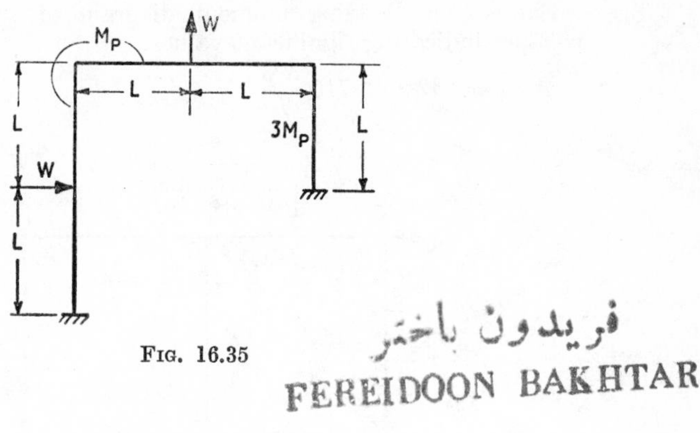

FIG. 16.35

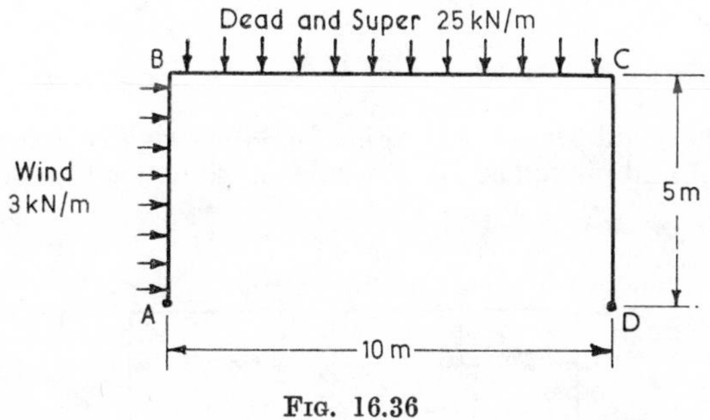

FIG. 16.36

be designed in mild steel by the plastic theory and is to be of constant section. The load factor for dead + superimposed loading is 1·75, and for dead + superimposed + wind loading 1·40. Calculate the minimum plastic modulus required for the frame section if the yield stress of the steel is 230 N/mm². Give a diagram showing the collapse mechanism for the frame for each loading condition considered. (*London*)

Answer. $1{\cdot}19 \times 10^6$ mm³ units.

12. The two-bay portal frame shown in Fig. 16.37 can be subjected to either of the following load systems:

(*a*) A horizontal load of $2W$ at B and a vertical load of $3W$ at the centre of span CD.

(*b*) A horizontal load of $2W$ at B and a vertical load of $3W$ at the centre of span BC.

If the plastic moment of the beams is $1{\cdot}5\ M_P$ and that of the columns M_P, calculate the minimum value of W which would cause collapse of the structure.

Sketch the bending moment diagram of the structure at collapse indicating significant values. (*Glasgow*)

Answer. $W = 2{\cdot}715\,\dfrac{M_P}{L}$.

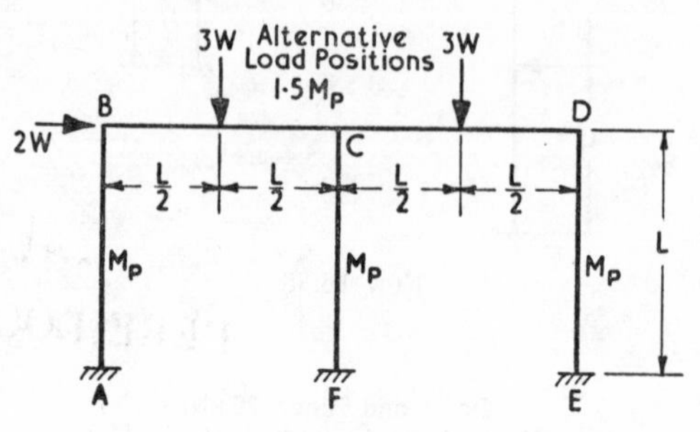

FIG. 16.37

13. The rigid frame AD (Fig. 16.38) is rigidly fixed at the support A and pinned at D. It carries working loads at the points

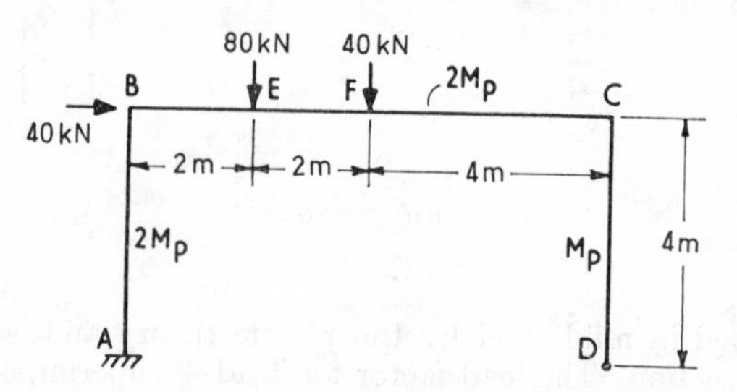

FIG. 16.38

B, E and F, as shown. Draw the bending moment diagram at collapse and calculate the plastic moduli of the members.

Load Factor = 1·75; Yield Stress = 225 N/mm².

(*Glasgow*)

Answer. AB, BC; $9{\cdot}68 \times 10^5$ mm³; CD: $4{\cdot}84 \times 10^5$ mm³.

14. The relative values of full plastic moments of resistance of the members of the two-bay portal frame shown in Fig. 16.39 are given in the circles. The stanchions are to be considered fixed at their feet and the foundations are able to develop the full plastic moments of the members. The frame is to carry the proportional working loads shown in the figure.

Using a plastic collapse method obtain the value of the elastic section modulus required for the stanchions to ensure a load factor of 1·75. The effects of instability and reduction in plastic moments of resistance due to axial forces can be neglected.

The shape factor of each cross-section may be taken as 1·15 and the value of the yield stress as 230 N/mm².

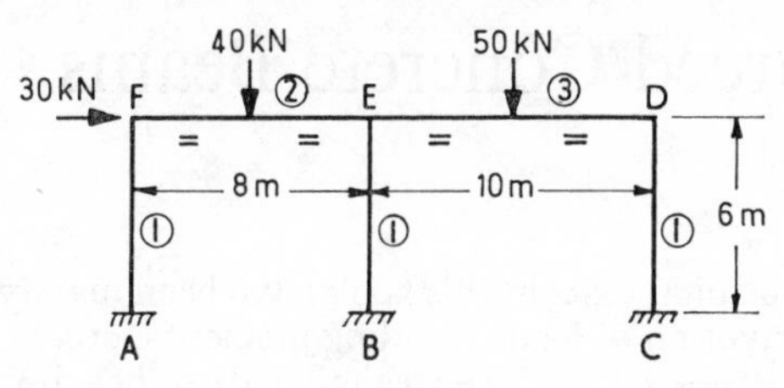

FIG. 16.39

Check your result by drawing the bending moment diagram for the structure at collapse. *(London)*

Answer. $2{\cdot}03 \times 10^5$ mm³ units.

15. Find the plastic collapse load of the ideally plastic structure shown in Fig. 16.40 by finding upper and lower bounds of the same value. *(London)*

Answer. $2{\cdot}67 \dfrac{M_P}{L}$.

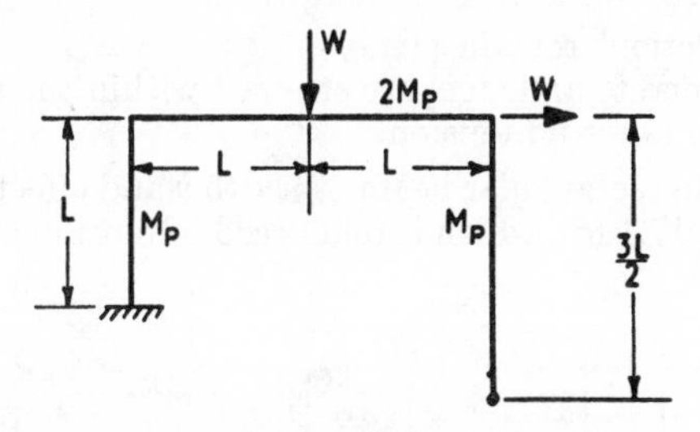

FIG. 16.40

16. A 375 mm × 125 mm R.S.J. is continuous over three spans, being simply-supported at A, B, C and D. AB = 6 m, BC = 12 m and CD = 6 m. Spans AB and CD each carry uniformly distributed loads of 50 kN/m while span BC carries a uniformly distributed load of 20 kN/m. The beam has an elastic section modulus of $9{\cdot}45 \times 10^5$ mm³ a shape factor of 1·17, and the yield stress of the material is 230 N/mm².

Calculate the load factor against plastic collapse of one or more of the spans, and design flange plating at suitable sections to increase the load factor by 25 per cent. *(London)*

Answer. Load Factor = 1·42; Thickness of flange plate 6 mm.

Chapter 17

Reinforced Concrete Beams

THE PREVIOUS CHAPTERS in this book have been mainly concerned with the derivation of loads, bending moments, etc. in structures, and where stresses have been calculated, it has been assumed that the member was monolithic. Reinforced concrete members do not, however, come within this class, and special methods have to be adopted when calculating the stresses in them. As with other structural materials, the members can be designed by the elastic or plastic methods, the former dealt with in this chapter, and the latter in Chapter 20.

Worked Examples on Singly Reinforced Rectangular Sections by Elastic Theory

The Elastic Theory as applied to reinforced concrete beams is based on the following three assumptions—

1. Plane sections remain plane.
2. Both concrete and steel are stressed within the elastic limit.
3. Concrete takes no tension.

Consider the rectangular beam of width b and effective depth d_1 shown in Fig. 17.1 (*a*), which is reinforced in tension only with A_{st}

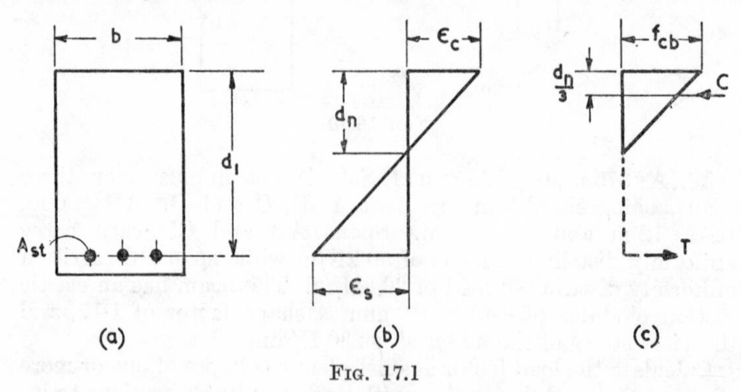

FIG. 17.1

units of steel. The first assumption means that the strain distribution across the section is linear, as shown in Fig. 17.1 (*b*), the maximum strain in the concrete in compression being ε_c and the strain in the steel (taken as uniform) being ε_s. The depth of the neutral axis is d_n. The second assumption means that the stress distribution in the compression section of the beam is triangular as shown in Fig. 17.1 (*c*), the fibre stress being taken as f_{cb} and the resultant compression acting at $\frac{d_n}{3}$ below the upper fibre. On the tension side the third assumption means that the total force is concentrated at the position of the steel reinforcement.

The following fundamental equations can be derived from the distribution of strains and stresses shown in Figs. 17.1 (*b*) and (*c*). From the straight line strain distribution—

$$\frac{\varepsilon_c}{\varepsilon_s} = \frac{d_n}{d_1 - d_n} = \frac{\frac{f_{cb}}{E_c}}{\frac{f_{st}}{E_s}} = \frac{mf_{cb}}{f_{st}} \quad . \quad . \quad . \quad (1)$$

where $m = \frac{E_s}{E_c}$ = modular ratio

f_{st} = tensile stress in steel.

Equating total compression to total tension gives

$$\tfrac{1}{2} \, . \, bd_n f_{cb} = A_{st} f_{st} \quad . \quad . \quad . \quad (2)$$

The moment of resistance of the section M_r is given by

$$M_r = \tfrac{1}{2} bd_n f_{cb} \left(d_1 - \frac{d_n}{3}\right) = A_{st} f_{st} \left(d_1 - \frac{d_n}{3}\right)$$

$$= \tfrac{1}{3} bf_{cb} \, d_n^2 + A_{st} f_{st} \, (d_1 - d_n) \quad . \quad . \quad . \quad (3)$$

The first two equations contain three possible unknowns—

$$d_n, \; A_{st}, \text{ and } \frac{f_{cb}}{f_{st}}.$$

Of these, d_n will generally be one and, since there are only two equations, this means that either A_{st} or $\frac{f_{cb}}{f_{st}}$ must be known. This is an important point for the student to remember. If the area of steel is given then it is the *stress-ratio* $\frac{f_{cb}}{f_{st}}$, which has to be determined. Thus, one can only specify the stress in either the steel or the concrete, not both stresses. If, on the other hand, the stresses in the materials are specified, then $\frac{f_{cb}}{f_{st}}$ is a known quantity and A_{st} is the unknown which must be determined from the equations.

If both materials are working at their permissible stresses, p_{cb} and p_{st}, this is known as "balanced design."

In dealing with problems of reinforced concrete the student must learn to use intelligently the three fundamental equations. The problems that occur in examinations can generally be solved only by the use of these equations, and they should be written down straight away at the start of a problem.

Example 17.1. The resistance moment of a singly reinforced concrete beam can be expressed as Qbd^2, where Q is a constant depending on the stresses and modular ratio, b = breadth of beam and d = effective depth.

Determine the value of Q when the stresses in the steel and concrete are 180 N/mm² and 10 N/mm² respectively, and $m = 15$.

If a beam 300 mm wide and 150 mm effective depth is subjected to a bending moment of 10 kN m and the stresses are limited to the values given above, which of the two materials is working at its limiting stress. *(St. Andrews)*

Solution. From the straight line strain relationship—

$$\frac{d_n}{d - d_n} = \frac{15 f_{cb}}{f_{st}} = \frac{15 \times 10}{180} = \frac{5}{6}$$

Hence $d_n = \frac{5}{11} d = 0{\cdot}455d$

$$M_r = \tfrac{1}{2} b d_n f_{cb} \left(d - \frac{d_n}{3}\right)$$

$$= \tfrac{1}{2} \times b \times 0{\cdot}455d \times 10\,(d - 0{\cdot}152d)$$

$$= 1{\cdot}93\, bd^2.$$

Hence, value of $Q = 1{\cdot}93$.

The answer to the second part of the question can be obtained from a study of this constant Q. It is clear that it depends on the neutral axis depth, d_n, and the fibre stress in the concrete f_{cb}.

If the stress ratio $\frac{f_{cb}}{f_{st}}$ is less than the value at balanced design then the depth of the neutral axis d_n will be less than its value at balanced design. Moreover, the concrete stress f_{cb} will also be less than its value at balanced design, consequently since f_{cb} and d_n are reduced the value of Q will also be reduced (it can be shown that the increase in the term $\left(d - \frac{d_n}{3}\right)$ is overbalanced by the reduction in d_n). Hence it can be stated that if Q is less than its value at balanced design the stress ratio $\dfrac{f_{cb}}{f_{st}}$ is less than its value at balanced design, and the steel will be fully stressed and the concrete understressed.

Conversely, if the stress ratio $\dfrac{f_{cb}}{f_{st}}$ is greater than its value at balanced design the depth of the neutral axis d_n will be greater than the value at balanced design and the value of Q will be greater. In this case the concrete will be fully stressed and the steel understressed.

In the example, the value of Q is given by

$$Q = \frac{10 \times 10^6}{150 \times 300 \times 300} = 0{\cdot}74.$$

This is less than the value at balanced design, 1·93, hence the steel is at its limiting stress.

Example 17.2. A singly reinforced rectangular concrete beam of effective span 5 m is required to carry a uniformly distributed load (including its own weight) of 15 kN/m run. The overall depth, D, is to be twice the breadth and the centre of the steel is to be at $0{\cdot}1D$ from the underside of the beam.

Find the dimensions of the beam and the area of steel required using working stresses of 8·5 N/mm² in the concrete and 140 N/mm² in the steel and taking $m = 15$. (*London*)

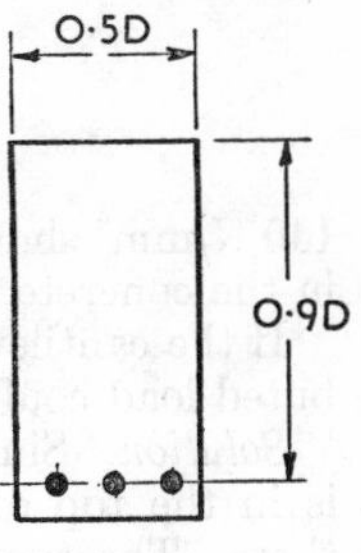

Fig. 17.2

Solution. A sketch of the beam is given in Fig. 17.2. The applied bending moment M is given by—

$$M = 15 \times 5^2 \times \frac{1}{8} = 47 \text{ kN m}$$

If d_n = depth to neutral axis and d_1 = effective depth = $0{\cdot}9D$. The straight line strain relationship gives

$$\frac{15 \times 8{\cdot}5}{140} = \frac{d_n}{d_1 - d_n}$$

$$\frac{d_n}{d_1 - d_n} = 0{\cdot}91, \text{ hence } d_n = 0{\cdot}476d_1.$$

The moment of resistance of the section M_r is given by—

$$M_r = \frac{8{\cdot}5}{2} \times 0{\cdot}476d_1 \times b \times 0{\cdot}841d_1 = 1{\cdot}70bd_1^2.$$

Substituting for b and d_1 in terms of D gives

$$1{\cdot}70 \times 0{\cdot}5D \times 0{\cdot}81D^2 = 47 \times 10^6$$

leading to

$$D = 410 \text{ mm}$$

The area of steel, A_{st}, is given by—

$$A_{st} = \frac{47 \times 10^6}{140 \times 0{\cdot}841 \times 0{\cdot}9 \times 410} = 1080 \text{ mm}^2$$

Example 17.3. The cross-section of a concrete step forming part of a flight which is cantilevered from a wall is shown in Fig. 17.3. Dimensions are in mm. Calculate what area of steel stressed to

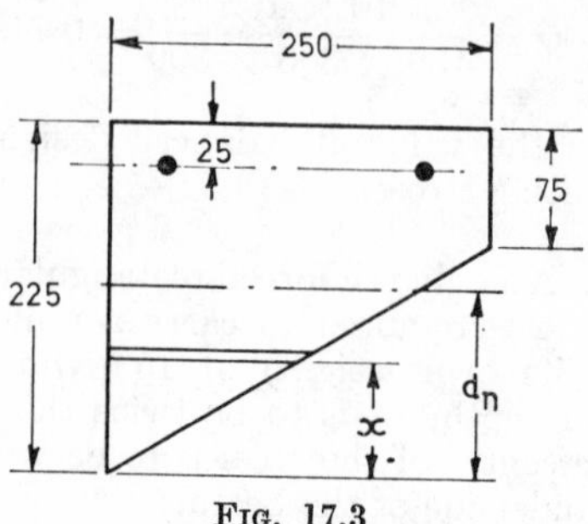

Fig. 17.3

140 N/mm² should be inserted in order that the maximum stress in the concrete shall be 7 N/mm.² Take $m = 15$.

If the cantilever is 1·4 m long, calculate what uniformly distributed load could be carried when the step is so reinforced.

Solution. Since the problem concerns a cantilever the tension is in the top of the section, hence the reinforcement is placed there. The compression part of the beam is triangular in cross-section, the position of the neutral axis being d_n from the most stressed fibre. At a distance x from this fibre the width of the section is $\dfrac{250x}{150}$. The stress at this point is $\left(\dfrac{d_n - x}{d_n}\right) f_{cb}$, since the stress distribution varies linearly from the neutral axis to the most stressed fibre.

Thus, the total compression

$$C = \frac{5\,f_{cb}}{3\,d_n}\int_0^{d_n} x\,(d_n - x)\,dx$$

$$= \frac{5\,f_{cb}}{3\,d_n}\left[\frac{d_n x^2}{2} - \frac{x^3}{3}\right]_0^{d_n}$$

$$= \frac{5}{18} f_{cb} \cdot d_n^2$$

The problem states that both materials are to be fully stressed, i.e. there is "balanced" design, and the depth of the neutral axis d_n can be obtained from the straight line stress distribution, giving

$$\frac{15 \times 7}{140} = \frac{d_n}{200 - d_n}$$

or $d_n = 85{\cdot}8$ mm.

If A_{st} = area of steel reinforcement, then since the total tension = total compression $A_{st} \times 140 = \dfrac{5 \times 7 \times 85{\cdot}8^2}{18}$

$$A_{st} = \frac{5 \times 7 \times 85{\cdot}8^2}{18 \times 140} = 102 \text{ mm}^2$$

The moment of the compressive forces about the neutral axis M_c is given by

$$M_c = \frac{5 f_{cb}}{3 d_n} \int_0^{d_n} x\,(d_n - x)^2\,dx = \frac{5}{36} f_{cb}\, d_n^3$$

The moment of the tensile force about the neutral axis M_t is given by

$$M_t = A_{st} f_{st} (d_1 - d_n)$$

and the total moment $M_r = M_c + M_t$.

Substituting the values for various quantities gives

$$M_c = \frac{5 \times 7 \times 85{\cdot}8^3}{36} = 612 \times 10^3$$

$$M_t = 102 \times 140 \times 114 = 1630 \times 10^3$$

$$M_r = 2242 \times 10^3 \text{ N mm}$$

If w = load per ft run on cantilever

$$M = w \times 1{\cdot}4^2 \times \frac{1}{2} \text{ N m}$$

Hence
$$w = \frac{2242 \times 2}{1{\cdot}4^2} = 2290 \text{ N/m run}$$

Examples on T-sections by Elastic Theory

Since the concrete below the neutral axis is not used for resisting bending, a T-section as shown in Fig. 17.4 is often used instead of a rectangular one. If the depth of the slab d_s is greater than the depth of the neutral axis d_n, then the beam is treated as a rectangular beam. In many problems, however, d_s is less than d_n

and the fundamental equations given for a rectangular beam have to be modified to allow for the compression which would be carried by the area shown shaded in Fig. 17.4 if the beam were rectangular.

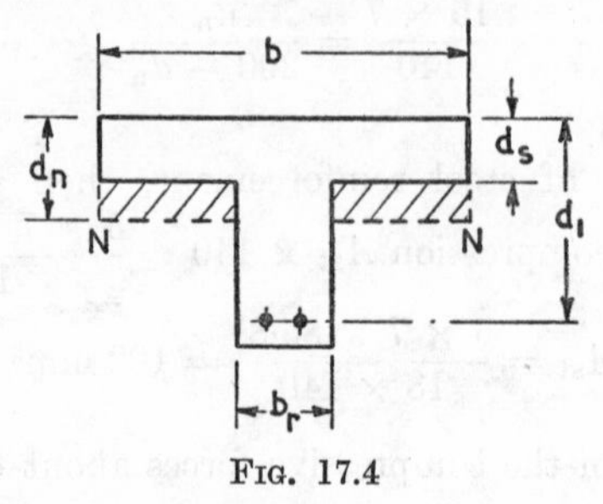

FIG. 17.4

The first equation derived from the straight-line strain relationship is unaltered. The stress at the underside of the flange is therefore—

$$\left(\frac{d_n - d_s}{d_n}\right) f_{cb}.$$

The second equation relating total tension to total compression now becomes—

$$A_{st} f_{st} = b d_n \frac{f_{cb}}{2} - (b - b_r) \frac{(d_n - d_s)^2}{d_n} \cdot \frac{f_{cb}}{2}$$

The moment of resistance of the section about the neutral axis, the third fundamental equation, is also changed and now becomes—

$$M_r = A_{st} f_{st} (d_1 - d_n) + \frac{b d_n^2}{3} \cdot f_{cb} - (b - b_r) \cdot \frac{(d_n - d_s)^3}{3\, d_n} \cdot f_{cb}$$

In cases where d_s is a small fraction of d_n almost the whole of the compression is carried by the flange and the small portion taken by the web can be neglected. Since the major part of the compression is carried by the flange it is often assumed for design purposes that the lever arm between the centres of the compressive and tensile forces is given by $d_1 - \frac{d_s}{2}$. If either of these assumptions is used in answering a question, this fact should be stated.

Example 17.4. The reinforced concrete T-section shown in Fig. 17.5 is subjected to a bending moment of 7·0 kN m. Derive an equation for the depth of the neutral axis in order that the stresses in the steel and concrete shall be 140 and 7 N/mm² respectively with $m = 15$.

Show that an approximate solution for this equation is $d_n = 86$ mm and calculate the amount of reinforcement required in the beam. (*St. Andrews*)

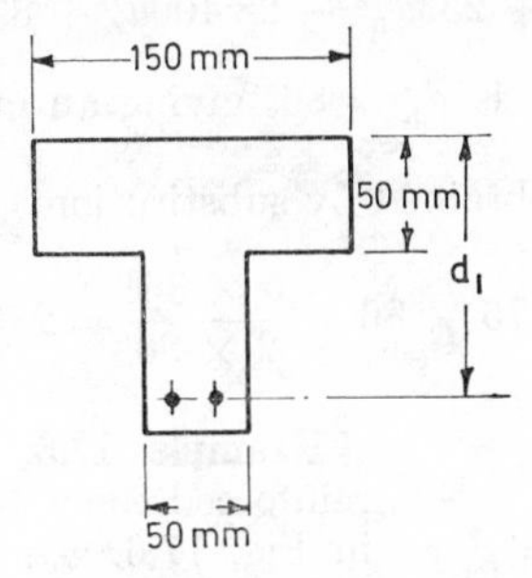

FIG. 17.5

Solution. From the straight-line strain relationship, since the stresses are known

$$\frac{15 \times 7}{140} = \frac{d_n}{d_1 - d_n}$$

hence $d_n = 0{\cdot}428d_1$ or $d_1 = 2{\cdot}33d_n$

Equating total tension to total compression gives

$$A_{st} \times 140 = 150d_n \times \frac{7}{2} - \frac{(d_n - 50)^2}{2d_n} \times 7 \times 100$$

which simplifies to

$$A_{st} = 3{\cdot}75d_n - \frac{5(d_n - 50)^2}{2d_n}$$

The moment of resistance of the section is given by

$$M_r = A_{st} \times 140\,(d_1 - d_n) + \frac{150d_n{}^2 \times 7}{3} - \frac{(d_n - 50)^3}{3d_n} \times 100 \times 7$$

This must be equal to the applied bending moment of 7×10^6 N mm. Substituting for A_{st} and d_1 gives—

$$7 \times 10^6 = \left\{3{\cdot}75d_n - \frac{5}{2}\frac{(d_n - 50)^2}{d_n}\right\} \times 140 \times 1{\cdot}33d_n + \frac{150d_n{}^2 \times 7}{3} - \frac{100(d_n - 50)^3 \times 7}{3d_n}$$

leading to the cubic equation

$$3{\cdot}49d_n^3 + 814d_n^2 - 99\,120d_n + 29\,1000 = 0$$

$$d_n^3 + 233d_n^2 - 28\,400d_n + 83\,500 = 0$$

a solution of which is $d_n = 86$, giving an effective depth d of 200 mm.

The steel area is obtained by substitution,

giving $$A_{st} = 3{\cdot}75 \times 86 - \frac{5 \times 36^2}{2 \times 86} = 285 \text{ mm}^2.$$

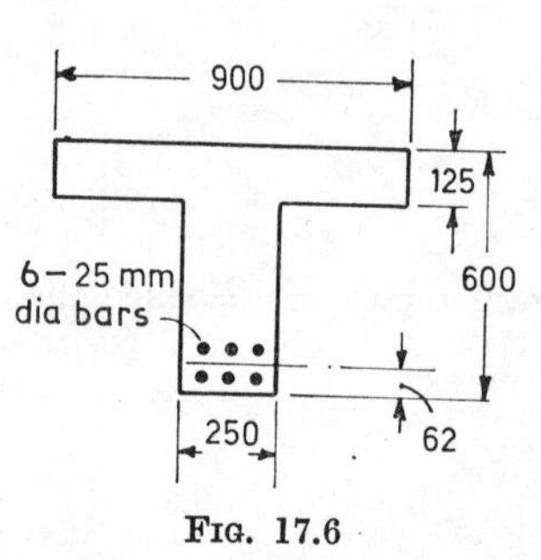

FIG. 17.6

Example 17.5. The section of a reinforced concrete T-beam is shown in Fig. 17.6. Dimensions are given in mm. Taking the modular ratio as 15 find—

(*a*) the position of the neutral axis,

(*b*) the moment of resistance of the beam for working stresses of 7 N/mm² compression in the concrete and 140 N/mm² tension in the steel,

(*c*) the maximum allowable uniformly distributed load, including the weight of the beam for a freely supported effective span of 8 m.

The beam is adequately reinforced against shear.

Solution. Let the depth to the neutral axis be d_n.
The straight-line strain relationship gives

$$15\frac{f_{cb}}{f_{st}} = \frac{d_n}{538 - d_n}$$

The total area of tension reinforcement $= 6 \times 0{\cdot}785 \times 25^2$
$= 2950 \text{ mm}^2.$

Equating total tension to total compression gives

$$2950f_{st} = 900d_n \,.\, \frac{f_{cb}}{2} - 650\,\frac{(d_n - 125)^2}{d_n}\,.\,\frac{f_{cb}}{2}$$

Eliminating $\frac{f_{st}}{f_{cb}}$ from these two equations gives

$$\frac{450d_n}{2950} - \frac{325(d_n - 125)^2}{2950d_n} = \frac{15(538 - d_n)}{d_n}$$

which leads to

$$450d_n^2 - 325(d_n^2 - 250d_n + 15\,600) = 15 \times 2950\ (538 - d_n)$$

$$d_n^2 + 1002d_n - 231\,000 = 0$$

the solution of which is $d_n = 195$ mm

$$\therefore \qquad \frac{f_{st}}{f_{cb}} = \frac{343}{195} \times 15$$

Hence, if $f_{st} = 140$, $f_{cb} = \frac{343}{195} \times \frac{140}{15} = 5{\cdot}3$ N/mm².

The moment of resistance is obtained by taking moments of compressive and tensile forces about the neutral axis—

$$\text{giving} \qquad M_r = 2950 \times 140 \times 343 = 142 \times 10^6$$
$$5{\cdot}3 \times 900 \times 195^2 \times 1/3 = 60{\cdot}3 \times 10^6$$

$$202{\cdot}3 \times 10^6$$

$$\text{less } 5{\cdot}3 \times \frac{70^3}{195} \times 650 \times 1/3 \qquad 2{\cdot}0 \times 10^6$$

$$\text{giving} \qquad M_r = 200 \text{ kN m}$$

If w = intensity of uniformly distributed load, then

$$w \times 8^2 \times \tfrac{1}{8} = 200$$

$$\text{leading to} \qquad w = 25 \text{ kN/m}$$

Example 17.6. A composite beam consists of a rolled steel joist 375 mm deep firmly connected to the underside of a concrete slab 750 mm wide and 150 mm deep. Determine the moment of resistance of this beam for maximum permitted stresses in the steel and concrete of 140 and 7 N/mm² respectively and a modular ratio of 15. The steel joist has a cross-sectional area of $9{\cdot}6 \times 10^3$ mm² and a maximum moment of inertia of $3{\cdot}12 \times 10^7$ mm⁴ units. (*London*)

Solution. Many problems in reinforced concrete can be solved by using what is known as the "transformed section." This consists of substituting for the steel an area of concrete equal to the steel area multiplied by the modular ratio and treating the beam as one composed entirely of concrete. Thus, the reinforced concrete beam shown in Fig. 17.7 (*a*) is transformed into the section shown in Fig. 17·7 (*b*) and the beam designed according to the standard methods on the assumption that it is a monolithic concrete beam capable of taking tension.

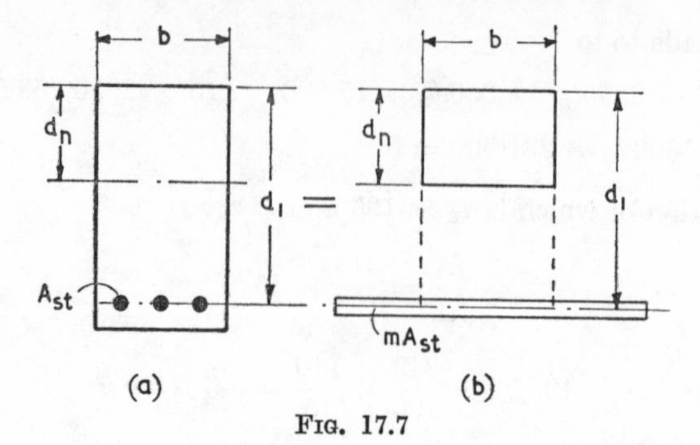

FIG. 17.7

The section used in the question is shown in Fig. 17.8 (*a*), the strain distribution being as shown in Fig. 17.8 (*b*). Dimensions are in mm.

In the previous problems in this chapter the nomenclature recommended by the British Standard Code of Practice has been used. The difficulty with this is the use of suffixes which often

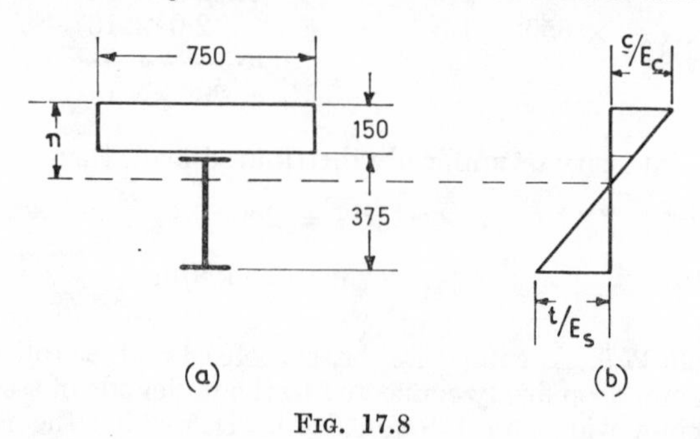

FIG. 17.8

lead to errors in answering questions. In this solution a simpler nomenclature is used with c and t the maximum fibre stresses in the concrete and steel respectively and n the depth to the neutral axis.

From the straight-line strain diagram

$\dfrac{15c}{t} = \dfrac{n}{525 - n}$. The stress at the underside of the concrete slab is $\dfrac{(n - 150)}{n} c$.

The total compression taken by the concrete

$$= 750 \times 150 \times \tfrac{1}{2}c\left(1 + \frac{n - 150}{n}\right) = 1{\cdot}125 \times 10^5 c\left(\frac{n - 75}{n}\right)$$

The total tension carried by the joist is the average tensile stress multiplied by the area

$$\text{Average stress} = \frac{t}{2} - \left(\frac{n - 150}{2n}\right) c$$

$$\text{Total tension} = 9{\cdot}6 \times 10^3 \times \tfrac{1}{2} \left\{ t - \frac{(n - 150)}{n} c \right\}$$

Equating total compression to total tension gives

$$1{\cdot}125 \times 10^5 c \left(\frac{n - 75}{n}\right) = 9{\cdot}6 \times 10^3 \times \tfrac{1}{2} \left\{ t - \left(\frac{n - 150}{n}\right) c \right\}$$

or $$23{\cdot}5 \frac{c}{t} \left(\frac{n - 75}{n}\right) = 1 - \left(\frac{n - 150}{n}\right) \frac{c}{t}$$

Substituting $\frac{c}{t} = \frac{n}{15(525 - n)}$ gives

$$23{\cdot}5\,(n - 75) = 15(525 - n) - (n - 150)$$
$$= 7850 - 15n - n + 150$$
$$39{\cdot}5n = 9770, \text{ or } n = 248 \text{ mm.}$$

The beam will now be transformed into a concrete beam. The second moment of area of the transformed beam about the neutral axis I_{NN} is given by

$$I_{NN} = \frac{750 \times 150^3}{12} + 750 \times 150 \times 173^2 + 15 \times 3{\cdot}12 \times 10^7$$
$$+ 15 \times 9{\cdot}6 \times 10^3 \times 89^2$$
$$= (2{\cdot}1 + 33{\cdot}7 + 46{\cdot}6 + 11{\cdot}4) \times 10^8$$
$$= 93{\cdot}8 \times 10^8 \text{ mm}^4 \text{ units}$$

The stress ratio $\frac{c}{t} = \frac{248}{15 \times 277}$

If the maximum fibre stress in the concrete is 7 N/mm² the corresponding stress in the steel $t = \frac{15 \times 277 \times 7}{248} = 117$ N/mm². This is less than the permissible, consequently the fibre stress in the concrete is the governing factor. The moment of resistance of the section is therefore calculated from the standard expression $M = \frac{fI}{y}$.

Substitution of the appropriate quantities gives

$$M_r = \frac{7 \times 93{\cdot}8 \times 10^8}{248} = 265 \text{ kN m}$$

Examples on Doubly Reinforced Beams by Elastic Theory

It has been pointed out in the examples on singly-reinforced beams that when the reinforcement is greater than that required for balanced design the concrete is fully stressed in compression but the steel in tension is understressed. From the practical viewpoint this is uneconomical, and in order that the steel in tension can be fully stressed the compression portion of the beam is strengthened by adding steel. The general case therefore is as shown in Fig. 17.9, where A_{sc} of steel have been placed above the neutral axis at a distance d_2 from the top fibre.

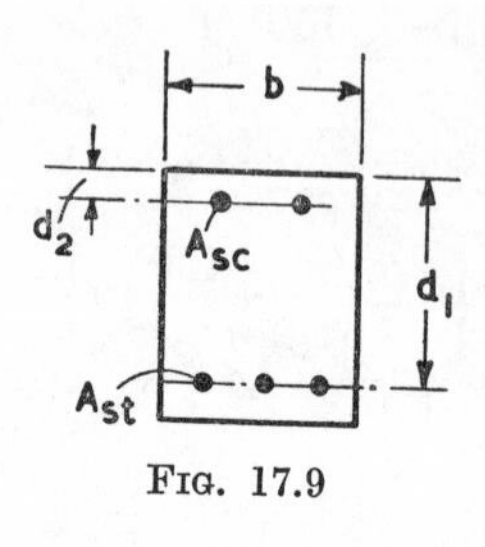

Fig. 17.9

The straight-line strain diagram shown in Fig. 17.1 (*b*) is unaffected by the addition of this steel, consequently the strain in the compression steel $= \dfrac{d_n - d_2}{d_n} \times$ fibre strain in concrete, and the stress is $mf_{cb}\left(\dfrac{d_n - d_2}{d_n}\right)$. The insertion of the steel, however, prevents an equivalent area of concrete from taking compression so that if the stress in the compression steel is taken as $(m - 1)\, f_{cb}\left(\dfrac{d_n - d_2}{d_n}\right)$ its true effect is obtained.

The moment of resistance of the beam is obtained by considering the fact that the steel in tension and concrete in compression are both fully stressed, consequently the neutral axis is in the same position as at balanced design conditions. The beam can, therefore, be imagined as a singly reinforced beam in balanced design with a steel area of $(A_{st} - A_{sc})$ together with a steel beam of top flange A_{sc} stressed to $(m - 1)\, f_{cb}\left(\dfrac{d_n - d_2}{d_n}\right)$, consequently the moment of resistance M_r is given by the expression

$$M_r = (A_{st} - A_{sc})\, f_{st}\left(d_1 - \frac{d_{nb}}{3}\right) + A_{sc}\,(m - 1)\, f_{cb}\left(\frac{d_{nb} - d_2}{d_{nb}}\right)(d_1 - d_2),$$

where d_{nb} = neutral axis depth at balanced design.

Example 17.7. A reinforced concrete beam 400 mm wide has to carry a bending moment of 300 kN m, the permissible stresses being 7 N/mm² in the concrete and 140 N/mm² in the steel. Calculate the relative economy in steel and concrete as between

(*a*) a singly reinforced section to balanced design, and
(*b*) a doubly reinforced section with effective depth restricted to 550 mm, both materials working at their full stress. Use $m = 15$. (*Glasgow*)

Solution. Let d_1 be the effective depth of the singly reinforced section. Let d_{nb} be the depth of the neutral axis at balanced design. Then for the stresses given—

$$\frac{15 \times 7}{140} = \frac{d_{nb}}{d_1 - d_{nb}}$$

leading to $$d_{nb} = 0{\cdot}428 d_1$$

and the lever arm is therefore $0{\cdot}857 d_1$.

The moment of resistance at balanced design M_{rb} is given by

$$M_{rb} = \tfrac{7}{2} \times 400 \times 0{\cdot}857 d_1 \times 0{\cdot}428 d_1{}^2 = 511 d^2 \text{N mm units}$$

and this equals the applied bending moment of 300×10^6 N mm.

Thus, $$d_1 = 10^3 \sqrt{\frac{300}{511}} = 765 \text{ mm}$$

$$A_{st} = \frac{300 \times 10^6}{140 \times 0{\cdot}857 \times 765} = 3280 \text{ mm}^2$$

With the effective depth restricted to 550 mm the depth of the neutral axis at balanced design is $0{\cdot}428 \times 550 = 235$ mm, and the lever arm is $0{\cdot}857 \times 550 = 470$ mm. The moment taken without the compression reinforcement $= \tfrac{7}{2} \times 400 \times 235 \times 470 = 155 \times 10^6$ N mm.

This requires a steel area of $\dfrac{155 \times 10^6}{140 \times 0{\cdot}857 \times 550} = 2360 \text{ mm}^2$

If the compression steel is placed at 60 mm below the top flange then the lever arm for this steel is 490 mm.

The moment to be taken by this steel is $(300 - 155)10^6 = 145 \times 10^6$ N mm. The area of tension steel to develop this moment at a lever arm of 490 mm is given by

$$A_{st_2} = \frac{145 \times 10^6}{140 \times 490} = 2110 \text{ mm}^2$$

Equivalent stress in compression steel $= 14 \times 7 \times \dfrac{175}{235} = 72{\cdot}8$ N/mm².

The area of compression steel $A_{sc} = \dfrac{145 \times 10^6}{72{\cdot}8 \times 490} = 4060 \text{ mm}^2$

Thus, with an effective depth of 550 mm the reinforcement required is—

$$\text{Tensile} = 2360 + 2110 = 4470 \text{ mm}^2$$

$$\text{Compressive} = 4060 \text{ mm}^2$$

giving a total area of 8530 mm², the increase being 160 per cent of the area of 3280 mm² with balanced design.

The saving of concrete $= (765 - 550) \times 400 = 8{\cdot}6 \times 10^4$ mm² representing 28 per cent of the area for a singly reinforced section.

Example 17.8. The sketch (Fig. 17.10) shows the section of a doubly reinforced concrete beam. Under test load on a simply supported span of 10 m strain gauges indicate zero strain at a depth of 360 mm below the top fibre. Taking this as the depth of the neutral axis calculate the maximum central load which could be applied so that the permissible stresses of 6·5 N/mm² in the concrete and 120 N/mm² in the steel are not exceeded. Take $m = 15$ and use the exact lever arm. Obtain the maximum shear stress immediately above the upper layer of tension steel.

(*Glasgow*)

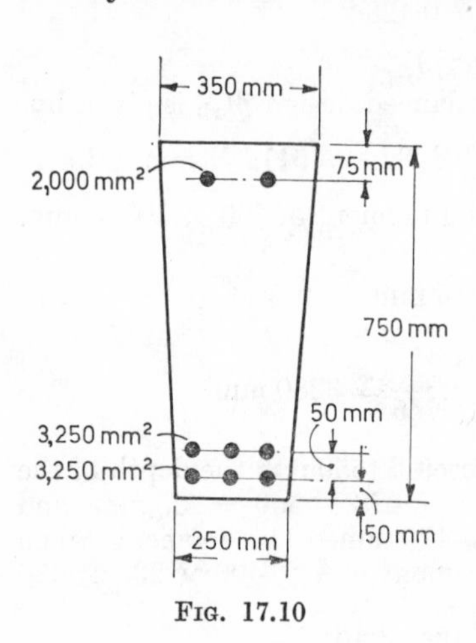

FIG. 17.10

Solution. Breadth of beam at neutral axis $= 350 - \left(100 \times \dfrac{360}{750}\right)$ $= 350 - 48 = 302$ mm. The second moment of area of the section about the neutral axis I_{NN} is given by

$$I_{NN} = \frac{350 \times 360^3}{3} - \frac{48 \times 360^3}{12} + 14 \times 2000 \times 285^2$$

$$+ 15 \times 3250 \times 340^2 + 15 \times 3250 \times 290^2$$

$$= (54{\cdot}5 - 0{\cdot}2 + 22{\cdot}8 + 56{\cdot}0 + 41{\cdot}1) \times 10^8$$

$$= 174{\cdot}2 \times 10^8 \text{ mm}^4 \text{ units.}$$

The stress in the steel, t, when the concrete is fully stressed is given by

$$t = 15 \times 6\cdot5 \times \frac{390}{360} = 107 \text{ N/mm}^2$$

This is less than the permissible stress, hence the concrete stress is critical and the moment of resistance of the beam, M_r, is given by

$$M_r = \frac{6\cdot5 \times 174\cdot2 \times 10^8}{360} = 315 \text{ kN m}$$

The applied bending moment $= \dfrac{W \times 10}{4} = 2\cdot5W$ kN m,

where W = central load. Hence, central load $= \dfrac{315}{2\cdot5} = 127$ kN.

In order to calculate the shear stress the value of the first moment of area G of the section above the neutral axis must be calculated. The value of G is given from the expression

$$G = \frac{350 \times 360^2}{2} + 2000 \times 14 \times 285 - \frac{48 \times 360^2}{6}$$

$$= 28\cdot6 \times 10^6$$

The shear stress at the neutral axis $= \dfrac{VG}{I_{NN}b}$

where V = shear force at section, b = width at section, and I_{NN} and G have the values previously defined.

Since concrete takes no tension this is also the shear just above the upper layer of steel.

The maximum shear $= \dfrac{127}{2} = 63\cdot5$ kN

$$\therefore \text{ maximum shear stress} = \frac{63\cdot5 \times 10^3 \times 28\cdot6 \times 10^6}{174\cdot6 \times 10^8 \times 302}$$

$$= 0\cdot35 \text{ N/mm}^2.$$

Example 17.9. Calculate the moment of resistance of the octagonal reinforced concrete beam section shown in Fig. 17.11 if the stresses in the steel and concrete are limited to 140 and 7 N/mm² respectively and $m = 15$. The bars are all of 650 mm² cross-sectional area. (*St. Andrews*)

Solution. Let n = depth to neutral axis and c and t the concrete and steel stresses respectively.

From the straight line strain relationship $\dfrac{mc}{t} = \dfrac{n}{450 - n}$.

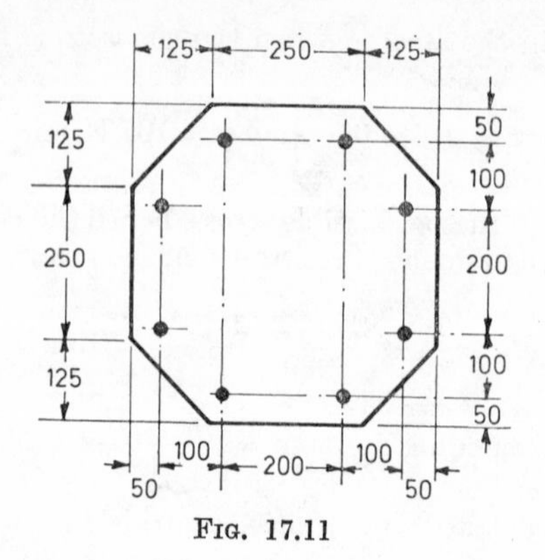

FIG. 17.11

Equating the total tension to total compression, remembering that the strain (hence stress) is proportional to the distance from the neutral axis, gives

$$\frac{500nc}{2} + 14 \times 1300\,\frac{(n-50)}{n}\,.\,c + 14 \times 1300\,\frac{(n-150)}{n}\,.\,c - 125\,\frac{(n-125/3)c}{n}$$

$$= 1300t + 1300t\left(\frac{350-n}{450-n}\right)$$

(The fourth term on the left-hand side represents the effect of the triangular corners and is obtained by integration as explained later.)

Eliminating $\frac{c}{t}$ gives the following quadratic in n

$$250n^2 + 18\,200\,(2n - 200) - 15\,620\left(\frac{n - 125}{3}\right) + 19\,500(2n - 800) = 0$$

which simplifies to $n^2 + 239n - 74\,360 = 0$
the solutions of which are $n = +180$ or -408, the latter being inadmissible.

If $t = 140$ N/mm² then the fibre stress in the concrete, c, is given by

$$c = \frac{140}{15}\cdot\frac{180}{270} = 6{\cdot}23 \text{ N/mm}^2$$

One difficult part in the solution of this question is to determine the effect of the triangular corners on the moment of resistance of the section.

Consider such a triangle as shown in Fig. 17.12 and measure the variable x from the bottom corner of the triangle. Then the stress on any element is $\dfrac{(n - d + x)}{n} . c$.

The width of the element is $\dfrac{x}{d} \times b$ and the moment of resistance of the element $\dfrac{(n - d + x)^2}{n} . \dfrac{x}{d} . b . c . \delta x$.

Thus, the moment of resistance of the triangular portion is given by—

$$\frac{cb}{nd}\int_0^d x\,\{(n - d)^2 + 2x\,(n - d) + x^2\} . dx$$

$$= \frac{cb}{nd}\left[(n - d)^2 . \frac{x^2}{2} + \frac{2x^3}{3}(n - d) + \frac{x^4}{4}\right]_0^d$$

which simplifies to $\dfrac{cbd}{n}\left(\dfrac{n^2}{2} - \dfrac{nd}{3} + \dfrac{d^2}{12}\right)$.

The moment of resistance of the octagonal section can be obtained by taking moments of the various forces about the neutral axis, assuming the concrete section to be rectangular and deducting the effect of the triangular portion.

Due to concrete in compression

$$500 \times 6{\cdot}23 \times \tfrac{1}{2} \times 180^2 \times \tfrac{2}{3} = 33{\cdot}6 \times 10^6$$

Due to steel in compression

$$14 \times 6{\cdot}23 \times 1300 \times \frac{130^2}{180} = 10{\cdot}6 \times 10^6$$

$$14 \times 6{\cdot}23 \times 1300 \times \frac{30^2}{180} = 0{\cdot}6 \times 10^6$$

Due to steel in tension

$$1300 \times 140 \times 270 = 49{\cdot}0 \times 10^6$$

$$1300 \times 140 \times \frac{170^2}{270} = 19{\cdot}4 \times 10^6$$

$$113{\cdot}2 \times 10^6$$

N mm units.

Less triangular portion of concrete

$$\frac{6{\cdot}23 \times 125^2}{180}\left(\frac{180^2}{2} - \frac{125 \times 180}{3} + \frac{125^2}{12}\right) = 6{\cdot}4 \times 10^6$$

Resultant moment of resistance $= 106{\cdot}8 \times 10^6$ N mm units.

$= 106{\cdot}8$ kN m

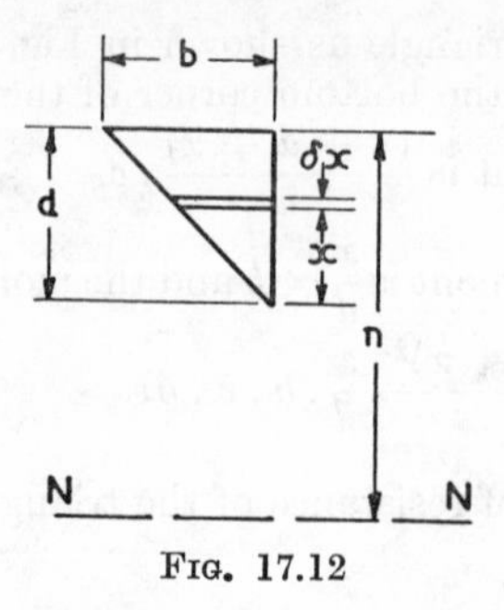

Fig. 17.12

Examples on Shear and Bond

The fundamental assumptions of the straight-line no-tension theory cause two main differences between reinforced concrete beams and standard monolithic beams when the problem of shear is considered. They are—

1. The distribution of the shear stress across the section.
2. The provision of reinforcement to resist the tensile stresses set up as a result of shear.

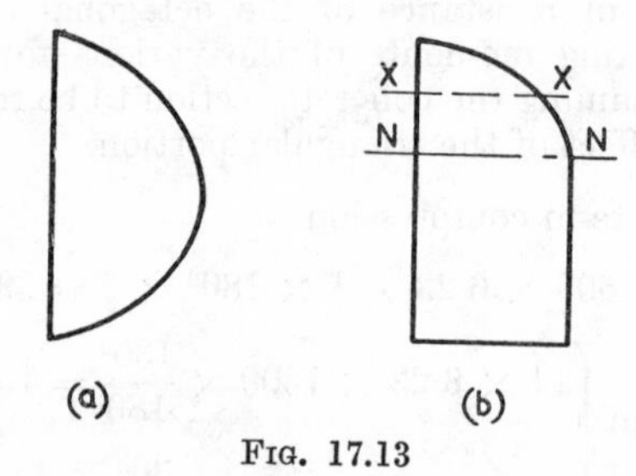

Fig. 17.13

In the case of a rectangular monolithic beam the distribution of shear stress across the section is parabolic, as shown in Fig. 17.13 (*a*). In a singly reinforced concrete beam the concrete below the neutral axis is unable to take bending stresses, consequently the shear stress from the neutral axis down to the reinforcement is uniform, the complete distribution across the section being as shown in Fig. 17.13 (*b*).

The maximum shear stress, v, is given by

$$v = \frac{V}{ab}$$

where V = total shear at section,
a = lever arm,
b = width of beam.

The intensity of shear v_x at any section x–x is given from the expression

$$v_x = \frac{VG_{xx}}{Ib_{xx}}$$

where G_{xx} = 1st moment of area of section above x–x about x–x

I = 2nd moment of area of transformed section about neutral axis

b_{xx} = width of section at x–x.

The second problem in connexion with shear concerns the reinforcement to be provided to resist the tensile stresses from shear.

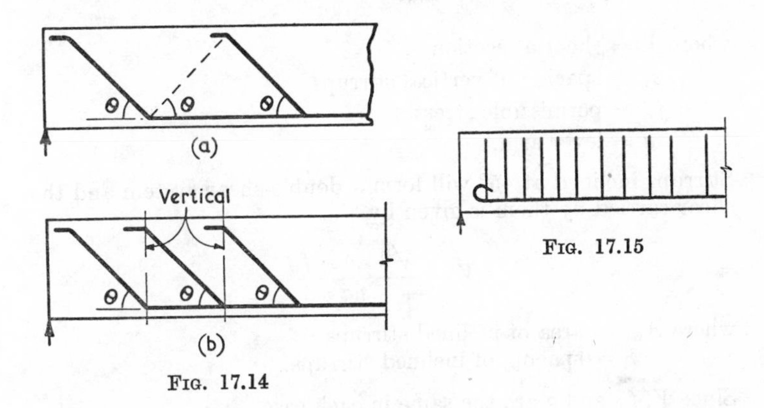

FIG. 17.14

FIG. 17.15

This is done either by (*a*) bent-up bars, (*b*) stirrups, or (*c*) a combination of both. In all cases the beam is assumed to behave as a trussed girder, the diagonal compression being provided by the concrete and the tension by the reinforcement.

Bent-up bars can, in general, be arranged in two ways, to give either a single shear system as shown in Fig. 17.14 (*a*) or a double shear system as in Fig. 17.14 (*b*). The shear carried by the beam being $A_{st} f_{st} \sin \theta$ in the first case, and $2A_{st} f_{st} \sin \theta$ in the second.

An elevation of a beam reinforced with stirrups is given in Fig. 17.15. The shear V_s carried by such a beam is given by

$$V_s = \frac{A_w f_{st} a}{s}$$

where A_w = area of stirrups

f_{st} = permissible tensile stress

a = lever arm

s = spacing of stirrups.

Example 17.10. Show that if anchorages could be neglected, the volume of shear reinforcement required for a length of concrete beam subject to shear is the same for vertical stirrups as for stirrups inclined at 45°.

A beam 200 mm wide and 400 mm deep to the centre of 400 mm² of steel reinforcement is subjected to a shearing force of 100 kN. The concrete has a modular ratio of 15 and the steel may be stressed to 140 kN/mm². Find the volume of shear reinforcement in the form of stirrups required per metre length of beam.

Solution. For vertical stirrups the area A_{wv} is given by

$$A_{wv} = \frac{Vs_v}{f_w a}$$

where V = shear at section
s_v = spacing of vertical stirrups
f_w = permissible stress
a = lever arm.

Stirrups inclined at 45° will form a double-shear system and the shear carried by them is given by—

$$V = \frac{\sqrt{2} \, . \, A_{wi} f_w}{s_i a}$$

where A_{wi} = area of inclined stirrups
s_i = spacing of inclined stirrups.

Since V, f_w and a are the same in each case

$$\frac{A_{wv}}{s_v} = \frac{\sqrt{2} A_{wi}}{s_i}$$

In a length of the beam, l, the number of stirrups is $\frac{l}{s_v}$ and $\frac{l}{s_i}$ for the vertical and inclined cases respectively. The length of an inclined stirrup (excluding anchorages) will, however, be $\sqrt{2} \times$ length of a vertical stirrup.

The volume of a stirrup is equal to its area multiplied by its length, therefore

$$\text{the ratio } \frac{\text{volume of inclined stirrups}}{\text{volume of vertical stirrups}} = \sqrt{2} \frac{A_{wi}}{A_{wv}} \cdot \frac{s_v}{s_i}$$

But from a previous equation $\frac{A_{wi}}{s_i} \cdot \frac{s_v}{A_{wv}} = \frac{1}{\sqrt{2}}$

$$\therefore \quad \frac{\text{volume of inclined stirrups}}{\text{volume of vertical stirrups}} = \sqrt{2} \, . \, \frac{1}{\sqrt{2}} = 1$$

For the second part of the problem the depth of the neutral axis must first be found. Using the transformed section and equating the first moments of area above and below the neutral axis gives (denoting the neutral axis depth by n)

$$200 \times n \times \frac{n}{2} = 15 \times 400\,(400 - n)$$

$$n^2 + 60n - 24\,000 = 0$$

giving
$$n = 127{\cdot}5 \text{ mm}$$

The lever arm $a = d_1 - \dfrac{n}{3} = 400 - \dfrac{127{\cdot}5}{3} = 358{\cdot}5$ mm

The shear stress $v = \dfrac{V}{ab} = \dfrac{100 \times 10^3}{358{\cdot}5 \times 200}$

$$= 1{\cdot}49 \text{ N/mm}^2$$

Take $s = 1000$ mm

Then $A_{\text{W}} = \dfrac{1{\cdot}49 \times 200 \times 1000}{140} = 2125 \text{ mm}^2$

The volume of reinforcement $= \dfrac{2125}{2}$ length of stirrup.

If it is assumed that the main bars are given 25 mm side cover and that the stirrup has a depth equal to the lever arm, then the length of the stirrup

$$= (2 \times 358{\cdot}5) + 150$$
$$= 867 \text{ mm}$$

Volume of stirrups per metre $= 867 \times \dfrac{2215}{2} = 0{\cdot}92 \times 10^6 \text{ mm}^3$ units.

Example 17.11. A reinforced concrete beam 225 mm wide and 300 mm deep to the centre of four 16 mm dia. steel rods has in one locality stirrups at 100 mm pitch. These stirrups are made from 10 mm dia. rod providng two vertical legs as shear reinforcement. Find the maximum allowable shear force which this locality may be subjected to if the allowable steel stress may not exceed 140 N/mm² and the modular ratio is 15. (*London*)

Solution. A sketch of the beam is given in Fig. 17.16. The area of reinforcement

$$= 4 \times \frac{\pi}{4} \times 16^2 = 805 \text{ mm}^2$$

Since the shearing force which can be carried by stirrups depends on the lever arm which, in turn, depends on the neutral axis the first step in the solution is to find the value of the neutral axis depth n.

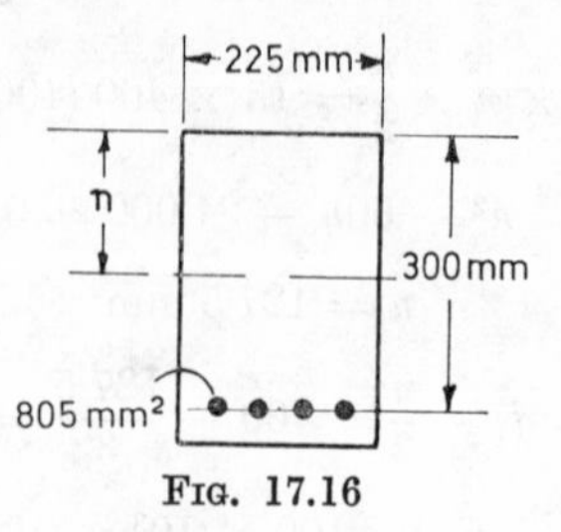

FIG. 17.16

Using the transformed section and equating moments above and below the neutral axis gives

$$225n \times \frac{n}{2} = 15 \times 805 \times (300 - n)$$

$$n^2 + 107n - 32\,100 = 0$$

the solution of which is $n = 133$ mm.

Hence the lever arm a is given by $a = 300 - \dfrac{133}{3} = 256$ mm.

Substituting in the formula for V gives

$$V = \frac{2 \times 78{\cdot}5 \times 140 \times 256}{100}$$

$$= 56\,200 \text{ N}$$

Example 17.12. The elevation of the shear reinforcement for a beam is shown in Fig. 17.17. The dimensions are in mm. If the stresses in the reinforcement are in no case to exceed 140 N/mm² calculate the shear resistance of the section. Assume $n = 0{\cdot}25d_1$.

If, after the beam has been constructed it is desired to increase its load-carrying capacity, would the addition of 75 mm of concrete on top of the flange have any effect on the shearing resistance? Take the permissible shear in plain concrete as 0·7 N/mm².

The bent up bars are 25 mm diameter and the stirrups 10mm diameter in pairs at 150 mm centres. *(Glasgow)*

Solution. The elevation of the beam is shown in Fig. 17.17 and it is clear from this that the bent up bars form a single shear system with tension and compression members inclined at 45° to the horizontal. Thus, if the triangle of forces is drawn for a lower panel point it is as shown in Fig. 17.18, where H, C and T represent the horizontal, compressive and tensile forces respectively, and it is clear that $H = \sqrt{2}T = \sqrt{2}C$.

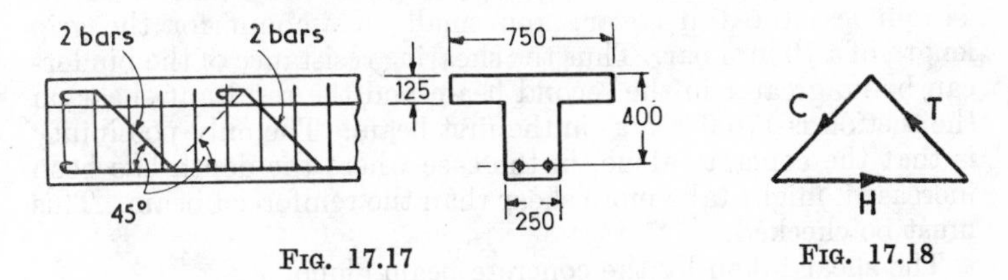

Fig. 17.17

Fig. 17.18

A difficulty with the question as set is in the amount of reinforcement available to resist the force H. If it can only be taken by two 25 mm dia. bars then the stress in these horizontal bars will be $\sqrt{2} \times$ stress in the inclined bars. It will be assumed that this is the case since no information is given; the error will be on the safe side. Consequently the stress in the inclined bars is $\frac{1}{\sqrt{2}} \times$ 140 N/mm². The shearing force resisted by the inclined bars is therefore

$$2 \times 490 \times \frac{1}{\sqrt{2}} \times \frac{1}{\sqrt{2}} \times 140 = 68{\cdot}6 \text{ kN}$$

The depth to the neutral axis $= 0{\cdot}25 \times 400 = 100$ mm. This is greater than the flange depth of the T-section, consequently the beam behaves as a rectangular beam and the lever arm is given by

$$a = 400 - \frac{100}{3} = 367 \text{ mm}$$

Thus the shear V_b taken by the binders is given by

$$V_b = \frac{4 \times 79 \times 140 \times 367}{150} = 108 \text{ kN}$$

The total shearing resistance of the section is therefore 68·6 + 108 = 176 kN.

When the depth of the beam is increased the lattice girder effect of the bent up bars cannot be increased since the position of the top flange of this hypothetical girder depends on the position of the bent up bars and this is unchanged.

The shear which can be carried by the binders depends on the grip length which they have from the point of maximum stress. This point is the neutral axis and in the first beam the binders will have a grip length of 100 mm less the top cover.

If, in the second beam, the shearing resistance assumed taken by the binders is to be increased, then the lever arm must be increased. If $n = 0{\cdot}25d_1$, as before, the position of the neutral axis is $0{\cdot}25 \times 475 = 119$ mm below the top flange. This means that the binders will have a grip length of only $(119 - 75) =$ 44 mm less the top cover: too small an amount for the grip length of a 10 mm bar. Thus the shearing resistance of the binders can be no greater in the second beam and the resultant shear on the section is 176·6 kN, as in the first beam. The only possibility is that the concrete alone, in the case where the depth has been increased, might take more shear than the reinforced beam. This must be checked.

The shear taken by the concrete beam alone

$$= 0{\cdot}7 \times \left(475 - \frac{119}{3}\right) \times 250$$

$$= 76 \text{ kN}$$

This is less than that taken by the original reinforced beam.

Example 17.13. A simply supported reinforced concrete beam of 375 mm effective depth and 500 mm wide is reinforced with four 25 mm dia. bars. It carries a uniformly distributed load, including its own weight of 20 kN/m run on a span of 6 m. The modular ratio is 15. Find from first principles the maximum intensity of bond stress. *(Oxford)*

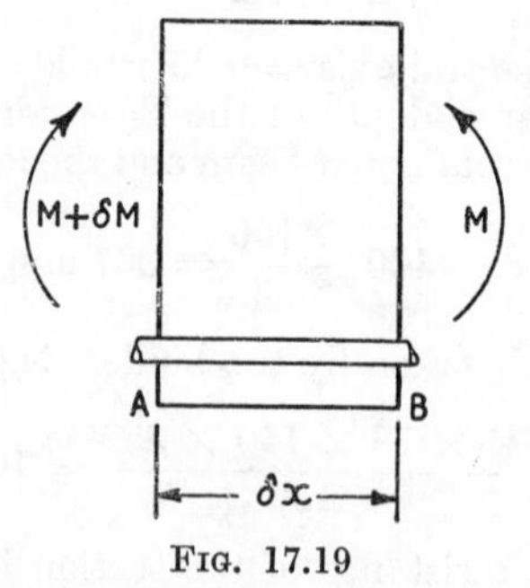

FIG. 17.19

Solution. Consider two sections of the beam distance δx apart and let the moments at the two sections be M and $(M + \delta M)$ respectively, as shown in Fig. 17.19.

The total tension at Section $A = \dfrac{M}{a}$, where a = lever arm.

The total tension at Section B $= \dfrac{M + \delta M}{a}$.

The difference between the two tensions $\frac{\delta M}{a}$ causes a bond stress to be developed between the reinforcement and the surrounding concrete.

If f_b is the intensity of this stress and o is the total perimeter of the bars forming the tensile stress, then $f_b \times o \times \delta x = \frac{\delta M}{a}$.

Hence $$f_b = \frac{\delta M}{\delta x} \cdot \frac{1}{ao}$$

But in the limit $\frac{\delta M}{\delta x} = V$, the shear at the section.

$\therefore$ the intensity of bond stress $f_b = \frac{V}{ao}$

This general expression can be applied to the example in the question when the lever arm a is known. This involves the determination of the neutral axis depth, n.

Using the transformed section and equating moments above and below the neutral axis gives

$$500 \times n \times \frac{n}{2} = 15 \times 4 \times 490\,(375 - n)$$

$n^2 + 118n - 44\,200 = 0$, the solution of which is $n = 159$ mm

The lever arm is therefore $375 - \frac{159}{3} = 322$ mm

The maximum shear force $= 60$ kN

Hence the maximum bond stress $= \frac{60 \times 10^3}{322 \times 4 \times \pi \times 25}$

$= 0{\cdot}59$ N/mm^2

Examples for Practice on Reinforced Concrete Beams by Elastic Theory

1. A rectangular reinforced concrete beam 600 mm deep and 200 mm wide has an effective span of 6 m, the ends being freely supported. The load is 18 kN/m run uniformly distributed, including tht weight of the beam. The reinforcement consists of four 22 mm diameter bars with centres 38 mm above the underside of the beam. Find the maximum compressive stress in the concrete and the maximum tensile stress in the steel, making $m = 15$.

Another beam of the same dimensions is required to carry a load of 25 kN/m run, including its own weight. Find the area of tensile and compressive reinforcement required, taking $c = 7$ N/mm² and $t = 140$ N/mm². The centre of the compression steel is to be at 38 mm from the top edge of the beam.

Answer. $c = 6{\cdot}03$ N/mm²; $t = 116{\cdot}5$ N/mm²; $A_t = 1622$ mm²; $A_c = 695$ mm².

2. Calculate the areas of tension and compression reinforcement required in the beam shown in Fig. 17.20, in order that the stresses in the concrete and reinforcement shall not exceed 7 and

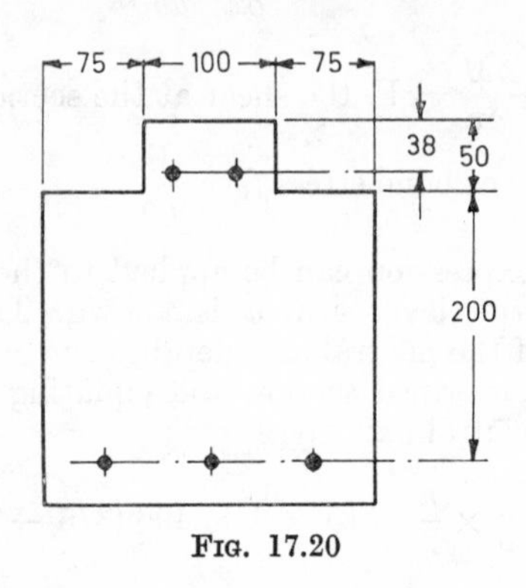

FIG. 17.20

140 N/mm² respectively when a moment of 15 kN m is applied. Take $m = 15$. The bottom portion of the beam is in tension. Dimensions are in mm. (*St. Andrews*)

Answer. $A_t = 509$ mm²; $A_c = 282$ mm².

3. A reinforced concrete beam of 300 mm effective depth and 250 mm width is subjected to a bending moment of 40 kN m. Calculate the total saving of reinforcement by using compression reinforcement centred at 38 mm below the top instead of tension reinforcement only. Take $c = 7$ N/mm², $t = 140$ N/mm² and $m = 15$. (*St. Andrews*)

Answer. Saving = 950 mm².

4. A concrete pile, 300 mm square in cross-section and 12 m long is reinforced with a 19 mm diameter bar in each corner, the bars being centred at 38 mm from each face. The pile is to be supported horizontally at two points so that the stresses are to be a minimum. Determine the position of the points of support and the stresses in the concrete and reinforcement. Take the weight of reinforced concrete as 23 kN/m³ and $m = 15$. (*St. Andrews*)

Answer. Support 2·49 m from ends; $c = 1{\cdot}65$ N/mm²; $t = 64{\cdot}8$ N/mm².

5. Determine the position of the neutral axis of the octagonal reinforced concrete section shown in Fig. 17·21. If the maximum stress in the concrete is 7 N/mm² calculate the stresses in each of the reinforcing bars which are 38 mm diameter. Take $m = 15$. (*Glasgow*)

Answer. $n = 145$ mm; 121, 62·3; 64·2, 9·5 N/mm².

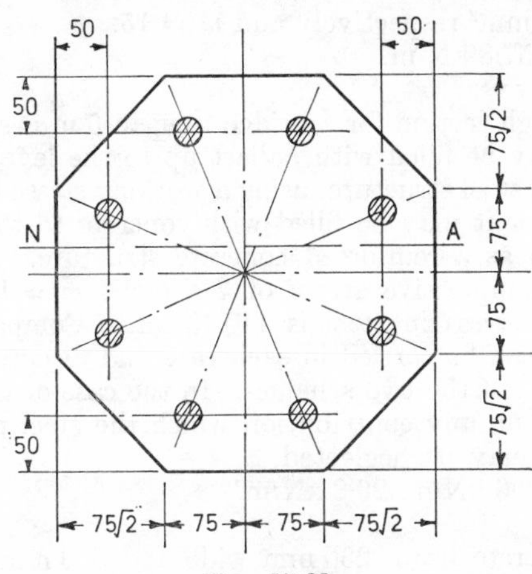

FIG. 17.21

6. The cruciform shown in Fig. 17.22 is reinforced with four similar bars. Calculate the area of these bars in order that when considering bending about a diagonal axis the neutral axis shall lie along the line N-N. What is the moment of resistance of the reinforced section about this axis? Take $m = 15$ and permissible stresses in steel and concrete as 140 and 7 N/mm² respectively. (*Glasgow*)

Answer. $A = 558$ mm²; $M = 41{\cdot}1$ kN m.

7. For the singly reinforced T-beam shown in Fig. 17.23 calculate the shear stress in the rib between the neutral axis and the upper layer of tension steel when the maximum shear on the section is 50 kN. If the permissible stresses in steel and concrete are 140 N/mm² and 7 N/mm² respectively and $m = 15$, obtain the moment of resistance of the section. Use the exact lever arm throughout. (*Glasgow*)

Answer. 0·69 N/mm²; $M = 143$ kN m.

8. The T-section shown in Fig. 17.24 is simply supported on a span of 10 m. If the stresses in the steel and concrete are not to exceed 140 and 7 N/mm² respectively and $m = 15$, calculate the maximum uniformly distributed load per metre run which can be carried by the section. Design any shear reinforcement which may be required. (*Glasgow*)

Answer. 27·7 kN/m; 10 mm dia. in pairs at 175 mm pitch.

9. Calculate the moment of resistance of the section shown in Fig. 17.25 if the stresses in the steel and concrete are not to exceed 140 and 7 N/mm^2 respectively and $m = 15$.

Answer. 557·8 kN m.

10. A trough section for a bridge to span 5 m is shown in Fig. 17.26. It may be filled with ballast up to the level shown, and designed as a steel structure, using a working stress in bending of 140 N/mm^2, or it may be filled with concrete to the same level and designed as a reinforced concrete structure. The concrete has a safe compressive stress of 7 N/mm^2, $m = 15$ and maximum tensile stress in steel is 140 N/mm^2. Compare the total loads which can be carried in each case and discuss very briefly the economics of the two schemes. In the case of the reinforced concrete section any contribution which the steel may make in compression may be neglected. (*Glasgow*)

Answer. 18·6 kN/m, 26·2 kN/m.

11. A concrete beam 250 mm wide and 500 mm deep to the centre of reinforcement is required to have a moment of resistance of at least 100 kN m. Only 25 mm diameter steel rods are available for reinforcement and any compression steel must be centred 50 mm from the top of the beam. The modular ratio is 16 and the allowable stresses for steel and concrete are 140 N/mm^2 and 6·5 N/mm^2 respectively. Find—

(*a*) the number of bars needed,

(*b*) the actual moment of resistance. (*London*)

Answer. 4 in tension, 3 in compression; 106·5 kN m.

12. A reinforced concrete beam of rectangular section is reinforced with steel bars on the tension side only.

(*a*) State the assumptions made in the elastic theory by which the bending stresses in such a beam are calculated.

(*b*) If $M = Qbd^2$ find from first principles the value of Q given the following design data: $t = 140$ N/mm^2, $c = 6·5$ N/mm^2, $m = 15$.

(*c*) Determine the position of the neutral axis for a beam 250 mm wide by 500 mm overall depth reinforced with two steel bars each 25 mm diameter and having 25 mm cover of concrete. Take $m = 15$.

Answer. $Q = 1·16$; 191 mm.

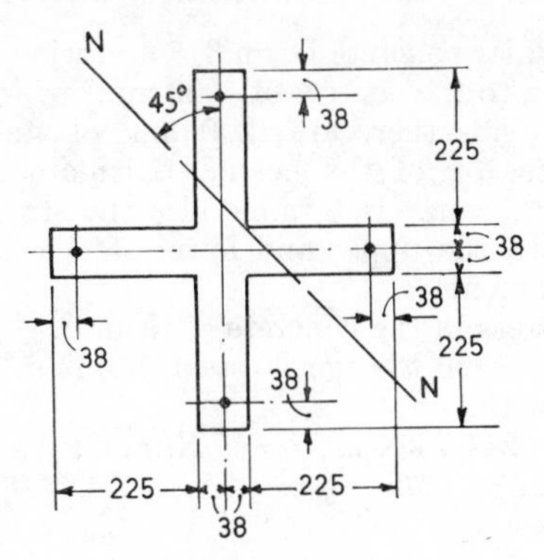

Fig. 17.22

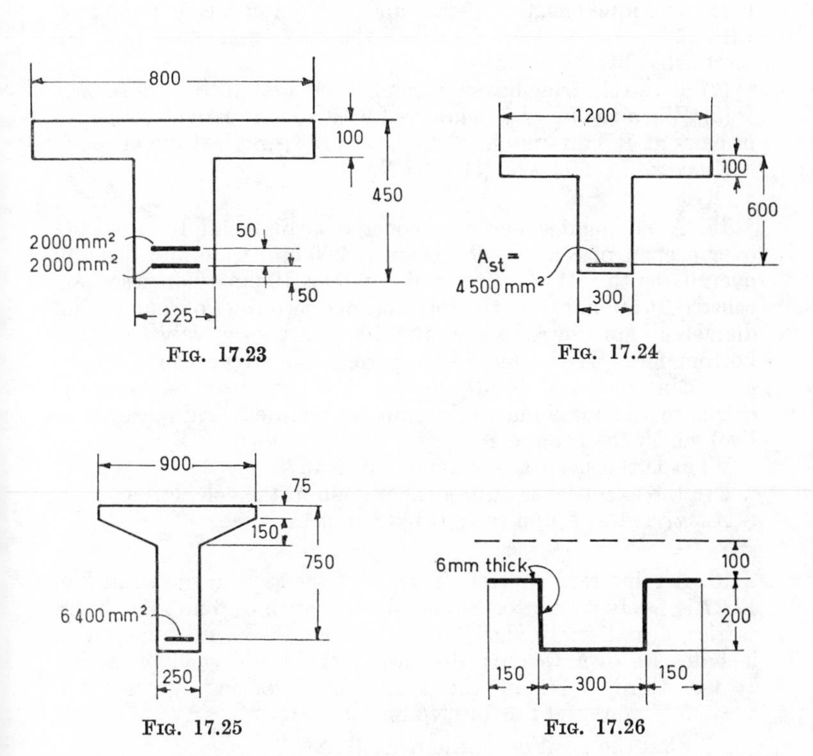

Fig. 17.23

Fig. 17.24

Fig. 17.25

Fig. 17.26

(Dimensions in the above Figures are in mm.)

13. A rectangular concrete beam 375 mm wide has an effective depth of 600 mm to the centres of 4000 mm^2 of tensile steel. On the compression side there are 1200 mm^2 of steel with centres 50 mm from the top of the beam. Calculate the moment of resistance of the beam. Determine also the stresses in the concrete and in both the upper and lower steel when the beam is carrying this moment.

Permissible stress in the concrete, 7 N/mm^2.

Permissible stress in the tensile steel, 140 N/m^2.

Modular ratio, 15.

Answer. $M = 244{\cdot}7$ kN m; $c = 7$ N/mm^2; $f_{sc} = 86{\cdot}5$ N/mm^2; $f_{st} = 118{\cdot}5$ N/mm^2.

14. Deduce the formula for the spacing of vertical stirrups when acting as shear reinforcement in a concrete beam. A rectangular beam 500 mm effective depth and 300 mm wide is reinforced with 1500 mm^2 of steel. Calculate the maximum shear that can be resisted by the beam—

(*a*) If the shearing stress in concrete is limited to 0·7 N/mm^2.

(*b*) When the beam is reinforced with 10 mm diameter stirrups in pairs at 150 mm pitch. Take $t_w = 140$ N/mm^2. Take $m = 15$.

Answer. (*a*) 90·3 kN, (*b*) 126 kN.

15. A rectangular reinforced concrete beam is simply supported over a span of 8 mm. The beam is 300 mm wide and 500 mm overall depth. It is reinforced by four 19 mm diameter bars centred at 50 mm from the top face, and two rows of four 19 mm diameter bars centred at 50 and 100 mm respectively from the bottom face. If the maximum permissible stresses in the steel and concrete are 140 N/mm^2 and 7·0 N/mm^2 respectively, calculate the maximum uniformly distributed load (dead plus live) which the beam can carry.

What is the shear stress under this load?

Use the exact lever arm of the section in the calculations.

Answer. 13·7 kN/m run; 0·468 N/mm^2.

16. A reinforced concrete T-beam of the section shown in Fig. 17.27 is freely supported on an effective span of 6 m. The beam carries a uniformly distributed load of 20 kN/m run (which includes its own weight) also two point loads each W acting at the third point of the span. The working stresses are $c = 7{\cdot}0$ N/mm^2 and $t = 140$ N/mm^2, $m = 15$.

(*a*) Find the position of the neutral axis.

(*b*) Find the maximum value of the point loads W.

(*c*) If at a section 1 m from one support one of the 25 mm diameter bars is bent up at 45° find what additional reinforcement in the form of vertical stirrups is required to resist the shear at this section.

Answer. 175 mm; 84·5 kN; 10 mm dia. single binders, 153 mm pitch.

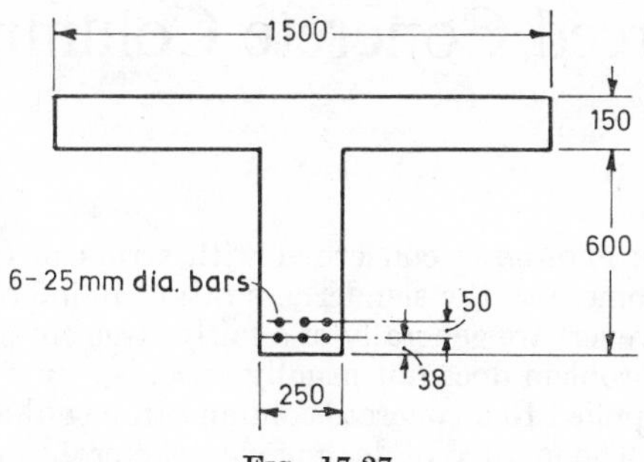

FIG. 17.27

17. A concrete beam 250 mm wide and 500 mm overall depth is to carry a uniformly distributed load of 10 kN/m run on a freely supported span of 8 m. The tensile reinforcement is centred at 50 mm from the lower side and any compression reinforcement at 38 mm from the top fibre.

Calculate the areas of reinforcement if the stresses in the steel and concrete respectively are not to exceed 140 and 7 N/mm^2 and $m = 15$. Calculate also the maximum shearing stress in this beam.

Answer. 1467 mm^2; 257 mm^2; 0·41 N/mm^2.

18. A reinforced concrete beam 225 mm wide and 300 mm deep to the centre of four 16 mm diameter steel rods has in one locality stirrups at 100 mm pitch. These stirrups are made from 10 mm diameter rod providing two vertical legs as shear reinforcement. Find the maximum allowable shear force to which this locality may be subjected if the allowable steel stress may not exceed 140 N/mm^2 and the modular ratio is 15. (*London*)

Answer. 56·5 kN.

19. A reinforced concrete beam 250 mm wide by 500 mm deep overall has 1300 mm^2 of steel centred 50 mm from its top face, and 650 mm^2 of steel centred 50 mm from its bottom face. If the working stresses in steel and concrete may not exceed 140 and 7 N/mm^2 respectively and the modular ratio is 15, determine the maximum hogging and sagging moments permissible. (*London*)

Answer. 60·3 kN m; 70·3 kN m.

Chapter 18

Reinforced Concrete Columns

MOST OF THE PROBLEMS concerned with struts in Chapter XII involved in some way the slenderness ratio. Reinforced-concrete columns, however, are generally of a fairly large cross-section and this type of problem does not usually arise.

The load applied to a concrete column can be either concentric or eccentric. The method of dealing with the problem depends on this. If the load is concentric then the load-carrying capacity of the column is derived from the study of its behaviour at failure. The formula given in the Code of Practice is based on this, and problems concerned with it are generally too simple to be set as examination questions.

If the load is eccentrically applied then the column is designed for direct-plus-bending stresses according to the straight-line no-tension theory outlined in the previous chapter. If the eccentricity is small, however, the combination of direct and bending stresses may not result in tension being developed. In this case the column is treated as monolithic, using the transformed section method.

The general type of problem involves a rectangular section, reinforced on each face with a load acting on one of the axes. Such a column is shown in Fig. 18.1.

If d_n denotes the depth to the neutral axis, then the stress in the concrete adjacent to the reinforcement on the compression face is $\left(\dfrac{d_n - d_2}{d_n}\right) \times c$, where c is the fibre stress in the concrete.

If t denotes the tensile stress in the reinforcement then the fundamental equations are—

$$mc/t = d_n/(d_1 - d_n)$$

$$\tfrac{1}{2}bd_nc + (m - 1)A_{sc}[(d_n - d_2)/d_n]c - A_{st}t = P$$

$$\tfrac{1}{2}bd_nc(d_1 - \tfrac{1}{3}d_n) + (m - 1)A_{sc}[(d_n - d_2)/d_n](d_1 - d_2)c = P(e + \tfrac{1}{2}d - d_2)$$

These three equations can be used to solve for three unknowns only and generally d_n will be one of these. If the details of the cross-section and the magnitude and position of the load are known then the equation for d_n is of the third degree, and in order to avoid the solution of this type of equation Wessmann's trial-and-error process is often used. This method can be employed in cases where the reinforcement is not symmetrical.

If in the above equations c or t are known and e is an unknown, the problem reduces to a quadratic in d_n.

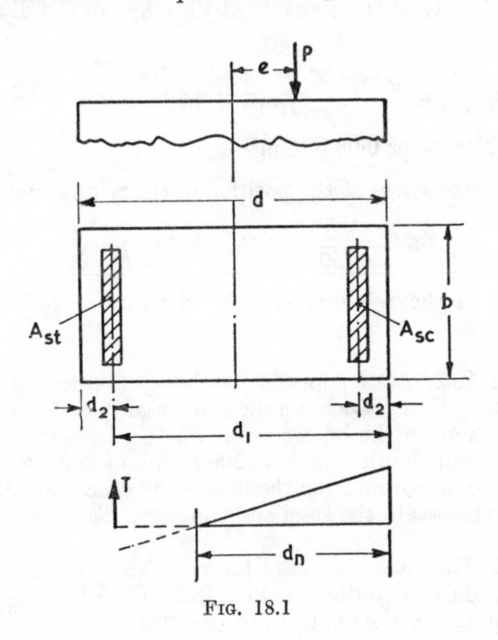

FIG. 18.1

Example 18.1. A concrete column 300 mm square is reinforced with a bar in each corner centred at 38 mm from each face, the bars all being 25 mm diameter. A load of 450 kN is placed at a point on one of the diagonals distant 40 mm from the centre of the column. Determine the maximum stresses in the concrete and reinforcement. Take $m = 15$. (*St. Andrews*)

Solution. The candidate is not told whether tension is developed. The eccentricity of load however is small, the point of application being less than $\frac{d}{6}$ (where d is the overall depth) from the centre, hence the stresses will be wholly compressive and the problem can be solved by use of the transformed section.

The equivalent area $A = 300^2 + 4 \times 14 \times 490 = 11{\cdot}75 \times 10^4$ mm^2.

The second moment of area of the transformed section about a diagonal, I, is given by

$$I = \frac{300 \times 300^3}{12} + 2 \times 14 \times 490 \times 112^2 \times 2$$

$$= (6{\cdot}75 + 3{\cdot}44)10^8 = 10{\cdot}19 \times 10^8 \text{ mm}^4 \text{ units.}$$

Thus the stresses in the concrete at the edges of the diagonal are given by

$$= \frac{450 \times 10^3}{11{\cdot}75 \times 10^4} \pm \frac{450 \times 10^3 \times 40 \times 150\sqrt{2}}{10{\cdot}19 \times 10^8} = 3{\cdot}83 \pm 3{\cdot}75$$

$$= 7{\cdot}58 \text{ N/mm}^2 \text{ or } 0{\cdot}08 \text{ N/mm}^2.$$

The concrete stress at the position of the reinforcement is

$$7{\cdot}58 - \frac{38}{150} \times 3{\cdot}75 = 6{\cdot}63 \text{ N/mm}^2$$

The stress in the reinforcement is therefore $15 \times 6{\cdot}63 = 99{\cdot}4$ N/mm^2.

Example 18.2. At a point in a reinforced concrete arch the thrust on the rib is 400 kN acting at a distance of 100 mm inside the section and on the longer axis. The rib is 300 mm wide by 550 mm overall depth and has 2000 mm^2 of tensile reinforcement centred at 50 mm from the face of the concrete. Obtain the maximum stresses in the steel and concrete. Take $m = 15$.

(Glasgow)

Solution. The cross-section of the rib is shown in Fig. 18.2 (*a*) and the strain distribution in Fig. 18.2 (*b*). Dimensions are in mm. Let n denote the depth to the neutral axis, c the maximum

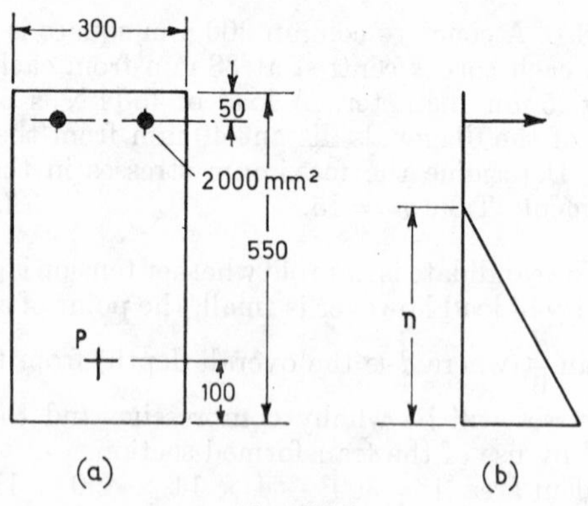

FIG. 18.2

compressive stress in the concrete and t the tensile stress in the steel. Then the straight line strain diagram gives

$$\frac{15c}{t} = \frac{n}{500 - n} \text{ leading to } \frac{t}{c} = \frac{15(500 - n)}{n}$$

Taking moments about the line of action of the load gives

$$\frac{300nc}{2}\left(\frac{n}{3} - 100\right) = 2000t(500 - 100) \text{ leading to } \frac{t}{c}$$

$$= \frac{3n(n/3 - 100)}{16\,000}$$

Equating gives $\dfrac{15(500 - n)}{n} = \dfrac{3n(n/3 - 100)}{16\,000}$

which leads to $n^3 - 300n^2 + 240 \times 10^3 n - 120 \times 10^6 = 0$ the solution of which (obtained by trial-and-error methods) is $n = 420$ mm.

The total compression P carried by the section is given by

$$P = \frac{300nc}{2} - 2000t$$

Substituting for t in terms of c gives

$$P = 150nc - 30\,000\,\frac{(500 - n)}{n}\,.\,c$$

But $P = 400 \times 10^3$ and $n = 420$, hence this equation becomes

$$400 \times 10^3 = c(63\,000 - 5700), \text{ giving } c = 7{\cdot}00 \text{ N/mm}^2,$$

and the tensile stress $t = 15 \times 7{\cdot}0 \times \dfrac{80}{420} = 20$ N/mm^2.

Example 18.3. Fig. 18.3 shows a cross-section of a reinforced concrete arch bridge and the detail of an element between the centre lines of two adjacent ribs. Dimensions are in mm. Under a particular loading condition the line of thrust normal to the section passes at a distance of 200 mm below the underside of the slab. Show that $n = 160$ mm. If the maximum compressive stress in the concrete under this loading is 6 N/mm^2 calculate the thrust per metre width of bridge. $m = 15$. (*Glasgow*)

Solution. Instead of building up the cubic equation as in Example 18.2, this example will be solved by the use of Wessmann's method, the basis of which is that the moments of the compressive and tensile forces about the load axis are equal.

Since the position of the load axis is known, while that of the neutral axis is not, it is more convenient to take moments about the load axis.

A position is assumed for the neutral axis, and using the transformed section the first and second moments F_p and I_p about the

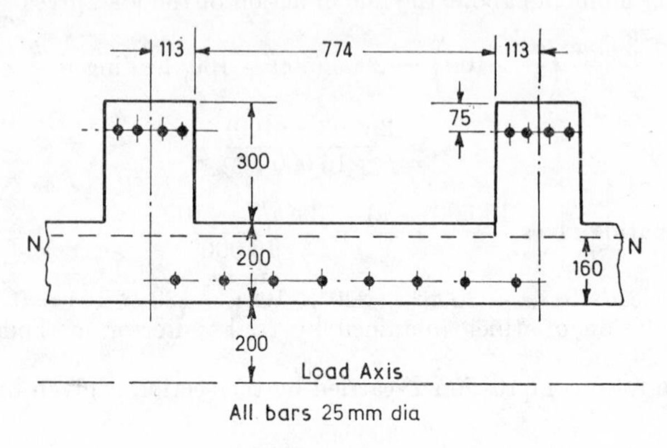

FIG. 18.3

load axis are obtained with this assumed position. Then the eccentricity of the load axis from the neutral axis, e, is given by $e = \frac{I_p}{F_p}$. If this agrees with the assumed position then this is correct. If it does not, then a fresh position is assumed for n and the process repeated until the assumed e equals the calculated $\frac{I_p}{F_p}$. Having obtained the position of the neutral axis the first moment of area of the transformed section about this axis, F_n, is obtained from the expression

$$F_n = F_p - Ae$$

where A = area of transformed section. The stresses are then obtained in the usual way.

Since in this example the question gives a clue to the position of the neutral axis the simplest assumption is to take it at 160 mm from the underside of the slab. The calculations are tabulated as in Table 18.1. The check is as near as one can expect with large quantities and slide-rule working. The first moment of area about the neutral axis, F_n is given by

$$F_n = 76{\cdot}9 \times 10^6 - 244{\cdot}2 \times 361 \times 10^3 = -10{\cdot}6 \times 10^6.$$

the negative sign arising because the load axis is in a negative direction from the neutral axis. Its significance can be neglected in the arithmetic.

The maximum stress $c = \frac{nP}{F_n}$ or $P = F_n \frac{c}{n}$

where P = total thrust on section.

Hence in the example, the thrust per metre width when the maximum compressive stress is 6·0 N/mm² is given by

$$\frac{6 \times 10{\cdot}6 \times 10^6}{160}\ \mathrm{N} = 397\ \mathrm{kN}.$$

TABLE 18.1

Section	Area	$\bar{x}$	F_p	I_p
Concrete	$1000 \times 160 = 160 \times 10^3$	280	$44{\cdot}8 \times 10^6$	125×10^8 $3{\cdot}4 \times 10^8 = \frac{1000 \times 160^3}{12}$
Steel (comp)	$14 \times 490 \times 8 = 54{\cdot}8 \times 10^3$	250	$13{\cdot}7 \times 10^6$	$34{\cdot}2 \times 10^8$
(tens)	$15 \times 490 \times 4 = 29{\cdot}4 \times 10^3$	625	$18{\cdot}4 \times 10^6$	$115{\cdot}0 \times 10^8$
	244×10^3		$76{\cdot}9 \times 10^6$	$277{\cdot}6 \times 10^8$

Calculated $e = \frac{27\,760}{76{\cdot}9} = 361$ mm

Assumed $e = 200 + 160 = 360$ mm

It is interesting to check this value of the thrust adopting the second of the fundamental equations given on page 372 taking $d_n = 160$ mm. With this value of d_n and $c = 6{\cdot}0$ N/mm² the tensile stress $t = 15 \times 6{\cdot}0 \times \frac{265}{160} = 14{\cdot}9$ N/mm². Thus the load carried is given by

$$6{\cdot}0 \times 10^3 \times 160 \times \tfrac{1}{2} = 480 \times 10^3$$

$$14 \times 6{\cdot}0 \frac{110}{160} \times 8 \times 490 = 226 \times 10^3$$

$$706 \times 10^3\ \mathrm{N}$$

$$\text{less } 4 \times 490 \times 14{\cdot}9 \quad 292 \times 10^3$$

$$414 \times 10^3\ \mathrm{N}$$

The load per metre width of section = 414 kN.

This, although not a perfect check, is reasonable.

13

Example 18.4. A rectangular reinforced concrete section of breadth 225 mm and overall depth 350 mm is to be subjected to an eccentric thrust of 70 kN. The ratio of the tension steel and compression steel areas is to be 2:1, the steel being centred at 50 mm inside the shorter face of the section. Calculate the steel areas which will ensure that under the given thrust the permissible stresses of 7 N/mm² in the concrete and 140 N/mm² in the tension steel will both be reached. Find also the eccentricity of the thrust. Use $m = 15$. *(Glasgow)*

Solution. The cross-section of the column is shown in Fig. 18.4 where n = depth to the neutral axis, A_t = area of steel on the

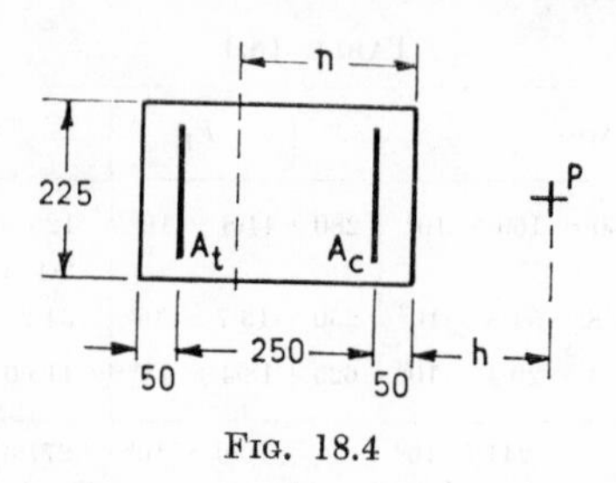

FIG. 18.4

tension face and A_c = area on the compression face. The load is at a distance h from the face. Dimensions are in mm.

From the straight-line strain relationship

$$\frac{15 \times 7}{140} = \frac{n}{300 - n}$$

leading to $$n = 128{\cdot}5 \text{ mm}$$

The total compression on the section P is given by

$$P = \frac{7}{2} \times 128{\cdot}5 \times 225 + 14 \times 7{\cdot}0 \times \frac{78{\cdot}5}{128{\cdot}5} \times A_c - 140A_t$$

Putting $P = 70 \times 10^3$ N and $A_t = 2A_c$ gives the following equation for A_c

$$70 \times 10^3 = 101 \times 10^3 + 60A_c - 280A_c$$

leading to $$A_c = \frac{31 \times 10^3}{220} = 141 \text{ mm}^2 \text{ and}$$

$$A_t = 282 \text{ mm}^2$$

Equating moments about the line of action of the load of the tensile and compressive forces gives

$$140 \times 282 \times (300 + h) = \frac{7{\cdot}0}{2} \times 225 \times 128{\cdot}5 \times \left(\frac{128{\cdot}5}{3} + h\right)$$
$$+ 14 \times \frac{78{\cdot}5}{128{\cdot}5} \times 7{\cdot}0 \times 141 \times (50 + h)$$

$$300 + h = 110 + 2{\cdot}56h + 10{\cdot}7 + 0{\cdot}21h$$

$$1{\cdot}77h = 179{\cdot}3$$

leading to $h = 101{\cdot}5$ mm

The eccentricity of the load is therefore

$101{\cdot}5 + 175 = 276{\cdot}6$ mm from the centre-line of the column.

Examples for Practice on Reinforced Concrete Columns

1. A reinforced concrete pipe of 400 mm internal diameter and 50 mm thick has longitudinal reinforcement equally spaced in the centre of its thickness consisting of 8 bars of 13 mm diameter. It is used as a column so that it is eccentrically loaded about a diametral axis containing two bars, and a concrete stress is developed which varies from 7 N/mm² compression at the farthest point of the face to zero at the centre. If $m = 15$ and no tension is taken by the concrete find the values of load and eccentricity. (The centroid of a semicircular plate is at $\frac{4r}{3\pi}$ from the diameter.) *(London)*

Answer. 122 kN; 274 mm.

2. The concrete column shown in Fig. 18·5 carries a load of 250 kN acting on the X–X axis. Dimensions are in mm. Determine the limits between which the load must act if the stresses

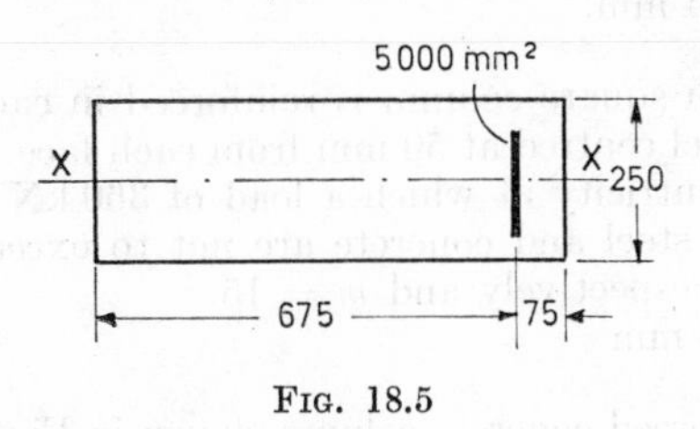

Fig. 18.5

in the concrete are not to exceed 7 N/mm² in compression and zero in tension. Take $m = 15$. *(St. Andrews)*

Answer. 212 and 569 mm from unreinforced edge.

3. Why is reinforcement placed in the corners of a concrete column? An axially loaded column 250 mm square has to carry a load of 400 kN. If the concrete stress is not to exceed 5·3 N/mm² calculate the area of reinforcement required, the steel stress being limited to 125 N/mm². If the reinforcement is centred at 38 mm from each face and a concrete stress of 7·0 N/mm² is allowed, determine the maximum eccentricity at which the load can act. Take $m = 15$. *(Glasgow)*

Answer. 276 mm² on each face; 9·6 mm.

4. A concrete column 300 mm square is reinforced with a bar in each corner centred at 38 mm from each face, the bars all being of the same size. A load is placed at a point on one of the diagonals, distant 40 mm from the centre of the column. Calculate the area of the bar necessary to prevent tension being developed at any point on the column. Take $m = 15$. *(St. Andrews)*

Answer. 400 mm².

5. The concrete column shown in Fig. 18.6 is reinforced with 2000 mm² of steel on each face. Calculate the maximum eccentricity about the X–X axis at which a load of 800 kN can act if

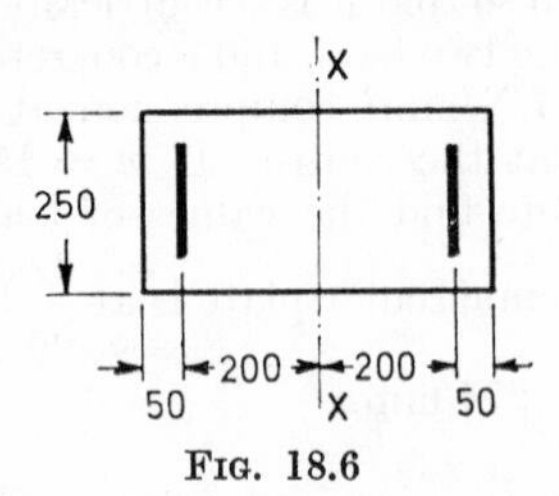

FIG. 18.6

the stresses in the steel and concrete are not to exceed 140 N/mm² and 7 N/mm² respectively and $m = 15$. *(St. Andrews)*

Answer. 62·4 mm.

6. A 300 mm square column is reinforced in each corner with 650 mm² of steel centred at 50 mm from each face. Calculate the maximum eccentricity at which a load of 360 kN can act if the stresses in the steel and concrete are not to exceed 140 N/mm² and 7 N/mm² respectively and $m = 15$. *(St. Andrews)*

Answer. 124 mm.

7. The reinforced concrete column shown in Fig. 18·7 carries a load of 450 kN at O. Dimensions are in mm. Determine the stresses in concrete and reinforcement. Take $m = 15$. *(St. Andrews)*

Answer. $c = 8{\cdot}06$ N/mm²; $t = 8{\cdot}06$ N/mm².

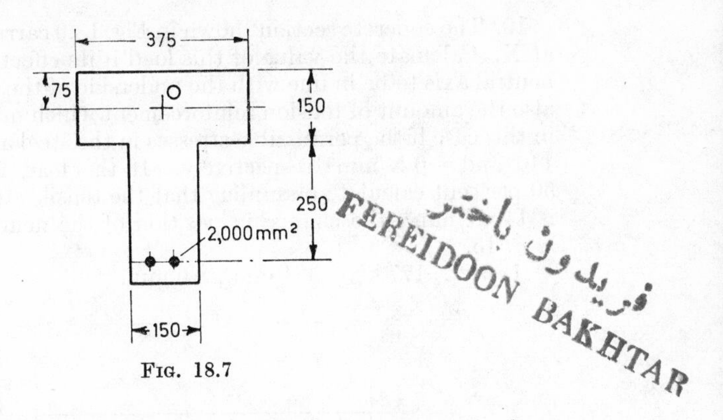

FIG. 18.7

8. A concrete column 300 mm sq. is reinforced with four 16 mm diameter bars centred at 38 mm from each face. Calculate the maximum stresses in the concrete and reinforcement when this column carries a load of 220 kN at a point on one of the diagonals distance 40 mm from the centre. The modular ratio is 15. *(Oxford)*

Answer. 4·40 N/mm²; 48·3 N/mm².

9. The reinforced concrete section shown in Fig. 18.8 carries a load of 1350 kN and a bending moment about X–X of 100 kN m.

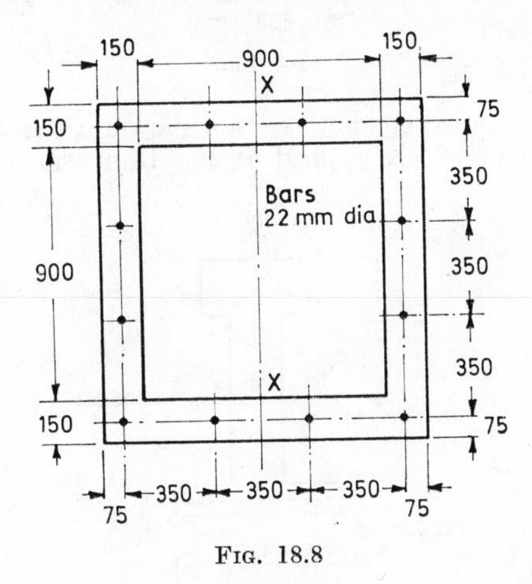

FIG. 18.8

Dimensions are in mm. Determine the maximum stresses in the steel and concrete. Take $m = 15$. *(Glasgow)*

Answer. 2·41 N/mm²; 35·3 N/mm².

10. The concrete section shown in Fig. 18.9 carries a load acting at X. Calculate the value of this load if its effect is to cause the neutral axis to be in line with the underside of the slab. Calculate also the amount of tension reinforcement which must be provided in this case if the permissible stresses in the steel and concrete are 140 and 7·0 N/mm² respectively. If the load is increased by 50 per cent calculate (assuming that the tensile stress in the steel is 140 N/mm²) the change in position of the neutral axis. Take $m = 15$. *(Glasgow)*

Answer. 177 kN; 564 mm², 8·0 mm.

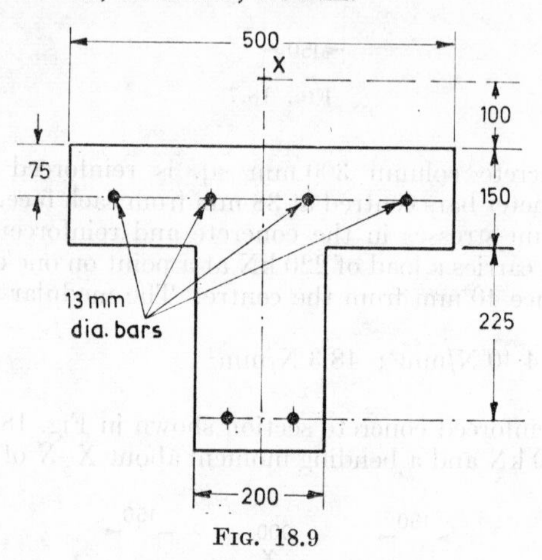

FIG. 18.9

11. The reinforced concrete section shown in Fig. 18.10 carries an axial load of 90 kN applied at X. Dimensions are in mm.

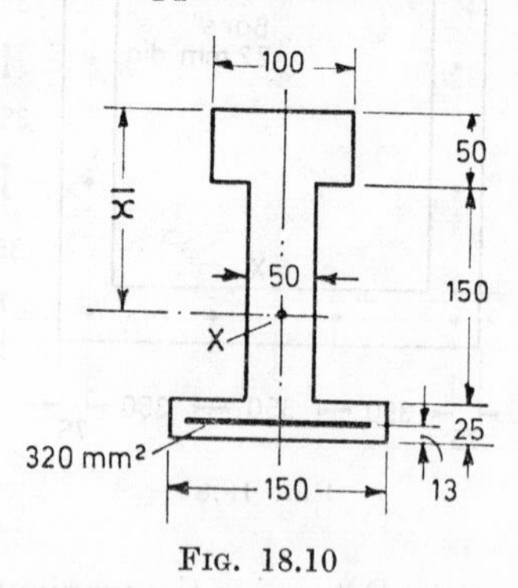

FIG. 18.10

What are the limiting values for $\bar{x}$ in order that tension shall not be developed in the column?

What is the maximum compressive stress under these conditions? (*Glasgow*)

Answer. 175 and 75 mm; 8·7 N/mm².

12. A reinforced concrete chimney has an internal diameter of 3 m and a constant wall thickness of 200 mm for its full height of 20 m. The reinforcement consists of an inner and outer concentric ring of steel mesh, each having a cross-sectional area of 1600 mm²/linear m. The rings are centred at 50 mm from the inner and outer surfaces respectively. Determine the minimum horizontal wind pressure which will just cause tension on a section at the base of the chimney, and find the compressive stress in the concrete for this condition.

Take density of concrete = 23 kN/m³ and $m = 15$. (*Glasgow*)

Answer. 1·25 kN/m; 0·920 N/mm².

13. A precast concrete beam 200 mm wide by 450 mm overall depth has two 13 mm dia. mild steel bars centred at 50 mm from the top face, and has a circular duct formed throughout its length, centred at 350 mm from the top. It is proposed to induce stresses in the beam by passing through the duct a 25 mm dia. high-tensile steel bar, threaded at its ends, and screwing up against bearing plates on the ends of the beam until the maximum compressive stress in the concrete is 12·5 N/mm². If E for mild steel is 200 kN/mm², for high tensile steel 185 kN/mm² and for concrete 13·3 kN/mm², calculate the extension which must be given to the threaded bar for a 6 m length of beam. Neglect the effect of the duct on the section modulus and the effect of dead load on the induced stresses. (*Glasgow*)

Answer. 26·1 mm.

14. A concrete column 450 mm × 225 mm is reinforced with 850 mm² of steel centred at 50 mm from each of the shorter faces.

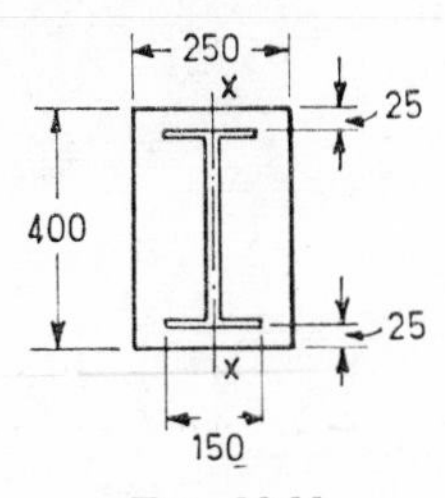

Fig. 18.11

It carries a load of 250 kN acting at an eccentricity of 125 mm from the shorter centre line.

Calculate the stresses in the concrete and reinforcement. Take $m = 15$.

Answer. $c = 4{\cdot}97$ N/mm², $t = 6{\cdot}30$ N/mm².

15. The carrying capacity of a steel column, of I section, made from 6 mm welded steel plates is increased by encasing it in concrete so that the combination can be considered to act as a reinforced concrete section. If the combined section which is illustrated in Fig. 18.11 carries a thrust acting on the X–X axis determine the magnitude and eccentricity of the thrust when the stresses in the concrete and steel are $7{\cdot}0$ N/mm² compression and $70{\cdot}0$ N/mm² tension respectively. The dimensions are in mm. Take $m = 15$. *(Glasgow)*

Answer. $P = 231$ kN; e about centre $= 250$ mm.

Chapter 19

Prestressed Concrete

ONE DIFFICULTY in using concrete as a structural material is that it has little resistance to tensile stresses. This can be overcome by inserting reinforcement where tensile stresses occur, and the examples in the two previous chapters have been worked out using such a method for dealing with the tensile forces.

An alternative method, however, is to prestress the concrete section so that the tensile stresses which are set up by the bending due to the applied load are neutralized by an applied compressive prestress. Thus, if Fig. 19.1 (*a*) represents the stress distribution

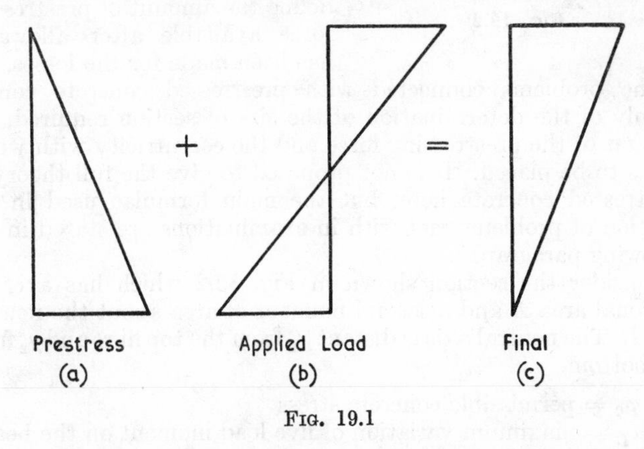

FIG. 19.1

due to prestress and Fig. 19.1 (*b*) that due to the applied bending moment, the resultant stress distribution is shown in Fig. 19.1 (*c*) and it is clear from this that tension has been eliminated.

Since little or no tension occurs, cracking is eliminated and the beam behaves as a concrete beam to which the ordinary theory of bending can be applied.

The prestress is applied to the concrete beam by tensioning steel cables. There are two distinct methods by which this is done. One is the "pretensioning" method and the other is "post-tensioning."

In pretensioning, the cable is first tensioned between end-posts and then the concrete cast around it. When the concrete is sufficiently hardened the cable is cut away from the end-posts thus transferring the load to the concrete and causing a prestress. When the load comes on the concrete, strain takes place due to the compression, and this is accompanied by an equal strain in the steel, causing some loss of prestress. Example 19.2 shows how this can be calculated.

In post-tensioning the concrete beam is cast with ducts running through it, into which cables can be inserted and stressed when the concrete has hardened. No loss due to transfer takes place with post-tensioned beams.

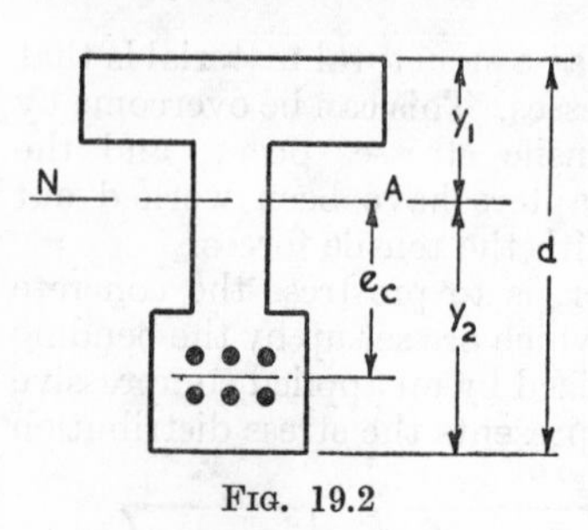

FIG. 19.2

The creep of concrete under load and the shrinkage on hardening cause strain in the steel, which reduces the stress in it, thus, after lapse of time there is some loss of prestress. The term "effective prestress" is used to define the amount of prestressing force available after allowance has been made for the losses.

The problems connected with prestressed concrete consist mainly of the determination of the size of section required, the amount of the prestressing force and the eccentricity with which it has to be placed. It is not proposed to give the full theory of prestressed concrete here, but the main formulae used in the solution of problems met with in examinations are stated in the following paragraphs.

Consider the section shown in Fig. 19.2 which has a cross-sectional area A and a second moment of area about the neutral axis I. The neutral axis is distant y_1 from the top fibre and y_2 from the bottom.

If p_c = permissible concrete stress
M_L = maximum variation of live load moment on the beam, and
Z = section modulus, then

$$M_L = Zp_c.$$

$Z = \frac{I}{y}$

This equation determines the section of beam required, and it is noted that it is dependent only on the live-load moment. It might appear from this that the dead load has no effect on the section required; this is not strictly true, since it does affect the cable eccentricity and this, in turn, may govern the size of the beam, as obviously, the cable eccentricity must never exceed the dimension y_2, less a reasonable allowance for cover.

The horizontal component, H, of the prestressing force is given from the equation

$$H = \frac{y_1}{d} A p_c$$

The cable eccentricity, e_c, at any section is given by

$$e_c = \frac{M_D}{H} - \frac{y_1}{d}\frac{M_1}{H} + \frac{y_2}{d}\frac{M_2}{H}$$

where M_D = dead load moment at the section

M_1 = maximum positive live load moment at the section

M_2 = " negative " " " " "

the signs for positive and negative live load moments being as used in Chapter 6.

The student should not, however, slavishly memorize these formulae for solving problems. In the majority of cases the solution can be obtained by the simple application of the fundamental equations for an eccentrically loaded column (which is how the member behaves during the prestressing process) and a beam subject to pure bending.

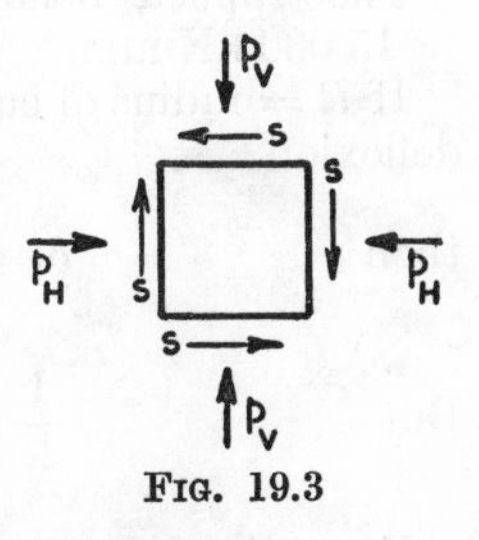

FIG. 19.3

Prestressing also has an effect on the tensile stresses set up as a result of shear. If there is an element of beam, as shown in Fig. 19.3, which in addition to being subjected to a shear stress of s, also has horizontal and vertical prestresses of intensity p_H and p_V respectively then the maximum tensile stress t_s on the element is given by

$$t_s = \tfrac{1}{2}\sqrt{(p_H - p_V)^2 + 4s^2} - \tfrac{1}{2}(p_H + p_V)$$

This will be less than the maximum tensile stress which would occur if there were no prestress, and if the intensity of prestress is sufficiently high, tension can be eliminated entirely.

Example 19.1. Compare briefly the two methods (sometimes called pre-tensioning and post-tensioning) which are used in making pre-stressed reinforced concrete beams. A concrete beam 6 m long, 200 mm wide and 300 mm deep overall is reinforced with a straight post-tensioned cable placed centrally 75 mm above the bottom of the beam. Estimate the amount by which the mid-point of the beam lifts off the form when the tension of 200 kN is applied to the cable. Take the weight of concrete as 23 kN/m³ and Young's Modulus as 13 kN/mm². Assume that the concrete does not crack in tension. *(Cambridge)*

Solution. During the application of the prestressing force the beam is subjected to a constant bending moment throughout its length (this is because the eccentricity of the cable remains constant) consequently it bends in a circular arc. The beam is sketched out in Fig. 19.4. The dimensions are in mm.

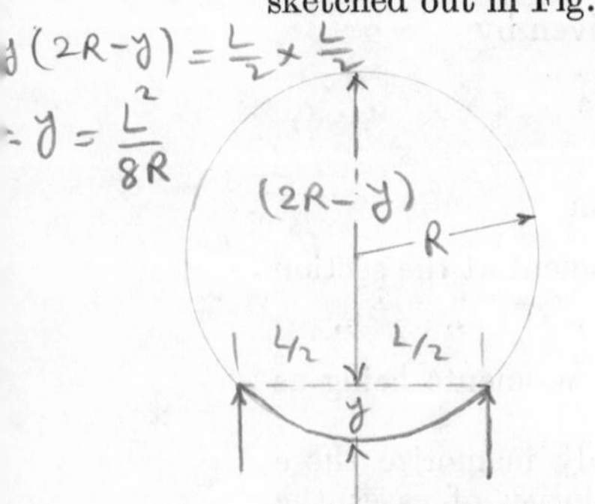

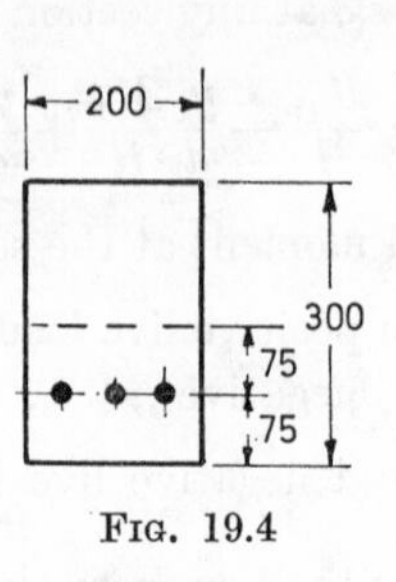

FIG. 19.4

The applied bending moment due to prestress = 200 × 75 = 15 000 kN mm.

If R = radius of curvature of the bent beam and δ_p the central deflexion

then $$\delta_p = \frac{L^2}{8R}, \quad \text{where } L = \text{span.}$$

But $$\frac{1}{R} = \frac{M}{EI}, \quad \text{hence } \delta_p = \frac{M}{EI} \cdot \frac{L^2}{8}$$

From the data supplied

$$L = 6 \text{ m}$$

$$E = 13 \text{ kN/mm}^2$$

Hence $$\delta_p = \frac{15\,000 \times 6^2 \times 10^6 \times 12}{13 \times 200 \times 300^3 \times 8} = 13{\cdot}5 \text{ mm}$$

When the beam lifts up off the form, however, its weight tends to pull it down again, thus the beam behaves as a beam carrying its own dead weight, the central deflexion under this being $\dfrac{5W_dL^3}{384EI}$, where W_d = total dead weight. For the dimensions give W_d = $0{\cdot}2 \times 0{\cdot}3 \times 23 \times 6 = 8{\cdot}28$ kN, and the central deflexion δ_w is given by $\delta_w = \dfrac{5 \times 8{\cdot}28 \times 6^3 \times 10^9 \times 12}{384 \times 13 \times 200 \times 300^3} = 4{\cdot}0$ mm

Thus the upward moment off the forms = 13·5–4·0
= 9·5 mm

Example 19.2. Pre-tensioned beams of the type shown in Fig. 19.5 are to be made by the long-bed process. The five wires are each loaded initially to 20 kN this load being released only after the hardening is complete. Determine the stresses at the top and bottom of a cross-section after transfer and estimate the load per

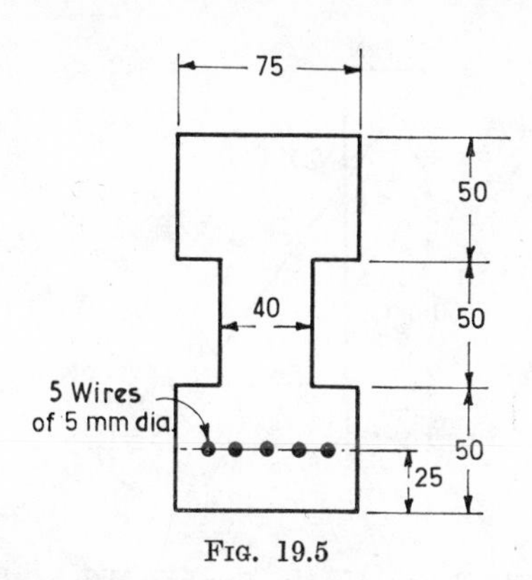

FIG. 19.5

ft which might safely be carried over a span of 6 m. Dimensions in Fig. 19.5 are in mm.

Weight of prestressed concrete 23 kN/m^3.

Modular ratio (allowing for creep effects) 15

Shrinkage 3×10^{-4}

Modulus of direct elasticity for steel 200 kN/mm^2

Allowable stresses in concrete 17 N/mm^2 compression, 1·5 N/mm^2 tension. (*Nottingham*)

Solution. The area of the section $A = (75 \times 150) - (50 \times 35) = 9500\ mm^2$.

The second moment of area about the neutral axis

$$I = 75 \times 150^3 \times \frac{1}{12} - 30 \times 50^3 \times \frac{1}{12} = 20{\cdot}74 \times 10^6\ mm^4$$

$$\text{Area of one wire} = \frac{\pi}{4} \times 5^2 = 19{\cdot}7\ mm^2.$$

The fibre stresses due to the prestressing force are given by

$$\frac{100}{9500}\left(1 \pm \frac{50 \times 75 \times 9500}{20{\cdot}74 \times 10^6}\right) = 28{\cdot}7 \text{ or } -7{\cdot}7,$$

the distribution of stress across the section being as shown in Fig. 19.6. The tensile stress of 7·7 N/mm² is an impossible value to attain but will actually never occur. It is seen from Example 19.1 that the beam will be lifted off the bed when the wires are released. As soon as this takes place the dead load comes into

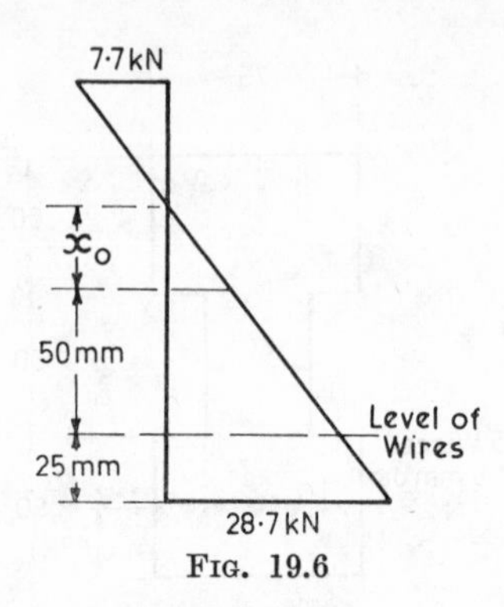

FIG. 19.6

effect and the actual stress is the combination of that due to dead load and prestress.

The distance x_0 of the point of zero stress from the centre-line of the section is given by

$$x_0 = \frac{20 \cdot 7 \times 10^6}{50 \times 9500} = 43 \cdot 5 \text{ mm}$$

The stress in the concrete surrounding the steel bars is therefore

$$\frac{93 \cdot 5}{118 \cdot 5} \times 28 \cdot 7 = 22 \cdot 6 \text{ N/mm}^2$$

When the stress is applied to the concrete it is strained and its length reduced. This reduction also means an equal reduction in length of the steel tendons causing a loss of stress

Let t_i = initial prestress in the steel.

t_l = loss of prestress in the steel on transfer.

$$\text{Final stress in concrete surrounding steel} = \left(\frac{t_i - t_l}{t_i}\right) \times 22 \cdot 6$$

$$\text{Compressive strain in this concrete} = \frac{22 \cdot 6}{E_c} \times \left(\frac{t_i - t_l}{t_i}\right)$$

The compressive strain in the steel is equal to this amount and therefore the loss of prestress in the steel $= m \times 22 \cdot 6 \left(\frac{t_i - t_l}{t_i}\right)$.

But by definition this loss $= t_l$

hence $$m \times 22{\cdot}6\left(1 - \frac{t_l}{t_i}\right) = t_l$$

$$15 \times 22{\cdot}6 = t_l\left(1 + \frac{15 \times 22{\cdot}6}{t_i}\right)$$

But $$t_i = \frac{20 \times 10^3}{19{\cdot}7} = 1015 \text{ N/mm}^2$$

Thus $$t_l = \frac{15 \times 22{\cdot}6}{1 + \dfrac{15 \times 22{\cdot}6}{1015}} = 254 \text{ N/mm}^2$$

Thus, the effective prestress in calculating the stresses at transfer $= 761$ N/mm².

Hence the fibre stresses at transfer are $\dfrac{761}{1015}(28{\cdot}7 \text{ or } -7{\cdot}7)$

$= 21{\cdot}5$ N/mm² compression at the bottom and $5{\cdot}77$ N/mm² tension at the top.

When calculating the load which can be carried by the beam a further reduction in prestress must be taken to allow for shrinkage which will occur with lapse of time.

Loss due to shrinkage $= 200 \times 10^3 \times 3 \times 10^{-4} = 60$ N/mm².

Thus the effective fibre prestresses are

$$\frac{701}{1015}(28{\cdot}7 \text{ or } -7{\cdot}7) = +19{\cdot}8 \text{ or } -5{\cdot}31.$$

The final stresses must not exceed $+17{\cdot}0$ at the top fibre or $-1{\cdot}5$ at the bottom.

Permissible change at top $= 17{\cdot}0 + 5{\cdot}31 = 22{\cdot}31$ N/mm².

Permissible change at bottom $= 19{\cdot}8 + 1{\cdot}5 = 21{\cdot}3$ N/mm².

The latter value is the smaller and therefore controls.

The maximum moment to which the prestressed beam can therefore be subjected

$$= 21{\cdot}3 \times \frac{20{\cdot}74}{75} \times 10^6 = 5{\cdot}87 \text{ kN m}$$

The load carried on a span of 6 m $= \dfrac{5{\cdot}87 \times 8}{6 \times 6}$

$= 1{\cdot}3$ kN/m run

The dead weight of the beam $= 218$ N/m run

$\therefore$ live load $= 1082$ N/m run.

Example 19.3. To carry a uniform live load of 7 kN/m it is proposed to use a post-tensioned concrete beam of symmetrical I-section having a cross-sectional area of $0{\cdot}36d^2$ mm² and a second moment of area of $0{\cdot}04d^4$ mm, d being the depth in millimetres. The permissible concrete stress is 14 N/mm², no flexural tension in the concrete is allowed and the cables can be located at a depth not exceeding $0{\cdot}9d$.

In such a beam the dead load stresses can, within certain limits, be neutralized by lowering the cables to an eccentricity greater than that required for live load alone. Calculate the limiting span for which full compensation of dead load stresses can be obtained in this beam. Disregard losses due to creep, shrinkage and friction and take the weight of concrete as 23 kN/m³. Check your solution by calculating independently the stresses when the span has its critical value. (*London*)

Solution. The section modulus $Z = 0{\cdot}08d^3$.
The live load moment $M_L = 14 \times 0{\cdot}08d^3 = 1{\cdot}12d^3$ N mm.
Since the beam is simply supported the maximum variation of live-load moment is equal to the maximum negative live-load moment, i.e. $M_2 = M_L$.
The dead load per m run $= 0{\cdot}36 \times 23d^2 \times 10^{-3}$ N.
Hence, if the span is L m, the maximum dead load moment M_D is given by

$$M_D = 1{\cdot}036d^2L^2 \text{ N mm.}$$

The horizontal prestress $H = \frac{1}{2} \times 14 \times 0{\cdot}36d^2 = 2{\cdot}52d^2$ N.
The maximum eccentricity of the cable $e_c = 0{\cdot}4d$.

But, generally, $e_c = \dfrac{M_D}{H} + \dfrac{M_2}{2H}$ for symmetrical sections, and substituting the values for this particular case gives

$$0{\cdot}4d = \frac{1{\cdot}036d^2L^2}{2{\cdot}52d^2} + \frac{1{\cdot}12d^3}{5{\cdot}04d^2}$$

$$0{\cdot}4d = 0{\cdot}411L^2 + 0{\cdot}222d$$

$$L^2 = 0{\cdot}431d$$

Hence the limiting span L is $0{\cdot}65\sqrt{d}$ where d is in millimetres and L in metres.

The actual stresses developed will now be checked using the standard beam and eccentrically loaded column theory.

Fibre stresses due to prestress

$$= \frac{2{\cdot}52d^2}{0{\cdot}36d^2}\left(1 \pm \frac{0{\cdot}4d \times 0{\cdot}5d \times 0{\cdot}36d^2}{0{\cdot}04d^4}\right)$$

$$= +\,19{\cdot}6 \text{ or } -\,5{\cdot}6 \text{ N/mm}^2$$

Fibre stresses due to live load $= \dfrac{1{\cdot}12d^3 \times d}{2 \times 0{\cdot}04d^4} = \pm\ 14{\cdot}0\ \text{N/mm}^2$

Fibre stresses due to dead load $= \dfrac{1{\cdot}036d^2 \times 0{\cdot}431d}{0{\cdot}08d^3}$

$= \pm\ 5{\cdot}6\ \text{N/mm}^2$

Thus the resultant stress at the top fibre $= -\ 5{\cdot}6 + 14{\cdot}0 + 5{\cdot}6$

$= +\ 14{\cdot}0\ \text{N/mm}^2$

(the + ve sign in this case denoting compression.)

The resultant stress at the bottom fibre $= +\ 19{\cdot}6 - 14{\cdot}0 - 5{\cdot}6 = 0$.

Hence the specified stress condition is satisfied and $L = 0{\cdot}21\sqrt{d}$ is the limiting span.

Example 19.4. Fig. 19.7 shows the cross-section of a prestressed concrete beam having an area of 20×10^4 mm² and an I of 20×10^9 mm⁴ units. This beam is designed to carry a central point load on a simply supported span of 16 m, the maximum permanent stresses being 14 N/mm² compression and zero tension.

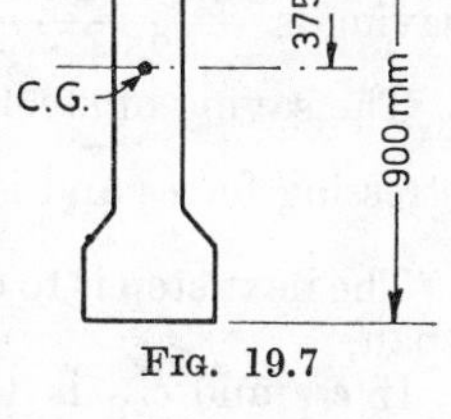

FIG. 19.7

(*a*) Determine the breadth of the rectangular section of a prestressed concrete beam designed for the same loading condition.

(*b*) Calculate the value of the point load.

(*c*) Calculate the saving in steel and concrete of the section shown compared with the rectangular section.

(*d*) Calculate the maximum compressive and tensile stresses in the section during transfer of prestress assuming an allowance of 16 per cent for creep and shrinkage losses in each beam.

(*e*) Draw a dimensioned sketch of the cable profile limits for each of the two beams. (*London*)

Solution. The minimum modulus of section

$$= \frac{20 \times 10^9}{525} = 3{\cdot}8 \times 10^7\ \text{mm}^3\ \text{units.}$$

In order to have the same load-carrying capacity the section modulus of the rectangular section must be the same as the I-section. If b = width of the rectangular section then

$$\frac{b \times 900^2}{6} = 3{\cdot}8 \times 10^7$$

giving $b = 282$ mm.
and a cross-sectional area of $25{\cdot}4 \times 10^4$ mm².

If W = central point load then

$$M_L = M_2 = W \times \frac{16}{4} = 4W \text{ kN m}$$

But $M_L = fZ = (14 \times 3{\cdot}8 \times 10^7)$ N mm.

hence $$W = \frac{3{\cdot}8 \times 10 \times 14{\cdot}0}{4} = 133 \text{ kN}$$

H_I and H_R, the prestressing forces for the I and rectangular sections respectively, are given by

$$H_I = \frac{375}{900} \times 20 \times 10^4 = 1170 \text{ kN}$$

$$H_R = \frac{450}{900} \times 25{\cdot}4 \times 10^4 = 1780 \text{ kN}$$

The saving of concrete when the I-section is used is given by the difference between the areas of concrete in each case. Thus the saving is $\frac{25{\cdot}4 - 20{\cdot}0}{25{\cdot}4}$ or 21·2 per cent.

The saving of steel is given by the difference between the prestressing forces and is $\frac{1780 - 1170}{1780}$ or 34·3 per cent.

The next step is to calculate the eccentricity of the cable at mid-span.

If e_{cI} and e_{cR} is the value for the I and rectangular section respectively then $e_{cI} = \frac{M_{DI}}{H_I} + \frac{y_2}{d} \cdot \frac{M_2}{H_I}$ and $e_{cR} = \frac{M_{DR}}{H_R} + \frac{M_2}{2H_R}$

M_{DI} = dead load moment for I-section

$$= \frac{23 \times 20}{100} \times 16^2 \times \frac{1}{8} = 147 \text{ kN m}$$

M_{DR} = dead load moment for rectangular section

$$= \frac{23 \times 25{\cdot}4}{100} \times 16^2 \times \frac{1}{8} = 187 \text{ kN m}$$

M_2 = live load moment = 532 kN m

Thus $$e_{cI} = \frac{147\,000}{1170} + \frac{525 \times 532 \times 10^3}{900 \times 1170} = 390 \text{ mm}$$

and $$e_{cR} = \frac{187\,000}{1780} + \frac{450 \times 532 \times 10^3}{900 \times 1780} = 254 \text{ mm}$$

In calculating the stresses at transfer the value for H is 1·16 times the final value, and the stresses are as follows, f_1 denoting the stress in the top and f_2 that in the bottom fibre.

I-*section*

$$f_2 = 1{\cdot}16 \times \frac{1170 \times 10^3}{20 \times 10^4}\left(1 + \frac{390 \times 525 \times 20 \times 10^4}{20 \times 10^9}\right)$$

$$= 20{\cdot}7 \text{ N/mm}^2$$

$$f_1 = 1{\cdot}16 \times \frac{1170 \times 10^3}{20 \times 10^4}\left(1 - \frac{390 \times 375 \times 20 \times 10^4}{20 \times 10^9}\right)$$

$$= -\,3{\cdot}12 \text{ N/mm}^2$$

Rectangular Section

$$f_2 = 1{\cdot}16 \times \frac{1780 \times 10^3}{25{\cdot}4 \times 10^4}\left(1 + \frac{6 \times 254}{900}\right) = 21{\cdot}9 \text{ N/mm}^2$$

$$f_1 = 1{\cdot}16 \times \frac{1780 \times 10^3}{25{\cdot}4 \times 10^4}\left(1 - \frac{6 \times 254}{900}\right) = -\,5{\cdot}6 \text{ N/mm}^2$$

The limits for the cable profile depend on the live and dead load moments at the section. The lower limit e_2 is given by

$$e_2 = \frac{M_D}{H} + \frac{I}{Ay_1}$$

The upper limit e_1 is given by

$$\left(\frac{M_D + M_2}{H} - \frac{I}{Ay_2}\right)$$

The live load moment M_2 is the same at any section for both beams but the dead load moment will vary for each beam.

For the rectangular section $\frac{I}{Ay_1} = \frac{I}{Ay_2} = d/6 = 150$ mm, but for the I-section these two values will vary $\frac{I}{Ay_1}$ being 266 mm and $\frac{I}{Ay_2}$, 190 mm. The cable profile limits are given in the tables on p. 396, Table 19·1 being for the I-section and Table 19.2 for the rectangular.

Examples for Practice on Prestressed Concrete

1. A reinforced concrete beam 250 mm wide and 600 mm deep is reinforced at a depth of 400 mm by four 14 mm diameter high tensile steel rods. Before casting the steel is stressed to 630 N/mm^2 and after the concrete has set the stressing device is removed. What is the maximum bending moment that can be applied to the beam without tension being developed in the concrete and what will the maximum stresses be then? The modular ratio is 15.

(*London*)

Answer. 78·0 kN m; 5·15 N/mm² and zero.

TABLE 19.1

Distance from centre, m	M_D	$\div H$	M_2	$\div H$	Σ	e_1	e_2
0	147	125	532	455	580	390	391
4	138	118	400	341	459	269	384
8	110	94	266	228	322	132	360
12	64	55	133	114	169	− 21	321

TABLE 19.2

Distance from centre, m	M_D	$\div H$	M_2	$\div H$	Σ	e_1	e_2
0	187	105	532	300	405	255	255
4	176	99	400	225	324	174	249
8	140	79	266	150	229	79	229
12	82	46	133	75	121	− 29	196

The cable profiles are sketched on Fig. 19.8

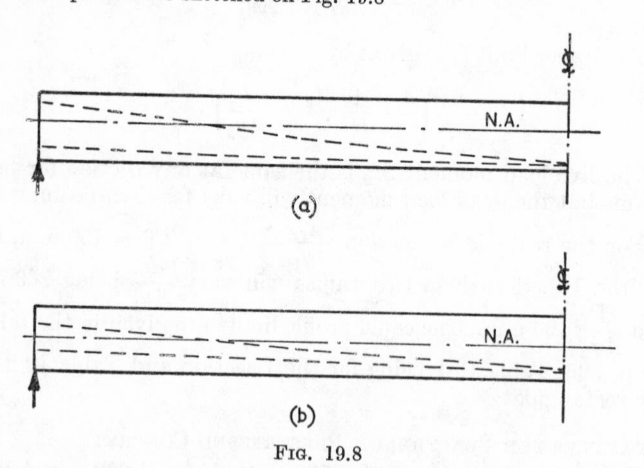

FIG. 19.8

2. A prestressed concrete beam has a rectangular section 150 mm wide by 375 mm deep. It spans between two supports 8 m apart and overhangs 2 m at each end. It is to be post-tensioned by two 19 mm diameter high-tensile steel rods which are to have an initial stress of 650 N/mm.^{2} An allowance is to be made for creep and shrinkage in the concrete amounting to a total strain of $6{\cdot}0 \times 10^{-4}$ and the modulus of elasticity of the steel is to be taken 185 kN/mm^2.

Determine the value of a single rolling load which this beam could carry if the permanent stresses in the concrete are not to exceed 14 N/mm^2 compression and zero tension. (*London*)

Show in a dimensional sketch the cable profile limits.

Answer. 16·5 kN; $e_c = 52{\cdot}3$ mm at mid-span and 62·5 mm at support.

3. The provisional design of a post-tensioned concrete beam is a symmetrical I-section, 900 mm deep with a cross-sectional area of $2{\cdot}07 \times 10^5$ mm^2 and an I of $1{\cdot}66 \times 10^{10}$ mm^4. The maximum practical eccentricity of the centroid of the tendons from the N.A. is 300 mm and the extreme fibre stresses due to self weight are 5·5 N/mm^2 and due to applied loading, 10·5 N/mm^2. The stresses allowed initially when prestressing are more severe than those permitted in the completed structure as follows—

	N/mm^2
Permissible initial compression during stressing	16·0
,, ,, tension ,, ,,	1·4
,, ,, compression under full load	12·5
,, ,, tension ,, ,, ,,	Nil.

Working from first principles and assuming a 10 per cent decline in prestress, determine the range of possible values of eccentricity and prestressing force and recommended specific values. (*London*)

Answer. 212–300 mm; 1512–1230 kN.

4. A prestressed concrete beam supported as shown in Fig. 19.9 (a) is to carry a uniformly distributed load which may cover the whole or part of the beam. Dimensions are in mm. Calculate

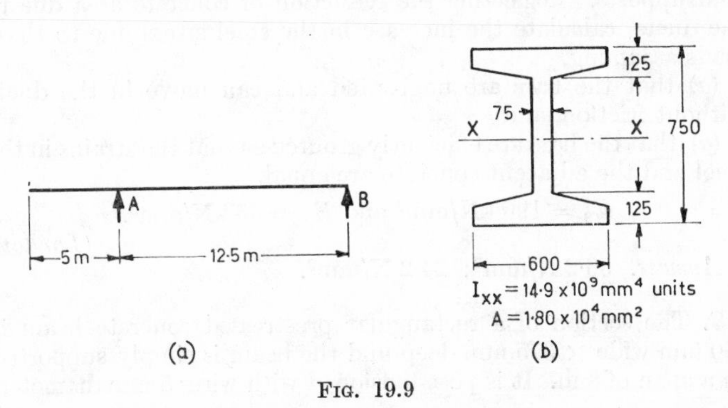

Fig. 19.9

the maximum value of this load if the cross-section of the beam is that shown in Fig. 19·9 (*b*) and the maximum permanent stress is limited to 14 N/mm^2, no tensile stress due to bending being allowed. Determine the shape of the cable and profile and the details of the cable section assuming reasonable values of safe steel stress. Calculate the maximum principal tensile stress due to shear. (*London*)

Answer. 21·6 kN/m run; $H = 1260$ kN; $A = 1480$ mm^2; 0·8 N/mm^2.

5. A rectangular pretensioned beam of 6 m span is 300 mm deep and 100 mm wide. Since, in handling, the beam is liable to be turned upside down, it is decided to limit the tensile flexural stress in this condition due to prestress and self weight to 1·4 N/mm^2. The compressive stress in the inverted position must not exceed 14 N/mm^2. Disregarding all time effects, determine the resultant eccentricity and the required force in the wires after release. Adopting the same limiting stresses, calculate the allowable applied loading in the normal position. Show by clear diagrams the stress conditions at the centre of the span both in the inverted and normal positions.

Describe very briefly how you would estimate the loss of stress in the wires and how you would dispose these in the section. Weight of concrete: 23 kN/m^3. (*London*)

Answer. $e = 44{\cdot}7$ mm; $P = 189$ kN; $w = 3{\cdot}74$ kN/m run.

6. The change of stress in the steel of a post-tensioned concrete structure due to the normal applied loading is sufficiently small to be disregarded in design. As an example, consider a post-tensioned slab spanning in one direction over 10 m span, 300 mm deep, with straight bars at a depth of 250 mm providing 1650 mm^2 of steel per metre width. The slab is subjected to two line loads of 45 kN/m applied along the third points of the span parallel to the supports. Neglecting the reduction of concrete area due to the ducts, calculate the increase in the steel stress due to these loads assuming,

(*a*) that the bars are ungrouted and can move in the ducts without friction, and

(*b*) that the bars are efficiently grouted so that the strains in the steel and the adjacent concrete are equal.

$$E_s = 190 \text{ kN/mm}^2 \text{ and } E_c = 35 \text{ kN/mm}^2.$$

(*London*)

Answer. 36·2 N/mm^2; 24·2 N/mm^2.

7. The section of a rectangular prestressed concrete beam is 150 mm wide × 375 mm deep and the beam is simply supported on a span of 8 m. It is post-tensioned with wires 5 mm diameter, initially stressed to 850 N/mm² and placed so that the initial tension in the concrete is zero and the maximum compression 14 N/mm². If the loss of steel stress is 15 per cent find the number of wires required and the uniform load per metre which can be applied to the beam, the final stresses being limited to the figures already quoted. No allowance need be made for loss of concrete due to the presence of cables and diagrams should indicate clearly the stresses at the various stages of loading. (*Aberdeen*)

Answer. 24, 5·22 kN/m run (including self weight).

8. The concrete beam shown in Fig. 19.10 is reinforced with 650 mm² of steel which is post-tensioned with an effective stress of 1030 N/mm² and not grouted up. The beam spans 12 m and carries a uniformly distributed load of 8 kN/m run and point loads of 30 kN at each of the third points. Calculate the stresses in the concrete due to prestress and applied loading. Take $m = 6$. (*St. Andrews*)

Answer. 20·6 N/mm²; − 9·1 N/mm².

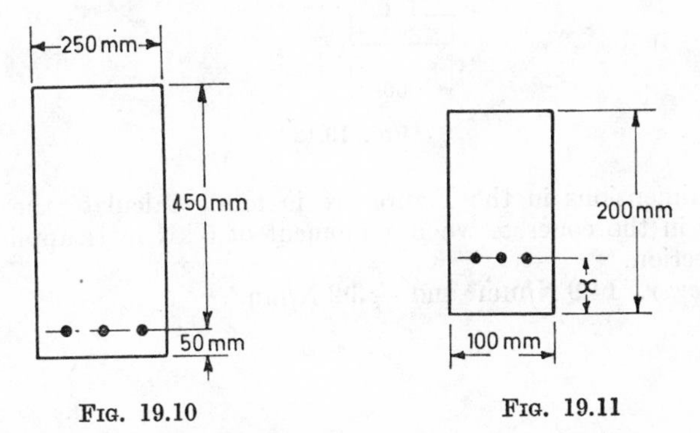

FIG. 19.10 FIG. 19.11

9. The concrete beam shown in Fig. 19.11 is reinforced with 60 mm² of steel which is given an effective prestress of 850 N/mm², the steel being bonded to the concrete. Calculate the dimension x and the maximum resistance moment which can be developed by the beam if the stresses in the concrete are not to exceed 2·5 N/mm² in tension and 40 N/mm² in compression. (*St. Andrews*)

Answer. 34 mm; 6·74 kN m.

10. A precast concrete beam 200 mm wide by 450 mm overall depth has two 13 mm diameter mild steel bars centred at 50 mm from the top face, and has a circular duct formed throughout its length, centred at 350 mm from the top. It is proposed to induce stresses in the beam by passing through the duct a 25 mm diameter high-tensile steel bar, threaded at its ends, and screwing up against bearing plates on the ends of the beam until the maximum compressive stress in the concrete is 12·5 N/mm². If E for mild steel is 205 kN/mm², for high-tensile steel 195 kN/mm² and for concrete 13·5 kN/mm², calculate the extension which must be given to the threaded bar for a 6 m length of the beam. Neglect the effect of the duct on the section modulus and the effect of dead load on the induced stresses. *(Glasgow)*

Answer. 15·3 mm.

11. The concrete beam shown in Fig. 19.12 is reinforced with 60 mm² of steel which is given an effective prestress of 700 N/mm².

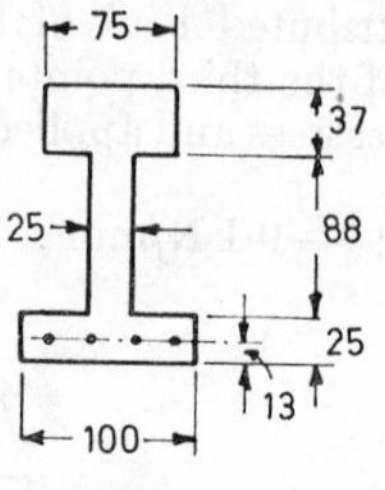

FIG. 19.12

The dimensions in the Figure are in mm. Calculate the fibre stress in the concrete when a moment of 5 kN m is applied to the section.

Answer. 14·9 N/mm² and − 3·2 N/mm².

Chapter 20

Reinforced Concrete Members Using Ultimate Load Methods

THE EXAMPLES given in Chapter 16 dealing with the application of plastic design methods to structural analysis could only be applied in the case of a ductile material having similar properties in tension and compression. They could not be applied without reservation to reinforced concrete. The reinforcement in the

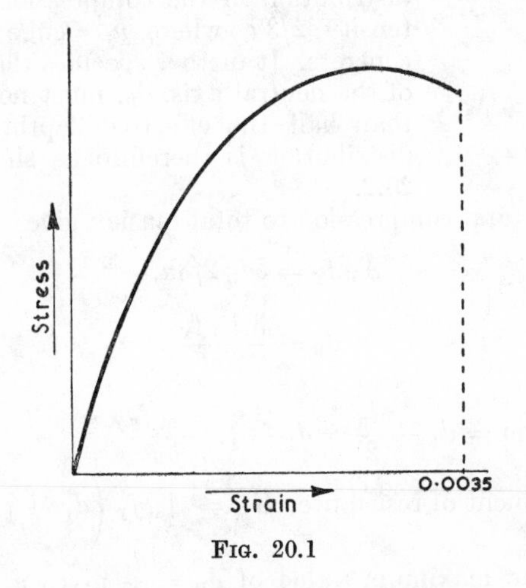

FIG. 20.1

tensile portion of the beam will have properties similar to those shown in Fig. 16.1 if mild steel is used. If a high-tensile steel is used the pronounced yield point will be absent but there will be a considerable strain (of the order of 0·020) before failure. On the other hand, the concrete in the compressive portion of the beam has stress-strain characteristics similar to that shown in Fig. 20.1, the features of which are the almost complete absence of any elastic portion and an ultimate strain very much lower than that in the reinforcement.

Two fundamental principles used in the straight-line no-tension theory will, however, apply to ultimate load theory. They are: (1) the straight-line strain distribution and (2) the fact that total tension equals total compression. The total tension is the area of reinforcement multiplied by the yield stress, and unless this total tension is very small failure of a reinforced concrete beam will generally take place by crushing of the concrete, since this material reaches its ultimate strain first.

Because it is impossible to give a general mathematical relationship for the stress-strain curve for concrete, a number of theories has been derived for the ultimate moment of resistance of a reinforced concrete beam depending on the assumptions made with regard to the stress-strain curve. It is not proposed to give these different theories here as most examples are based on the British Standard Code of Practice 114. The theory therein given is based on a uniform stress distribution in the compression face of intensity $2/3u$, where $u =$ cube strength of concrete. It further specifies that the depth of the neutral axis, d_n, must not be greater than half the effective depth. The force distribution is therefore as shown in Fig. 20.2.

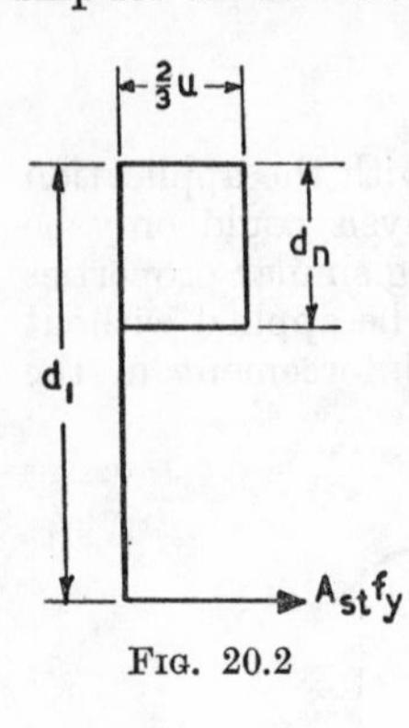

FIG. 20.2

Equating total compression to total tension gives

$$A_{st} f_y = b d_n \, 2/3u,$$

leading to

$$d_n = \frac{3A_{st} f_y}{2bu}$$

$$\text{The lever arm} = d_1 - \frac{d_n}{2} = d_1 - \tfrac{3}{4}.\ \frac{A_{st} f_y}{bu}$$

giving a moment of resistance, $M_R = A_{st} f_y \left(d_1 - \frac{3}{4} \frac{A_{st} f_y}{bu} \right)$

Taking the maximum value of $d_n = \frac{d_1}{2}$ gives a moment of resistance $= \frac{b d_1^2 u}{4}$. If the factored applied moment exceeds this then the compressive resistance of the beam must be increased by using compression steel. The straight-line strain diagram gives the value of the stress in this steel as $760 \left(\frac{d_n - d_2}{d_n} \right)$ or the yield stress, whichever is the smaller. The recommended value of the load factor is 1·8 and the Code also recommends that the value to be taken for u is 2/3 of the actual cube strength.

Instead of designing to the ultimate load and using a load factor of 2 the Code also permits designing for working load moments and resisting these by a stress distribution similar to that shown in Fig. 20.2, the uniform stress in the concrete being $2/3p_{cb}$, i.e. 4·65 N/mm^2 for a nominal mix 1:2:4 concrete. (This forms the basis of Example 20.2 for practice.)

The collapse mechanisms, given in the examples following Example 16.3 in Chapter 16, may not apply to reinforced concrete structures because in order for them to develop, considerable rotation may be necessary at some of the hinges. The low value of the ultimate strain in concrete does limit the hinge rotation in a reinforced concrete member, and this may prevent the collapse mechanism from forming. The permissible hinge rotation is a matter on which no generally accepted decision has been made and problems on this are unlikely to arise in examinations.

The problems are mainly concerned with the ultimate moment of resistance of beams and with eccentrically loaded columns taking a rectangular stress block on the compression face.

Example 20.1. Define the term "load factor." State the formula for deriving the moment of resistance of a rectangular beam using the load-factor method of design. Compare the results obtained with this method and with the classical method for a beam with the following dimensions: $b = 300$ mm $d_1 = 500$ mm, steel percentage (not economic) 1·35 per cent. (*London*)

Solution. It is presumed that by the classical method is meant the straight-line no-tension theory. No values are given for the stresses, and although a comparison might be made in general terms the preamble to the paper in which the question occurs gives the following values to be taken for the stresses—

Steel 140 N/mm^2. Concrete 8·5 N/mm^2. $m = 15$. These values will be taken in the solution of the problem and based on them the following additional information—

Yield strength of steel 230 N/mm^2.

Cube strength of concrete 25·5 N/mm^2.

Straight-line no-tension theory

$$\frac{mc}{t} = \frac{n}{500 - n} \text{ giving } \frac{c}{t} = \frac{n}{15(500 - n)}$$

$$At = 150nc \text{ giving } \frac{c}{t} = \frac{A}{150n}$$

where
$$A = \frac{1{\cdot}35 \times 300 \times 500}{100}$$

Hence
$$\frac{n}{15(500 - n)} = \frac{1{\cdot}35 \times 300 \times 500}{150 \times 100n}$$

This leads to the following quadratic in n

$$n^2 + 202n - 101\,000 = 0$$

the solution of which is $n = 233$ mm.

Hence

$$\frac{c}{t} = \frac{233}{15 \times 267}$$

If $t = 140$; $c = \dfrac{140 \times 233}{15 \times 267} = 8{\cdot}14$ N/mm²

This is less than the permissible stress, hence the steel stress of 140 N/mm² is critical, and the moment of resistance is given by

$$M_{\mathrm{R}} = 140 \times \frac{1{\cdot}35 \times 300 \times 500}{100} \times \left(500 - \frac{233}{3}\right)$$

$$= 121 \times 10^6 \text{ N mm} = 121 \text{ kN m}$$

With the ultimate-load method, using a load factor of 2 and taking the cube strength as 2/3 × actual cube strength, the calculations are as follows—

$$2/3 \times 25{\cdot}5 \times 300 \times n = \frac{230 \times 300 \times 500 \times 1{\cdot}35}{100}$$

giving

$$n = 91{\cdot}5 \text{ mm}$$

and the lever arm $= 500 - \dfrac{91{\cdot}5}{2} = 454$ mm

The ultimate moment of resistance $= \dfrac{230 \times 1{\cdot}35 \times 300 \times 500}{100} \times 454$

$$= 212 \text{ kN m}$$

Hence the safe moment of resistance = 106 kN m.

If the cube strength is taken equal to the actual value of 25·5 N/mm²

then

$$n = \frac{230 \times 300 \times 500 \times 1{\cdot}35}{100 \times 300 \times 25{\cdot}5} = 60{\cdot}6 \text{ mm.}$$

and the lever arm = 470 mm, giving the ultimate moment of resistance as 220 kN m and the safe moment as 110 kN m.

Example 20.2. A rectangular reinforced concrete beam is to be 300 mm wide and 600 mm deep to the centres of steel bars.

(*a*) Using the elastic theory with a modular ratio of 15 and working from first principles, determine the required area of steel bars and the total inclusive uniformly distributed load the beam can carry over a simply supported span of 8 m. The permissible stresses in the steel and concrete are assumed to be reached simultaneously and are 140 and 7·0 N/mm² respectively.

(*b*) The concrete cube strength at 28 days is 21 N/mm² and the yield strength of the steel 230 N/mm². Stating clearly your assumptions regarding stress distribution at failure, determine the ultimate collapse load for a rectangular beam 300 mm wide and 600 mm deep to the centres of 1930 mm² of steel.

(*London*)

Solution

Part (*a*)

Let n = depth to neutral axis, then

$$\frac{n}{600 - n} = \frac{15 \times 7}{140} \text{ leading to } n = \frac{3}{7} \times 600 = 257 \text{ mm}$$

Equating total tension to total compression gives

$$A \times 140 = 300 \times 257 \times \frac{7}{2}$$

leading to $A = 1930$ mm².

The moment of resistant M_R is given by

$$M_R = 1930 \times 140 \times \left(600 - \frac{257}{3}\right) = 139 \text{ kN m}$$

If w = uniformly distributed load on span of 8 m then

$$w \times 8^2 \times \frac{1}{8} = 139$$

leading to $$w = 17{\cdot}4 \text{ kN/m run.}$$

Part (*b*)

If n = depth to neutral axis then, equating total tension to total compression with a rectangular stress block of intensity $2/3u$ gives $1930 \times 230 = 300 \times n \times \frac{2}{3} \times 21$

hence $$n = 106 \text{ mm}$$

and the lever arm $= 600 - \dfrac{106}{2} = 547$ mm.

The ultimate moment of resistance $= 1930 \times 230 \times 547$
$= 243$ kN m.

Example 20.3. The concrete used for a rectangular beam may be assumed to have a parabolic stress-strain curve showing a strain of 0·001 at 30 N/mm² and 0·002 at 40 N/mm² at which point it may be assumed to fail suddenly. The steel has a stress-strain curve which is linear to the yield point at 1000 N/mm² and 0·005 strain, stress remaining constant for further strain.

Find the steel area and the moment of resistance at failure of a balanced design in terms of the beam breadth b and the effective depth d_1 with no pre-stress in the steel. (*London*)

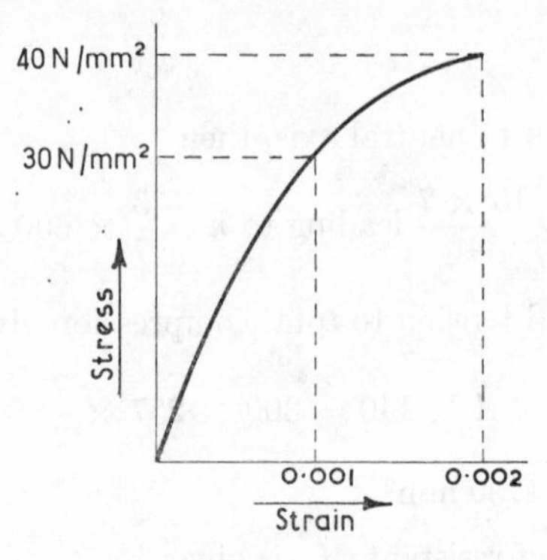

FIG. 20.3

Solution. The stress-strain curve for the concrete is as shown in Fig. 20.3. If f represents the stress and ε the strain, then the relationship between stress and strain is given by

$$f = k\varepsilon\,(0{\cdot}004 - \varepsilon)$$

In order to determine k the data gives $\varepsilon = 0{\cdot}002$ when $f = 40$. Substituting gives $k = 1 \times 10^7$, hence the stress-strain relationship is $f = 1 \times 10^7\ \varepsilon(0{\cdot}004 - \varepsilon)$ N/mm².

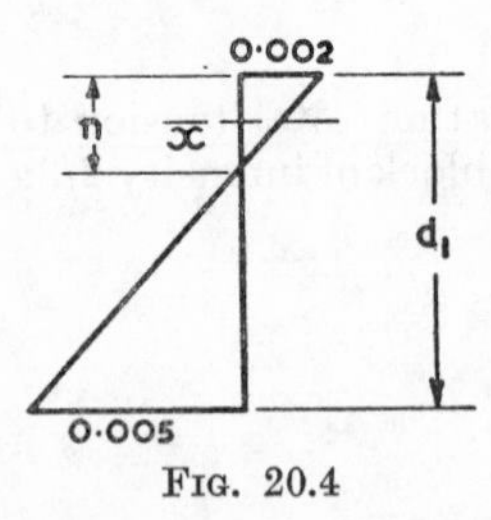

FIG. 20.4

Balanced design implies simultaneous failure of the concrete and steel, hence the strain distribution across the section at failure is as shown in Fig. 20.4, the fibre strain in the concrete being 0·002 and the strain in the steel 0·005. If n denotes the depth to the neutral axis, then

$$\frac{n}{d_1 - n} = \frac{0{\cdot}002}{0{\cdot}005} \text{ giving } n = 0{\cdot}286d_1$$

The next part of the solution consists in determining the total compression which can be carried by the concrete.

At a distance x from the neutral axis the strain ε_x is given by

$\varepsilon_x = \frac{x}{n} \times 2 \times 10^{-3}$ and the stress f_x by

$$f_x = 1{\cdot}0 \times 10^7 \times \frac{x}{n} \times 2 \times 10^{-3} \left(4 \times 10^{-3} - \frac{x}{n} \times 2 \times 10^{-3}\right)$$

or $f_x = 20 \frac{x}{n}\left(4 - \frac{2x}{n}\right)$

Hence the total compression C is given by

$$C = 20 \frac{b}{n} \int_0^n x \left(4 - \frac{2x}{n}\right) dx$$

$$= 20 \frac{b}{n} \left[\frac{4x^2}{2} - \frac{2x^3}{3n}\right]_0^n$$

$$= 80/3bn.$$

The total tension $T = 1000A_{st}$

where A_{st} = steel area = $\frac{pbd_1}{100}$, where p = percentage of reinforcement.

Equating total tension to total compression gives

$$1000 \frac{pbd_1}{100} = \frac{80b}{3} \times 0{\cdot}286d_1$$

leading to $$p = \frac{8 \times 0{\cdot}286}{3} = 0{\cdot}764$$

The moment of resistance of the section is, in this case, best obtained by taking moments of the compressive and tensile forces about the neutral axis. The moment of the compressive forces M_{CN} is given by

$$M_{CN} = 20 \frac{b}{n} \int_0^n x^2 \left(4 - \frac{2x}{n}\right) dx$$

$$= \frac{20b}{n} \left[\frac{4x^3}{3} - \frac{2x^4}{4n}\right]_0^n = 16{\cdot}67bn^2$$

Substituting $n = 0{\cdot}27_1\, 6d$gives

$$M_{CN} = 1{\cdot}27bd_1^2$$

The moment of the tensile forces about the neutral axis M_{TN} is given by

$$M_{TN} = 1000A_{st}\,(d_1 - n)$$

Substituting for A_{st} and d_1 leads to

$$M_{TN} = 5{\cdot}54bd_1^2$$

Hence the moment of resistance of the section $= 6{\cdot}81bd_1^2$

Examples for Practice on Reinforced Concrete Members Using Ultimate-Load Methods

1. Comment briefly on the use of the plastic method for designing a reinforced concrete section subjected to bending.

Whitney's theory states that the ultimate moment of resistance of a reinforced concrete beam

$$= 0{\cdot}333bd^2\,\mu \text{ or } bd^2pt_y\left(1 - 0{\cdot}59\,\frac{pt_y}{\mu}\right)$$

the appropriate formula being used if the beam is over- or under-reinforced. Comment on the value to be used for μ in this formula.

A reinforced concrete slab of 106 mm effective depth carries a dead load of 3·5 kN/m² and a live load of 9·0 kN/m² on a simply supported span of 4 m. Using load factors of 1·5 on the dead load and 2·5 on the live load calculate the amount of reinforcement required if the concrete used has a cube strength of 21 N/mm² at 28 days. The yield stress in the steel is 240 N/mm². Compare this with the area of reinforcement using ordinary design methods. (*St. Andrews*)

Answer. Whitney 2520 mm²/m width; Elastic 1980 mm²/m.

2. A reinforced concrete beam of uniform depth and 300 mm width supports a 3 m width of slab and is continuous over two equal spans of 6 m. The dead load on the slab is 7 kN/m² (including an allowance for the weight of the beam itself) and the superimposed load is 5 kN/m². Design the beam by the Load Factor method described in C.P. 114 1957 and calculate the number of 25 mm diameter bars required over the centre support and in the middle of the spans. Take the permissible stress in the steel as 140 N/mm² and the permissible compressive stress in the concrete in bending as 7 N/mm². (*Nottingham*)

Answer. Effective depth, 555 mm; Support 6 bars; Mid-span, 3 bars.

3. A rectangular reinforced concrete beam is 250 mm wide and 500 mm deep to the centre of three 25 mm diameter steel bars.

(*a*) Using the elastic theory with a modular ratio of 15 and permissible stresses in the steel and concrete of 140 N/mm² and 7·0 N/mm² respectively, determine the total inclusive uniformly distributed load the beam can carry over a simply supported effective span of 6 m.

(*b*) The concrete cube strength for the above beam at 28 days is 21 N/mm² and the yield strength for the steel 250 N/mm². Stating clearly your assumptions regarding stress distribution at failure, determine the ultimate load the beam can carry uniformly distributed over the same span. (*London*)

Answer. Elastic Theory, 18·3 kN/m run; Ultimate, 37·0 kN/m run.

4. Calculate the maximum uniformly distributed load which can be carried per metre width by a singly reinforced concrete floor slab of 188 mm effective depth and 6 m span if the concrete has a cube strength of 21 N/mm². Calculate also the area of tension steel of yield stress 250 N/mm² which must be provided to develop the necessary resistance moment. State fully the assumptions you make. (*London*)

Answer. 41·3 kN/m run; $A_{st} = 5270$ mm².

5. A reinforced T-section of 150 mm effective depth has a flange 300 mm wide and 25 mm deep and a web 50 mm thick. Calculate the area of reinforcement of yield stress 275 N/mm² if the section is to have an ultimate moment of resistance of 15 kN m when the concrete has a cube strength of 27·5 N/mm². (*London*)

Answer. 400 mm².

6. A reinforced concrete beam 250 mm wide and 450 mm effective depth has to be designed for an ultimate moment of resistance of 350 kN m. The cube strength of the concrete is 21 N/mm² and the yield stress in the steel is 240 N/mm². Calculate the amount of reinforcement required centering any compression reinforcement at 50 mm below the top of the beam. (*London*)

Answer. $A_{st} = 4165$ mm²; $A_{sc} = 940$ mm².

7. A concrete column 250 mm × 400 mm is to carry an ultimate load of 450 kN acting on the longer axis of symmetry with an eccentricity of 300 mm. The column is symmetrically reinforced with steel centred at 50 mm from each of the shorter faces. If the concrete has a cube strength of 21 N/mm² and the steel a yield stress of 275 N/mm² calculate the area of reinforcement required.

Answer. 1790 mm².

8. A short reinforced concrete column is 300 mm square and has 12 bars of 25 mm diameter as main reinforcement with centre-lines in the form of a 225 mm square and bars at 75 mm centres. It is well reinforced laterally. What load, at what eccentricity, would just cause failure in compression if one face is at zero stress? The stress-strain curve for concrete is assumed parabolic, with failure at 40 N/mm² at a strain of 0·002 where it fails suddenly, the general equation of the curve being (stress) $= 1{\cdot}0 \times 10^7$ (strain) $\times$ (0·004 — strain) up to that point. The stress-strain curve for steel is linear with $E = 200$ kN/mm² up to a stress of 280 N/mm² at which point it yields at constant stress. (*London*)

Answer. 3240 kN; 42 mm.

9. A short reinforced concrete column has the section shown in Fig. 20.5, the steel bars being each of 25 mm dia. All dimensions

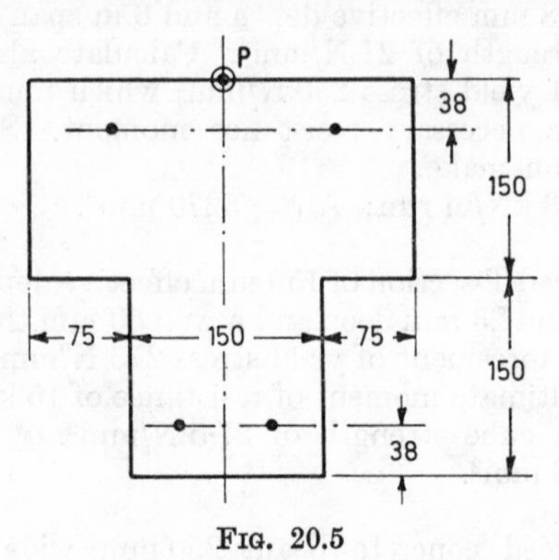

FIG. 20.5

in the Figure are in mm. The force and moment on the column are equivalent to a load P acting at the mid-point of the 300 mm side.

(*a*) If P is 200 kN calculate the critical stresses assuming that elastic conditions prevail and that m is 15.

(*b*) Calculate the load P which would cause collapse taking the cube strength of the concrete as 28 N/mm², the yield stress of the steel in tension, and compression as 280 N/mm² and the limiting strain in the concrete as 0·0033.

Answer. (*a*) 6·8 N/mm² (compression); 51·0 N/mm² (tension). (*b*) 835 kN.

Index

Fundamentals of Fortran Programming

by McCracken